FISH AND SHELLFISH PROCESSING

FISH AND SHELLFISH PROCESSING

M.T. Gillies

NOYES DATA CORPORATION
Park Ridge, New Jersey London, England
1975

Library of Congress Catalog Card Number: 75-2942
ISBN: 0-8155-0574-4
Printed in the United States

Published in the United States of America by
Noyes Data Corporation
Noyes Building, Park Ridge, New Jersey 07656

FOREWORD

The detailed, descriptive information in this book is based on U.S. patents since 1960 relating to the large scale processing of fish and shellfish. Where it was necessary to round out the complete technological picture, some earlier, but very relevant patents were included.

This book serves a double purpose in that it supplies detailed technical information and can be used as a guide to the U.S. patent literature in this field. By indicating all the information that is significant, and eliminating legal jargon and juristic phraseology, this book presents an advanced, commercially oriented review of modern methods of seafood processing as depicted in U.S. patents.

The U.S. patent literature is the largest and most comprehensive collection of technical information in the world. There is more practical, commercial, timely process information assembled here than is available from any other source. The technical information obtained from a patent is extremely reliable and comprehensive; sufficient information must be included to avoid rejection for "insufficient disclosure." These patents include practically all of those issued on the subject in the United States during the period under review, there has been no bias in the selection of patents for inclusion.

The patent literature covers a substantial amount of information not available in the journal literature. The patent literature is a prime source of basic commercially useful information. This information is overlooked by those who rely primarily on the periodical journal literature. It is realized that there is a lag between a patent application on a new process development and the granting of a patent, but it is felt that this may roughly parallel or even anticipate the lag in putting that development into commercial practice.

Many of these patents are being utilized commercially. Whether used or not, they offer opportunities for technological transfer. Also, a major purpose of this book is to describe the number of technical possibilities available, which may open up profitable areas of research and development. The information contained in this book will allow you to establish a sound background before launching into research in this field.

Advanced composition and production methods developed by Noyes Data are employed to bring our new durably bound books to you in a minimum of time. Special techniques are used to close the gap between "manuscript" and "completed book." Industrial technology is progressing so rapidly that time-honored, conventional typesetting, binding and shipping methods are no longer suitable. We have bypassed the delays in the conventional book publishing cycle and provide the user with an effective and convenient means of reviewing up-to-date information in depth.

The Table of Contents is organized in such a way as to serve as a subject index. Other indexes by company, inventor and patent number help in providing easy access to the information contained in this book.

15 Reasons Why the U.S. Patent Office Literature Is Important to You –

1. The U.S. patent literature is the largest and most comprehensive collection of technical information in the world. There is more practical commercial process information assembled here than is available from any other source.

2. The technical information obtained from the patent literature is extremely comprehensive; sufficient information must be included to avoid rejection for "insufficient disclosure."

3. The patent literature is a prime source of basic commercially utilizable information. This information is overlooked by those who rely primarily on the periodical journal literature.

4. An important feature of the patent literature is that it can serve to avoid duplication of research and development.

5. Patents, unlike periodical literature, are bound by definition to contain new information, data and ideas.

6. It can serve as a source of new ideas in a different but related field, and may be outside the patent protection offered the original invention.

7. Since claims are narrowly defined, much valuable information is included that may be outside the legal protection afforded by the claims.

8. Patents discuss the difficulties associated with previous research, development or production techniques, and offer a specific method of overcoming problems. This gives clues to current process information that has not been published in periodicals or books.

9. Can aid in process design by providing a selection of alternate techniques. A powerful research and engineering tool.

10. Obtain licenses — many U.S. chemical patents have not been developed commercially.

11. Patents provide an excellent starting point for the next investigator.

12. Frequently, innovations derived from research are first disclosed in the patent literature, prior to coverage in the periodical literature.

13. Patents offer a most valuable method of keeping abreast of latest technologies, serving an individual's own "current awareness" program.

14. Copies of U.S. patents are easily obtained from the U.S. Patent Office at 50¢ a copy.

15. It is a creative source of ideas for those with imagination.

CONTENTS AND SUBJECT INDEX

INTRODUCTION

It was really only a few years ago that people suddenly became aware of the constant, inexorable, geometric rise in the world population. Nutritionists and economists from industrialized and developing countries alike began to search for ways in which food, and particularly proteinaceous food, could be produced in quantities great enough to feed the ever-growing population of this planet of ours, whose food resources were suddenly recognized by more and more people as being less than infinite.

It became fashionable for a while to point out that the 71% of the earth's surface, which is ocean, produces from 6 to 12% of the world's current protein supply, although the seas receive more than twice as much of the solar energy, which is the source of all biological activity, as does the land. Surely, it was argued, the potential harvest from the sea could fully supply the earth's protein needs if more efficient harvesting methods could be developed.

Only five years ago some experts were estimating the potential seafood catch at 2,000 million metric tons per year. Now, however, recent hard-headed, scientific examination of ocean resources has reduced our expectations considerably, and many experts now believe that, using current technology, a figure of 100 million metric tons a year is more realistic, a rise of only about 42% over the current catch of 70 million metric tons.

It is now clear that many varieties of fish and shellfish have been overfished in many places, and that pollution has taken a heavy toll, so that frightening drops in the yearly catch of many species have been suffered. There are, for instance, few herring left in the North Sea and along the European coast; sardines have virtually disappeared from the California coast; commercial haddock fishing is forbidden off the East Coast of the U.S. because of scarcities; and, most gravely, the huge anchovy catch off the Peruvian coast, which amounted to 12.3 million metric tons in 1970, dwindled so severely (1.8 million tons in the first 6 months of 1973) that in mid-1973 the Peruvian government prohibited anchovy fishing in order to preserve the species. Although a contributory cause of the drop in the anchovy harvest was undoubtedly a shift in the Humboldt current, the experience of the anchovy fishermen is one which may well be repeated again and again, as the fishermen of all nations take more and more fish with more and more sophisticated equipment and larger and larger factory ships.

It becomes obvious now that the world's need for protein from the sea can at this point best be served by research into methods of sea farming and also by development of uses for species of fish and shellfish not presently accepted in the world market. Other ways to contribute to an increase in the availability of seafood are within the purview of this book—improving the economy of fish and shellfish processing, better methods of preserving seafood, less wasteful butchering and canning techniques, and, perhaps most important of all, the introduction of edible products which use some of the incredible numbers of fish which are caught and thrown away as trash in the process of netting desirable and salable seafood items. (This last subject is touched on in the chapter on fish protein concentrates.)

This book reviews patent activity in the seafood field since 1960. The first chapter deals with the preservation of fish, suggesting methods for bringing fish or shellfish to the market or cannery in an organoleptically and bacterially satisfactory state even though they may have been on board ship for some days. The second chapter reviews patents in the huge and important tuna fish industry, and deals primarily with ways for improving the economy of processing tuna fish from catch to can. Methods of processing other varieties of fish are treated in Chapter 3.

A large part of the expense in producing seafood products like frozen shrimp and crab, scallops and clam chowder, etc. comes from the traditional hand labor involved in removing the inedible parts of such shellfish. Many of the patents in Chapters 4 and 5 are concerned with methods and apparatus for such problems that have proved their usefulness in commercial application.

Chapter 6 reviews patents in the field of animal foods, and includes such interesting and unusual applications as dolphin food, mink food, and food for tropical fish.

Probably the most interesting development is discussed in Chapter 7, on fish protein concentrates, which many experts feel are the most promising substances available from the sea in contributing an answer to some of the world's food problems.

SEAFOOD PRESERVATION AND STRUVITE PREVENTION

REFRIGERATION ON BOARD SHIP

Refrigeration is of particular importance as regards fish, since in contrast to the tissues of warm-blooded animals, the tissues of fish are not sterile. They contain large quantities of seawater bacteria, which in conjunction with enzymes develop an intensive decomposing activity immediately after biological death has occurred. The autolytic processes usually proceed with extraordinary rapidity, because the conditions existing on board ship after the fish have been caught differ only to a slight degree from natural living conditions. Even if the autolysis initially does not produce any harmful decomposition products it does cause an unpleasant and penetrating fish odor due to formation of organic amines, for example, trimethylamine.

These phenomena can be counteracted only if intensive refrigeration is carried out as soon as the fish are on board. Many refrigerating processes have become used, such as, for example, the immediate freezing by means of a Birdseye deep-freeze apparatus, but these processes are only suitable for large fishing trawlers, or the like. In small vessels, the fishermen are restricted to adding what ice they can carry with them, which hardly lowers the fish to an appreciable degree as compared with their normal body temperatures of from 4° to 10°C.

Moreover, oxygen, which promotes the development of the seawater bacteria and causes oxidation of the fatty substances, has unrestricted access to the fish, and pathogenic bacteria find an excellent nutrient medium in the water from melting ice enriched with albumin. This deterioration, which takes place at ever-increasing speeds, makes it necessary for the fishing vessel to make its return journey often before it is nearly full, and causes its economy to be jeopardized.

H. Beckman; U.S. Patent 2,987,404; June 6, 1961 developed a process for conserving and refrigerating fresh-caught fish in a fishing ship by subjecting them to a vacuum, the selected degree of which is such that the fish cools approximately to freezing temperature by the vaporization of the moisture on its surface, but is not frozen.

Therefore, the considerable heat of evaporation of the water, which amounts to 583 cal/g at room temperature, is used for cooling the food which is to be kept fresh, in that the evaporation of the water is produced by applying a high vacuum. Since the heat of evaporation of the water at 0°C is somewhat greater than at room temperature, namely 597 cal/g, an even better effect is obtained when fresh or salt water ice is added to the food to be conserved in the containers.

The process is based on the following biological and physical principles; the fish must be killed immediately after being caught; the treatment must be carried out within 4 to 6 hours; the fish must be deaerated and degasified, after carrying out the gutting operation. By this means, the absorption of those gases which promote the decomposition process is prevented. Furthermore, there is also removed thereby the oxygen which promotes the development of seawater bacteria, and oxidation processes within the cells are prevented.

The process is carried out by storing the fish, after they have been gutted, in vacuum tanks, from which air can be evacuated by a steam jet pump, to a pressure of 2 mm of mercury. At this pressure, the water adhering to the fish is evaporated down to −10°C, and the fish is cooled by a thin layer of ice formed, which is immediately liquefied again by the heat flowing from the fish, so that the evaporation continues without interruption.

To accelerate the process, ice can be added to the tank in an amount of approximately 10% of the weight of the fish and a stirrer used. The tank may be filled all at once, or gradually, depending on its size. When it is completely filled it is left without any special supervision, except that the temperature and vacuum are checked periodically.

By practical experience in using the process it has been found that fish that have not been butchered suffer damage in their cell structure by decreasing the ambient pressure to about 4 mm Hg, because gases forming in the fish body during storage do not disappear quickly enough. Thus a swelling of the fish occurs, especially in the region of the belly, because the extremely aggressive digestive enzymes in the digestive tract effect a strong gas formation. At a very high vacuum, these gases may therefore lead to damage in the structure of the fish.

To avoid this difficulty, a further refinement is made by *H. Beckman; U.S. Patent 3,097,099; July 9, 1963.* It was found that if the fish were exposed during their storage to a higher pressure, say from 80 mm Hg upwards, the swelling then disappeared without hindering the diffusion of the gases.

The pressure to be applied during stoarge depends on the sort of fish. Thus, this increase of pressure is effected only after the fish has been degasified and deaerated by applying the high vacuum and has been cooled down to freezing temperature by vaporizing the moisture present on its surface, but has not been frozen. This pressure may be generated by means of vapors, gases or liquids. It is of course understood that the pressure-increasing means must be fully neutral and may not effect any chemical reactions, deposits or other transformations and contain neither air nor oxygen.

So any desired pressure increase may be achieved by introducing vapors of cooling means, for instance frigene 12, i.e., difluorodichloromethane; or gases, for instance nitrogen; or a liquid, for instance sterilized water. By introduction of frigene 12, herrings, which are very delicate may be preserved in an unobjectionable condition for about 3 weeks at a pressure of 100 mm Hg.

CHEMICAL METHODS

5-Amino-Hexahydropyrimidine

R.C. Sanders; U.S. Patent 2,963,374; December 6, 1960; assigned to Warner-Lambert Pharmaceutical Company has found that the storage life of fresh fish can be greatly enhanced if the fish is treated with a 5-amino-hexahydropyrimidine of the following formula:

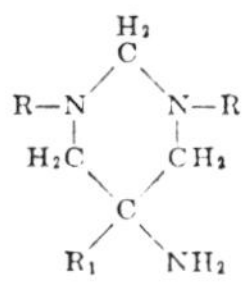

where R is an alkyl, aryl, aralkyl, alkaryl, hydroxyalkyl, aminoalkyl or cycloalkyl radical and R_1 represents hydrogen or a lower alkyl or hydroxymethyl radical, or with a salt of the 5-amino-hexahydropyrimidine compound. The fish so treated are markedly resistant to supporting the growth of microorganisms and are able to undergo the usual forms of storage for periods of time substantially in excess of those experienced when they are placed in storage in an untreated form. Undesirable changes in taste, form and appearance due to bacterial action are greatly minimized. The compounds may be applied in the form of an aqueous solution or dispersion during various stages of processing.

Satisfactory results are obtained employing a solution of the 5-amino-hexahydropyrimidine containing up to about 1.0% by weight. It is preferred that not more than about 0.1% by weight of the 5-amino-hexahydropyrimidine be placed on the food being treated and less than that amount, for example, as low as 0.01% or even less is satisfactory. These 5-amino-hexahydropyrimidines have a rather low toxicity, both acute and chronic, and can be tolerated in appreciable amounts.

Fresh fish in whole or in fillet form may be treated with excellent results. The perishability of fish or other seafood in transit is ordinarily avoided by icing the fish to maintain the temperature below that at which bacterial growth or multiplication is favored. Advantageously, the ice employed for this purpose may be prepared by freezing an aqueous solution containing up to about 0.1% by weight of the desired 5-amino-hexahydropyrimidine and employing the ice so formed for icing purposes.

The 5-amino-hexahydropyrimidines exhibit a pronounced affinity for protein materials and subsequent icings need only be conducted with ordinary forms of ice. No impairment in the degree of protection obtained is observed where ordinary ice is employed to supplement the original icing. Among the 5-amino-hexahydropyrimidine compounds which have been found effective may be mentioned:

1,3-bismethyl-5-methyl-5-amino-hexahydropyrimidine
1,3-bis(beta-ethylhexyl)-5-methyl-5-amino-hexahydropyrimidine
1,3-bis(dicyclohexyl)-5-methyl-5-amino-hexahydropyrimidine
1,3-bisbenzyl-5-methyl-5-amino-hexahydropyrimidine

Example 1: A preservative bath of the following composition is prepared:

	Parts by Weight
1,3-bis(beta-ethylhexyl)-5-methyl-5-amino-hexahydropyrimidine	0.1
Polyoxyethylene sorbitan monooleate (Tween 80)	1.4
Water, q.s. 1,000.0	

Freshly prepared fish fillets are immersed in this bath for two minutes and then iced for shipment. This treatment is found to extend the shipping range materially and to reduce the frequency of reicing in transit while maintaining the appearance and acceptability unchanged.

Example 2: A 0.005% aqueous solution of 1,3-bis(beta-ethylhexyl)-5-methyl-5-amino-hexahydropyrimidine trihydrochloride is frozen into the form of flake ice and used for the icing of fresh shrimp in transit. This treatment effectively controls bacterial action and maintains the product in acceptable market condition not only in transit but subsequently with only a minimum of reicing.

Sorbic Acid and Antioxidant

J.T.R. Nickerson and L.D. Starr; U.S. Patent 2,933,399; April 19, 1960 developed a process for the preservation of fresh fish fillets and lightly smoked or salted fish, such as finnan haddie, which are susceptible to bacterial decomposition and oxidative rancidity even when refrigerated.

Insofar as oxidative rancidity is concerned antioxidants have been effective, both the fat-soluble type such as butylated hydroxy anisole (BHA), butylated hydroxy toluene (BHT), catechin, quercetin, and 2,6-dimethyoxyphenol (DMP), and the water-soluble or synergistic type, such as ascorbic acid, citric acid and phosphoric acid, either singly or in combination. However, it has been shown that such antioxidants have little or no effect in inhibiting the growth of those decomposing and putrefaction-producing bacteria that are mainly responsible for the short storage or shelf life of processed tissue; and it has been found that bactericidal agents while inhibiting the growth of putrefaction-producing bacteria often facilitate oxidative rancidity since they prevent the production of reducing substances formed by bacterial growth.

It was discovered that when a relatively small amount of sorbic acid, and/or a water-soluble sorbic acid salt is applied to the exposed surface of fresh animal tissue, together or in conjunction with an antioxidant, the growth of decomposing and putrefaction-producing bacteria is inhibited at refrigerator temperatures above freezing. Sorbic acid and its water-soluble salts, unlike certain bactericidal agents such as chlorine and hydrogen peroxide, do not interfere with the action of antioxidants used in preventing oxidative rancidity, or in any way adversely affect the appearance, color or palatability of the processed tissue. In other words, antioxidants and sorbic acid, its water-soluble salts and esters, when used in conjunction, are synergetic in that neither interferes with the effectiveness of the other and both cooperate to inhibit deterioration of mechanically processed fresh animal tissue.

In accordance with the process sorbic acid, and/or a water-soluble salt or ester, or a mixture thereof, together with an antioxidant, are applied to, mixed with, or otherwise incorporated in the processed tissue so as to come in contact with the exposed surfaces. The amount of sorbic acid compound may vary from a minimum of approximately 0.03%, based on the weight of the processed animal tissue under treatment, to a practical maximum which need not exceed 0.3%.

The preferred procedure for treating filleted fish comprises either applying to the exposed surfaces or immersing the cut tissue in an aqueous medium or solution containing from 0.1 to 5.0% of a sorbic acid compound and up to 0.2% of an antioxidant, which may be either a water- or fat-soluble type, for a period of from 2 minutes to 0.1 minute.

Where a synergistic or water-soluble antioxidant is to be used in conjunction with one or more fat-soluble types, an aqueous dispersion may be prepared by dissolving the antioxidants in an emulsifier of the sorbitan derivative type, or in propylene glycol, then adding the solution to water containing the sorbic acid compound and dispersing.

The amount of fat-soluble antioxidants, i.e., diphenols or compounds of similar electron configuration, may vary from a minimum of the order of 0.005% based on the weight of animal tissue, to the limit of common or practical usage which is of the order of 0.05%; and the same is true for the water-soluble or synergistic type antioxidants, although as a practical matter it might be desirable to use the latter type in somewhat higher concentrations, e.g., up to 0.2%, but when the fat-soluble and water-soluble types are used together, lesser amounts of each are generally employed.

In all cases the incorporation of the sorbic acid compound and antioxidant is not only effective to inhibit both the growth of decomposing and putrefaction-producing bacteria and oxidative rancidity, but also produces no adverse effect on the palatability and color of the tissue. As the result the shelf or storage life at refrigerator temperatures above freezing is considerably prolonged. The following example will illustrate the process.

Example: A fish fillet from one side of a haddock was washed and placed in storage at 36° to 40°F to serve as a control and the fillet from the other half of the haddock was immersed for 30 seconds in a dispersion comprising 0.25% of sodium sorbate, 0.25% of sorbic acid and 0.2% of ascorbic acid as the antioxidant.

Both control and treated samples were held at 36° to 40°F for a period of eight days. Bacterial counts and organoleptic observations were made after three, five and eight days of storage, and the results are indicated in the table on the following page.

	Time Held, days	Odor	Standard Plate Count, bacteria/gram, Average
Plain fillet	3	Fishy	3,800,000
Fillet treated with sorbic acid and sodium sorbate, antioxidant mixture	3	Bland	850,000
Plain fillet	5	Stale	15,000,000
Fillet treated with sorbic acid and sodium sorbate, antioxidant mixture	5	No off odor	2,400,000
Plain fillet	8	Very stale	57,000,000
Fillet treated with sorbic acid and sodium sorbate, antioxidant mixture	8	Slightly stale	16,000,000

2-Haloacetoacetic Acids

W.E. Rader and M. Legator; U.S. Patent 3,044,884; July 17, 1962; assigned to Shell Oil Company have successfully used the lower alkyl esters of 2-haloacetoacetic acids to effectively inhibit and prevent growth and development of microorganisms which cause deterioration of foods. These esters have the formula:

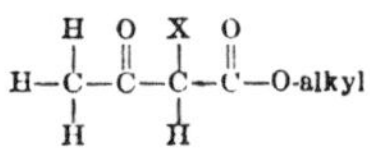

where alkyl represents the lower alkyl group and X represents halogen. Because they appear to be the most effective biocides, the esters of the kind where the alkyl group contains up to 4 carbons, and X represents either bromine or chlorine, are preferred.

These microbiocidal esters also possess the other requirements for a useful food preservation additive, that is, they are not toxic to the human or animal consuming the food at the concentrations required to preserve the food and they do not adversely affect the palatability of the food. Further, these esters are volatile, so that they will be removed completely from the food during cooking. Still further, they are believed to be converted in the metabolic processes of humans and animals to innocuous products, which are excreted with other wastes. These esters are essentially neutral compounds and do not modify significantly the pH of the food products treated with them.

These 2-haloacetoacetic acid esters are highly active against a broad spectrum of microorganisms, including bacteria, yeasts, fungi and actinomycetes, and retain this activity in the presence of organic materials such as foods.

These esters of haloacetoacetic acids are soluble to some extent in water, and can in many cases be most conveniently and effectively applied in the form of solutions in water. They readily disperse in water, so that a solution-dispersion of the esters also can be used. If desired, suitable edible emulsifying agents can be added. These esters also are soluble and/or readily dispersible in other materials, such as vegetable oils, fats and the like, and can be applied in the form of solutions or dispersions in these media. The esters also can be incorporated in gelatin, algae, pectins or the like, or used in wrapping materials.

The efficacy of the microbiocides of this process for preserving shrimp was demonstrated as follows. Fresh shrimp were placed in the bottoms of Petri dishes, the test chemicals were added to glass planchets placed next to the shrimp and the dishes were homogenized and examined for organoleptic changes. The results were as follows:

Test Compound	Concentration in parts per million by weight	pH of shrimp at end of test	Odor Rating	Reduction in number of bacterial colonies from number in nontreated control
Methyl 2-chloro-acetoacetate	400	7.5	Normal	Some reduction.
	600	7.5	do	Substantial reduction.
	800	7.5	do	Marked reduction.
None (control)		8.0	Foul	

No unfavorable odors were imparted to the shrimp by the chemicals.

Polyphosphate Dip to Prevent Thawing Drip

An extremely difficult problem in the fish industry has been the fact that frozen fish is subject to large moisture losses on thawing and again on cooking. The lost moisture, known as thawing drip, carries with it a large portion of the soluble protein, minerals, and other nutrients. The amount of moisture loss on thawing may be as high as 20% of the fish weight following prolonged frozen storage and as high as 14% after only 1 month of frozen storage.

After cooking a portion of frozen fish in a closed vessel, the housewife frequently finds a tough, fibrous, concentrated portion of fish surrounded by a pool of liquid containing soluble protein, minerals and vitamins. Where the fish are fried or baked, the exuded fluid may be evaporated, leaving the soluble nutrients as a dried film on the cooking vessel. Because of the loss of the fluid, the cooked fish is dry, fibrous, devoid of much of its natural tenderness and flavor, and, of course, less nutritious.

J.H. Mahon; U.S. Patent 3,036,923; May 29, 1962; assigned to Hagan Chemicals & Controls, Incorporated has described a process to minimize moisture and nutrient loss on thawing and cooking frozen fish.

It was found that the sodium and potassium molecularly dehydrated phosphates having a molar ratio of alkali metal oxide to P_2O_5 of about 1:1 to about 2:1 will significantly reduce thawing drip without any undesirable side effects. Throughout the description when molecularly dehydrated phosphates are referred to, both crystalline and glassy forms within the specified range of ratios of alkali metal oxide to P_2O_5 are included.

It is a common practice in the fish industry to dip scaled and filleted or skinned and filleted fish in a sodium chloride brine prior to freezing. It is thought that dipping the fish in a sodium chloride brine reduces the moisture loss on thawing, enhances the palatability of the fish, deters the growth of bacteria, and reduces the thawing drip following frozen storage. However, the primary reason for the brine dip or wash is to remove particles of scale and slimy material from the fillets. The process may have evolved from the custom of washing fish in ocean water.

Dipping fish in salt alone does not significantly reduce the thawing drip unless very high salt concentrations are used. High salt dips are not desirable because they cause rapid color and flavor changes including rancidity. The use of certain molecularly dehydrated phosphates alone or together with salt will result in a tastier, juicier fish with a more appetizing appearance. Moreover, fish treated according to this process have a significantly reduced thawing drip, reduced tendency to turn yellow and are less susceptible to the development of rancidity. No ammonia is produced upon cooking fish treated in this way.

The following table demonstrates the effectiveness of the process over long periods of frozen storage. The fish fillets used were air-expressed from Boston to Pittsburgh under continual refrigeration. After dipping in the specified solution for 2 minutes the fillets were drained for 15 seconds, sealed in polyethylene bags and held at 0°F. It is well to note that the 4% salt controls exhibited the greatest thawing drip in each case.

Fillets Frozen and Stored at 0°F

All Dip Solutions Contain 4% Salt	7 Days, percent	18 Days, percent	39 Days, percent	85 Days, percent
12.5% $Na_5P_3O_{10}$[1]	1.3	0.9	1.2	2.1
10% Na 60% P_2O_5 glass	1.4	2.1	1.5	2.0
20% Na 60% P_2O_5 glass	1.3	0.9	1.3	1.4
30% Na 60% P_2O_5 glass	1.2	0.9	1.6	1.3
Control—4% salt alone	7.8	8.1	9.4	8.5
30% $(NaPO_3)_6$	2.6	1.3	4.3	2.3
10% Na trimetaphosphate	3.1	3.4	4.2	9.0
15% Na trimetaphosphate [1]	3.5	2.0	3.3	5.7
10% Na tetrametaphosphate [1]	3.4	6.0	8.5	7.2
Control (4% salt)	5.8	8.2	11.6	9.5

[1] Effective saturation with 4% salt at 40° F.

Chelating Agents

In the past, freshly caught fish have generally been preserved for the market by storage in ice at about 31° to 33°F. If no steps other than icing were taken, fish stored in this manner spoiled after 3 or 4 days storage. After this time, the fish developed an unacceptable odor and flavor, which rendered them unfit for human consumption.

Various steps have been taken in an attempt to increase the storage time of fresh fish packed on ice. The most widely employed method until recently consisted in dipping fresh fish for about 20 seconds in an aqueous bath containing approximately 10 ppm of an antibiotic such as chlorotetracycline or oxytetracycline. This method usually resulted in increasing the storage life of the fish fillets to about 10 to 20 days when packed in ice.

Castell and other members of the Canadian Fisheries Research Board found that such chelating agents as ethylenediaminetetraacetic acid, its sodium salts, sodium citrate, and the like, increased the storage time of fish. For instance, about 250 ppm of ethylenediaminetetraacetic acid, when deposited upon fish, were found to increase the permissible storage time of the fish to approximately 6 days.

Another chelating agent, diethylenetriaminepentaacetic acid, when applied in similar amounts to fresh fish fillets, provides approximately 8 to 9 days protection. Combinations of chelating agents and antibiotics, were used in the processing of fresh fish into various food products obtainable from the fish by a digestive process. The preservative agent such as oxytetracycline or chlorotetracycline, was added to prevent bacteria growth, to control undesirable loss of nutritive value of the fish products made by breaking the fish down through an enzymatic digestive process. In this instance, the chelating agents are employed to maintain the preservative action of the known preservatives, oxytetracycline and chlorotetracycline.

Because of the potential undesirable consequences possible by the ingestion of uncontrolled quantities of antibiotics by humans, the FDA has banned the application of such antibiotics to fish.

T.E. Furia; U.S. Patent 3,563,770; February 16, 1971; assigned to Geigy Chemical Corporation has found that the storage life of fresh fish can be increased substantially over the time of storage previously possible by use of chelating agents alone, and without antibiotics. According to this process, the storage time of iced fresh fish is greatly enhanced by applying to the fish a composition comprising from about 500 to about 999 parts of (a) a compound of the formula:

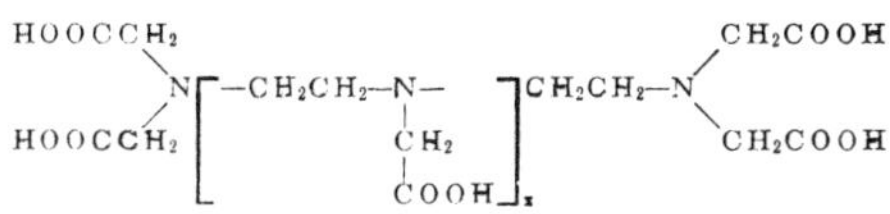

where x is an integer of from 0 to 6, or a water soluble, alkali metal, or ammonium, or amine salt thereof; and from about 1 to about 500 parts by weight of (b) a compound of the formula:

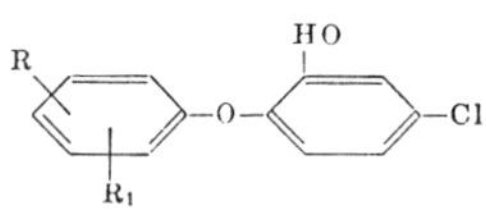

where R is a halogen atom, and R_1 is either chlorine or hydrogen. Particularly useful compounds falling within the scope of (a) are ethylenediaminetetraacetic acid and ethylenetriaminepentaacetic acid, and the alkali metal, ammonium, and amine salts thereof. In the compound of (b), R is preferably chlorine and R_1 is preferably chlorine.

Specifically preferred compounds within the scope of (b) are: 2-hydroxy-3',4,4'-trichlorodiphenyl ether, and 2-hydroxy-4,4'-dichlorodiphenyl ether, particularly the former. One of the advantages in employing the compounds of (b) in this composition is that they generally volatilize at relatively low temperatures and thus under most cooking conditions will be removed from the fish.

A preferred method for utilizing the process comprises dipping the fish for 20 seconds in an aqueous solution of about 0.10 part 2-hydroxy-3',4,4'-trichlorodiphenyl ether and about 0.90 part sodium salt of ethylenediaminetetraacetic acid.

The graph, Figure 1.1, demonstrates the greatly increased storage time imparted to fresh fish fillets treated according to this process. The left hand side of the graph shows the number of days that fish fillets are preserved when stored in ice at 31° to 33°F. At the bottom of the graph the various percentages of 4 representative components of the compositions of this process are indicated. In each instance, the total of solution components is 1.0%.

FIGURE 1.1: PRESERVATION OF FRESH HADDOCK FILLETS

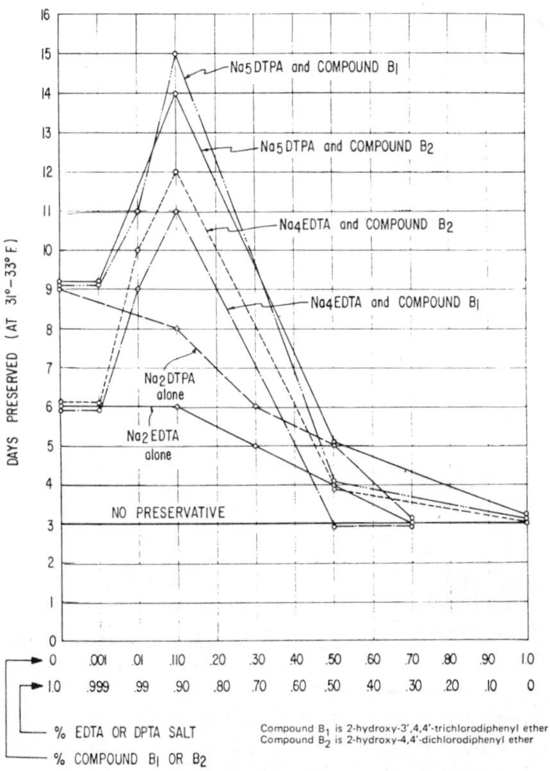

Source: T.E. Furia; U.S. Patent 3,563,770; February 16, 1971

Nitrous Oxide

It is known that preserved foods are subject to two kinds of deterioration: biological and chemical. Chemical alterations are due to the action of the oxygen of the air and to internal reactions of the constituents or substances of the food in question. These alterations take place unaffected by the sterilization of the substances, and unaffected also by the type of the wrapping within which the food is packed, and whether or not the wrapping is more or less hermetically closed.

Chemical changes occur in various forms, a few of which are harmful, while the others, although being harmless, give a disagreeable savor to the food, or modify its properties so as to render the food unacceptable for human consumption. The alterations of this kind occur particularly in those substances which include double bonds.

C. Balestra; U.S. Patent 3,335,014; August 8, 1967; assigned to Nitrox SA, Switzerland has developed a process for the preservation of foods by submerging them in a protecting atmosphere constituted totally or in part by nitrous oxide.

An important feature of the nitrous oxide is that its protective action against oxidation and chemical reactions of the foodstuffs is independent of temperature, and not at all dependent on the pressure. Foodstuffs can be preserved at room temperature, thus avoiding the cost of freezing or refrigeration. The fact that the action of the nitrous oxide is not dependent on any absolute pressure allows the utilization of any type of receptacle, from the smallest bag to the largest silo. The only requirement is that the receptacles be impervious to the passage of the nitrous oxide. Thus the container may be made not only of metal, but also of other laminated materials, plastic and/or derivatives, or paper impregnated with synthetic solutions.

Example: An experiment was carried out with sardines purchased from the trade. The contents of a commercial tin of sardines preserved in oil were divided into two portions. The sardines were separated from the oil by draining, and enclosed in two receptacles which were hermetically closed; one of them was saturated with nitrous oxide at 1.1 atm absolute pressure, after evacuation to 160 mm of Hg, while the other receptacle with the second portion of sardines was used for comparison. Both of the hermetically closed receptacles were subjected to simultaneous sterilization in a vacuum boiler.

After having left the receptacles for 120 days at a temperature varying between 25° and 28°C, they were opened and the following results were found. The nitrous oxide-treated sardines had maintained their orginal color, odor and savor. The sardines of the other receptacle had assumed a color tending to orange color, had a very heavy odor of cod-liver oil, and were not edible.

Ammonia

The process described by *F.W. Mitchell; U.S. Patent 3,442,661; May 6, 1969; assigned to W.R. Grace & Co.* relates to the preservation of fish by treatment with ammonia in amounts greater than 0.1% by weight of the fish.

Ammonia has already been used as a fish preservative. However, this process as practiced, requires that fish be first eviscerated, immersed in ammonia solution, and enclosed in an air-tight storage compartment under a concentrated ammonia atmosphere. Clearly, evisceration is a formidable deterrent to many small fishing vessels, especially when a catch contains many small fish, such as anchovies. For instance, a small fishing vessel is one whose capacity is 100 to 200 tons. At an average weight of 30 grams per anchovy, literally millions of fish are stored; if evisceration is required to preserve each fish, it is much cheaper to take the spoilage loss.

It is noted that it is known in the art that evisceration tends to prevent decomposition of the fish, even without preservatives. This is due to the removal of the stomach and gut enzymes having rapid digestive activity. The eviscerated fish has none of these enzymes present, and clearly will not decompose as fast as a fish which is not eviscerated. It is a matter of common knowledge that whole fish are observed to initiate rapid decomposition in the gut area. Besides the purely practical impossibility of requiring evisceration of a fish catch, as noted above, many end uses of the fish require the use of the whole fish.

It has been found, that, when using ammonia for a fish preservative, the weight of ammonia used in relation to the weight of the fish must be closely controlled. When there is 0.1 to 1% by weight ammonia used to treat the fish, no evisceration is required to keep the fish in good condition for a reasonable length of time. Preferably, about 0.3 to 1% by weight ammonia is used.

It must also be noted in connection with the preservation of fish by a basic agent that only ammonia can be used. For instance, when an equivalent amount of sodium hydroxide is used, the fish decays rapidly after 2 to 3 hours. This effect of a strong base is probably due to hydrolysis of the amide linkages in the amino acid constituents of fish protein.

When this process is followed, the fish are sprayed with the preferred amount of ammonia, and then stored normally in the ship's hold or fishbox without further treatment or special equipment. The following illustrates the procedure.

Example: The weight of freshly caught fish is estimated, and about 1% by weight of anhydrous ammonia is sprayed over the fish preferably in an enclosed space, i.e., below deck. The fish are then kept in the hold or fishbox, and are preserved for about 2 weeks or more. At the end of this time, the fish are in excellent condition: firm, yet not tough or gelatinized. The treated fish were not soft and needed no further treatment to restore firmness to the flesh. This shows particularly the criticality of the amount of ammonia used: when greater quantities of ammonia are used with whole (noneviscerated) fish, the fish become soft, and remain soft even after being washed.

After storage, the fish are removed and, for many purposes, such as the manufacture of fish meal, may immediately be used without further treatment. However, for fish of the finest quality, with no taste or residue of ammonia discernible, they may be washed with water or seawater. The thus-preserved fish can then be processed by ordinary means into fish meal, fish protein concentrate or other high-protein fish products.

Propionic and Benzoic Acids Buffered with Their Salts

The use of such organic acids as formic, acetic, propionic and benzoic acids and their alkaline salts has long been recommended for the preservation of fish during storage and distribution.

B. Gonthier and J. Mocotte; U.S. Patent 3,600,198; August 17, 1971; assigned to Progil, France have found that a synergistic activity of the biocidal properties of these materials can be obtained if a mixture of propionic and benzoic acids buffered with their corresponding alkali metal or magnesium salts is used.

The compositions are mixtures of a buffered mixture of propionic acid/metal propionate and of a buffered mixture of benzoic acid/metal benzoate the total pH of which is lower than 7 and preferably between 4 and 5.

The respective ratios of alkali metal or magnesium salt to the organic acid in the buffered systems depend upon the desired pH; to obtain a pH within the limits indicated, the molar ratio of salt/corresponding acid may vary from 0.1/1 to 100/1. The quantities of each buffered system in the mixture may be chosen within a large range, the volume ratios of propionic acid and propionate to benzoic acid and benzoate being preferably between 0.5/1 and 99/1.

Further, the efficient quantities of the biocidal mixtures employed to obtain the desired protection and to extend fish preservation time depend upon several factors, such as: fish type (from seawater or fresh water); time chosen for treatment (immediately after fishing or N days after it); application mode (incorporation in ice or fish washing water); duration, etc.

It may be said by way of example that when the described mixture is intended to be incorporated in the ice used for fish storing, it is advantageous to use 0.5 to 10 grams of mixture per liter of ice. When the bactericidal composition is used during fish treatment by soaking, aqueous soaking baths may contain 0.5 to 200 grams of active product per liter of water, the optimal quantity varying according to immersion time. For a very short time, for example, 2 to 10 minutes, an aqueous solution containing 50 to 200 g/l of product will be satisfactory. For much longer times, such as, for example, 10 to 20 days, active substance concentration will preferably range between 0.5 and 5 g/l.

These biocidal mixtures are excellent agents for the preservation of seawater and fresh water fishes. They have the advantage of not changing the aspect and taste of the aquatic vertebrates with which they have been contacted. Moreover, they are not injurious since the constituents composing them have been accepted by the European Economical Community as preserving agents for foodstuffs.

The method used for determining and measuring fish deterioration, in the following tests is the recommended one, known by the experts in this field, that is the analytical determination of nitrogen-containing bases produced by autolysis and fish bacterial degradation, after the stage of cadaveric stiffness. In this well-known method, the whole of the steam-volatile bases is determined under the form of total volatile basic nitrogen. In the following examples, the results are expressed under the shortened term TVBN (total volatile basic nitrogen), in mg of nitrogen per 100 grams of fish.

Example 1: A germicidal ice was made by freezing a water to which had been added 2 g/l of a mixture A having 95 parts (by volume) of buffered propionic acid/sodium propionate of pH 4.5 and 5 parts (by volume) of buffered benzoic acid/sodium benzoate of pH 4.5.

Simultaneously, for purposes of comparison, there were prepared 2 other germicidal ices from a 2 g/l aqueous solution, containing either the buffered mixture B propionic acid/sodium propionate of pH 4.5 or the buffered mixture C benzoic acid/sodium benzoate of pH 4.5.

Whitings eviscerated and stored for 4 to 6 days after being caught were divided into 4 lots which were stored as follows: the first lot in ordinary ice, free from bactericide, the second lot in ice containing the mixture A according to the process, the third and fourth lots in ice containing respectively product B and product C.

After 6 days of preservation, the fish lots were separated from their ice and measurements of total volatile basic nitrogen were made as described hereinabove. Fishes stored in ordinary ice (blank) presented a TVBN of 36 mg/100 grams of fish and the ones preserved in ices B and C gave respectively the numbers of 23 mg/100 grams and 32 mg/100 grams. However, fishes kept in the germicidal ice A presented a TVBN of only 18. Moreover, fish meat of this last lot was firm, and tasting proved that it had kept all its flavor, while the fishes of the reference group were in a decomposition state and the ones of groups B and C were consumable only with difficulty. During a continuation of the tests for a further month, the following TVBN numbers were obtained: blank, 70; ice B, 40; ice C, 67; ice A, 21.

Example 2: Pieces of whiting muscles caught 8 days before were immersed for 2 days in aqueous solutions containing 2 g/l of the following products at a temperature of 25°C. For group D the products were a buffered system of propionic acid-sodium propionate at pH 4.5 and for group E the products were propionic acid-sodium propionate plus benzoic acid-sodium benzoate with pH 4.5; the ratio by volume of the two constituents being 70/30. After 48 hours of immersion, the following results were obtained by TVBN measurement (in mg/100 grams of fish):

Blank (nontreated water)	430
Water treated with D	33
Water treated with E	27

Example 3: Whole whitings, caught 4 days previously, were immersed for 5 minutes at a temperature of 25°C, in aqueous solutions containing 50 g/l of the products D and E identical to Example 2. After immersion, fish were left for 48 hours and TVBN was measured. The results were as follows (in mg/100 grams of fish):

Blank (nontreated water)	131
Water treated with D	36
Water treated with E	27

Tert-Butyl Hydroperoxide

J.J. Cavallo and R.A. Reynolds; U.S. Patent 3,622,351; November 23, 1971; assigned to Atlantic Richfield Company have discovered that tert-butyl hydroperoxide (TBHP), having the chemical structure

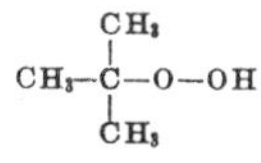

is an unusually effective bactericidal and/or bacteriostatic agent for the control of bacteria growing on fish.

A recent review of current chemical methods being used and investigated by the fishing industry to prevent the microbial degradation of fresh fish is presented by J.M. Shewan and G. Hobbs, *Progress in Industrial Microbiology,* 6, 169 to 208 (1967). Chemical agents such as boric acid, salicylic acid, potassium nitrate, formaldehyde, sodium hypochlorite, carbon dioxide, chlorine, hydrogen peroxide, acetic acid, hexamethylenetetramine, sulfur dioxide, benzoic acid, and sodium and potassium nitrites have been used in an attempt to control bacterial growth on fresh fish.

More recently, the antibiotics, chlortetracycline, oxytetracycline, and chloromycetin have been tested and found to be effective. Public health organizations, however, have not entirely endorsed the use of these antibiotics in the preservation of food in the fear that this practice may result in the production of pathogenic microorganisms resistant to these compounds.

There are two basic problems involved in using the aforementioned types of biocides: cost-effectiveness and stability. Inhibitory concentrations of these compounds usually range between 1 ppm to 100 ppm. Even at these low concentrations the use of some of the more expensive compounds becomes prohibitively costly. Tert-butyl hydroperoxide, however, is less expensive and its effective concentration range extends from about 5 ppm to 50 ppm.

In order for a compound to maintain long term biocidal effectiveness, the integrity, of the chemical structure, or that portion of the chemical structure that is antimicrobial, must be retained. Biocides such as quaternary ammonium compounds lose their antimicrobial activity in the presence of certain ions. Benzoic and salicylic acid are relatively stable, chemically, but many microorganisms are capable of enzymatically degrading them. Most of the other compounds listed above have a deleterious effect upon the quality of the fish. None of these compounds, therefore, are used to any great extent for fish preservation.

Tert-butyl hydroperoxide has been shown by Milas and Harris, *Journal of the American Chemical Society,* 60, (8) 2434, (1938), to be very stable under ordinary conditions and in the presence of enzymes capable of degrading hydrogen peroxide and alkyl hydroperoxides. These workers were also able to show that tert-butyl hydroperoxide retained its peroxidic activity after being exposed to palladium black catalyst and 10% sodium hydroperoxide solution for several months. Other investigators have shown that tert-butyl hydroperoxide retains 50% of its activity after exposure to 100°C for 165 hours in dilute benzene solutions.

Aqueous solutions of this compound may be sprayed on both the fresh fish and fish fillets. The fish may also be dipped into appropriate aqueous solutions of the compound. The most effective way of utilizing this compound for fresh fish preservation is believed to be by incorporating the compound into ice used for shipboard storage.

TBHP in concentrations of from 5 to 50 ppm can economically and effectively be used in preventing the growth of bacteria associated with low-temperature (4°C) spoilage of fish. The preferred range is from 20 to 30 ppm dissolved in water. The following examples describe the use of TBHP to prevent growth of bacteria responsible for the spoilage of fish.

Example: A fillet of halibut was aseptically divided into three approximately equal parts (30 to 40 grams). One part served as a control and was placed unaltered into a petri dish and stored at 4°C. The remaining two parts were dipped in a 20 ppm aqueous solution of TBHP, for 1 minute and 5 minutes, respectively. These samples were also placed in petri dishes and stored at 4°C. Viable cell counts were made on these samples at zero time, 4 days, and 12 days. The procedure used for obtaining the viable counts was as follows:

(1) A portion of each sample was removed aseptically, weighed (approximately 1.0 gram) and placed in a sterile mortar and pestle.

(2) The fish muscle was then ground into very fine particles.

(3) 5 ml of sterile saline solution (0.85%) was added and the mixture blended thoroughly.

(4) The resultant solution was poured into a centrifuge tube and centrifuged for 15 minutes at 1,500 rpm.

(5) The supernatant was transferred to a sterile graduate and diluted with additional sterile saline solution to a volume of 10 ml.

(6) Serial 1:10 dilutions were made and plated on Tryptone Glucose Extract Agar.

(7) After 2 days incubation at room temperature, the plates were counted.

The data clearly showed that tert-butyl hydroperoxide is a good chemical agent for controlling bacterial growth associated with fish spoilage. Further, organoleptic examination of the fillet samples as to odor, color and texture left little doubt that the tert-butyl hydroperoxide had effectively controlled the growth of the active bacterial spoilers. For example, while the control had a very strong and obnoxious rotten fish odor, the TBHP-treated samples retained the fresh fish odor originally present. The texture of the control was poor, the flesh being quite soft. The 1-minute-dipped sample still had its firm original texture. The 5-minute-dipped sample was soft, indicating that the dip time was too long.

POLYPEPTIDE HADDOCK FRACTION FOR PREVENTING BOTULISM

For some time, the outgrowth of spore-forming bacteria has been prevented in canned foods by heating the food in a sealed can at a temperature of about 250°F for varying periods of time to kill any *Clostridium botulinum* and other spore-forming bacteria present in the food. If these bacteria are not killed, and the food is ingested, botulism will result with its high incidence of death and if other bacteria are not killed the food spoils. While heating for different periods of time is an effective method for killing these bacteria, the process is undesirable in that heating causes the food to lose flavor and to become partially deteriorated.

In the course of work performed for the Atomic Energy Commission, *J.T.R. Nickerson and J.M. Zak; U.S. Patent 3,765,908; October 16, 1973; assigned to Massachusetts Institute of Technology* discovered that a polypeptide fraction isolated from fresh haddock prevents the outgrowth of *Clostridium botulinum* type E. The polypeptide fraction is obtained by blending the meat of fresh haddock with distilled water and recovering the polypeptide fraction by extraction and separating the extract on the basis of molecular weight. The polypeptide fraction that retards the growth of *Clostridium botulinum* type E has a molecular weight between about 1,100 and 1,400.

This fraction, when added to foods not containing proteolytic enzymes that break down the polypeptides, reduces the heat needed to kill the *Clostridium botulinum* type E present. Sufficient heating to effect the desired results can be from about 1.5 to 30 minutes at a temperature between about 200° and 270°F. Unpackaged food such as fresh fish is treated under milder conditions than packaged foods to retain food flavor without food deterioration. Thus, unpackaged food is treated at a temperature between about 200° to 220°F for short periods while packaged foods can be treated at any temperature within the broad range.

The polypeptides are isolated by first isolating a protein fraction of haddock extract obtained from haddock meat which is either fresh or has been maintained at low temperatures to prevent degradation of the polypeptides by the proteolytic enzymes present in the haddock. The haddock extract is obtained by comminuting the haddock meat in the presence of distilled water maintained at a relatively low temperature below about 5°C. The water and comminuted haddock are mixed until homogenization is attained. The resulting homogenate then is separated into a liquid fraction and a solids fraction such as by centrifugation, with the desired fish extract comprising a liquid fraction. The liquid fraction either can be separated to obtain the polypeptides as described below or can be stored at low temperatures about -18°C until used in further processing.

The protein fraction is isolated from the fish extract by any desirable means. A particularly useful means is described by Alden in *Methods in Enzymology*, Volume 1, (1955) in an article entitled "Protein Fractionation on the Basis of Solubility in Aqueous Solutions of Salts and Organic Solvents."

In this method, a specificed concentration of ammonium sulfate aqueous solution is added to the extract and intimately mixed therewith at a relatively low temperature. A suspension is obtained which then is centrifuged to recover a precipitate. The precipitate can be further purified by dialysis against distilled water and the dialyzed protein fraction can be concentrated such as by freeze-drying.

The protein fraction is separated into its components on the basis of molecular weight by any desirable means such as gel filtration carried out on Sephadex columns according to the methods described by Determan in *Gel Chromatog-*

raphy (New York: Springverlag, 1968). Separation of the protein fraction conveniently is separated on a G-25 Sephadex column.

When employing gel filtration, the protein fraction is dissolved in a buffer to form a protein solution and then is passed through the gel column. A suitable solvent for use in fractionation comprises an 0.01 molar phosphate buffer of pH 6.8 to 7.0 which can be prepared from a mixture of sodium dibasic phosphate and potassium monobasic phosphate.

The polypeptide fraction comprises a mixture of polypeptide having a molecular weight within the range of about 1,100 to about 1,400 which is obtained for the low-molecular-weight protein fraction of haddock. The fraction comprises predominately polypeptide having a molecular weight of about 1,200 and about 1,350.

The isolated polypeptide fraction is added to the food to be processed in amounts of at least about 500 mg/kg of food, preferably between about 1,000 mg and 10,000 mg/kg of food. Obviously, more polypeptide can be employed but no significant improvements are obtained with amounts added above the stated range. The food to which the polypeptide fraction is added should contain no proteolytic enzymes. Otherwise, the polypeptide would be broken down in the food and would be rendered useless for its intended purpose.

Suitable foods to which the fraction can be added include vegetables, soups, fish (preheated), or meat (preheated). The fish and meat are preheated to eliminate the proteolytic enzymes therein. In a preferred embodiment the food to be processed and the polypeptide fraction are mixed and the mixture is sealed in a container such as a glass jar or metal can. The sealed container containing the mixture is heated, usually with steam, to the desired temperature for the desired period and then cooled. In this manner, food deterioration and loss of flavor are minimized.

APPARATUS FOR COATING SEAFOOD

The practice of treating fish fillets and other seafood products with liquid treating agents is well-known. For example, fish fillets and other seafood are treated with solutions of molecularly dehydrated phosphates prior to freezing to reduce the loss of soluble proteins as freezer drip on thawing. However, the actual mechanics of treatment have been relatively crude and wasteful. Generally the seafood has been soaked in or otherwise treated with a solution of treating agent either while stationary or on a conveyor. This results in large losses of treating chemicals and often times less than satisfactory treatment. Similarly, seafood is frequently treated with a brine solution during processing for market.

M.V. Satz, L.A. Edwards and T. Livsey; U.S. Patent 3,806,612; April 23, 1974; assigned to Calgon Corporation have developed a seafood processing machine which eliminates the problems of treating agent waste and of incomplete or unsatisfactory treatment. This machine automatically treats seafood with a liquid treating agent with the minimum of losses and with the maximum of efficiency.

Preferably a reservoir of treating agent, a rotatable drum mounted in a framework, means for rotating the drum, a plurality of spaced flexure members on the interior of the drum, a perforate infeed chute, such as a wire-rod chute, at one end of the drum receiving product to be treated and discharged into the interior of the drum, means for spraying treating agent at high pressure on opposite sides of the product on the infeed chute and an outfeed chute at the other end of the drum receiving the treated product from the drum and delivering it to a means for removing the product from the treating apparatus are provided.

Preferably the drum is provided with flexure members in the form of longitudinal baffles fixed to the interior of the drum. These baffles or any part of them, or the drum itself, may be inclined so as to assist in moving the product axially through the drum. Wire scoops are preferably provided adjacent the discharge end of the drum for collecting product and discharging it to the outfeed chute.

Figure 1.2a is an isometric view of an apparatus according to this process; Figure 1.2b is a generally schematic illustration of the treating agent spray system of Figure 1.2a; Figure 1.2c is a generally schematic illustration of the operational steps carried out in the machine of the process; Figure 1.2d is an end elevation, partially cut away of the drum, of the apparatus of Figure 1.2a; and Figure 1.2e is a section on the line **V–V** of Figure 1.2d.

Referring to the drawings the following are illustrated: A frame and housing **10** carrying a cylindrical drum **11** having an input end **11a** and a discharge end **11b** mounted in the frame for rotation about a generally horizontal axis. The inclination of the axis of the drum can be varied by inclination screw **12** which raises and lowers yoke **13** carrying support rollers **14** at the discharge end of the drum. The drum is provided on its inner periphery with spaced longitudinal baffles **15** extending from one end to the other and with discharge scoops **15a** adjacent the discharge end.

An infeed chute **16** is provided at the input end of the drum. The chute is preferably formed of spaced steel rods or wires to form a perforate member. Fish fillets **17** or other seafood products to be treated are discharged from a feed conveyor **18** onto the infeed chute. Feeler members **19** connected to a switch **20** extend above the surface of the infeed chute adjacent the entry end to sense the presence of product to be treated.

Treating solution is prepared in a tank **21** equipped with an agitator **21a** and pumped by pump **22** through solenoid valve **23** to supply tank **24** equipped with a level sensing device (not shown) which operates the solenoid valve to maintain a supply of treating agent. A pump **25** receives treating agent from the supply tank and delivers it through solenoid valve **26**, and lines **27** having pressure gauges **28** and manual valves **27a** to bifurcated mist sprayers **29**, which have one arm above and one below the perforate infeed chute so as to spray both sides of product entering on the chute.

A perforate outfeed chute **30**, preferably of the same form as the infeed chute, extends into the discharge end of the drum and receives the treated product from the discharge scoops. The outfeed chute carries the product to an outfeed conveyor **31** or other mechanism for removing the treated product to a packing area.

FIGURE 1.2: MACHINE FOR APPLYING TREATMENT TO SEAFOODS

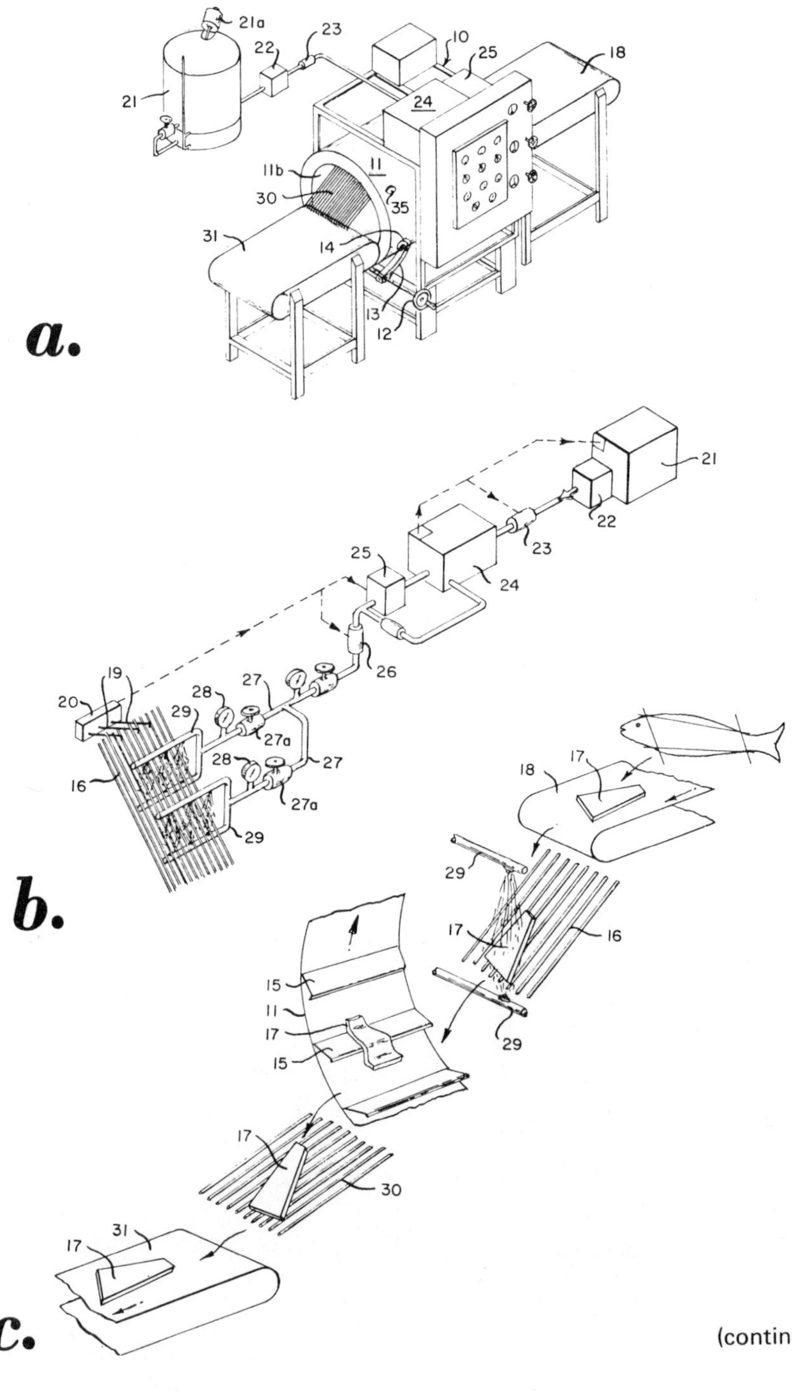

(continued)

FIGURE 1.2: (continued)

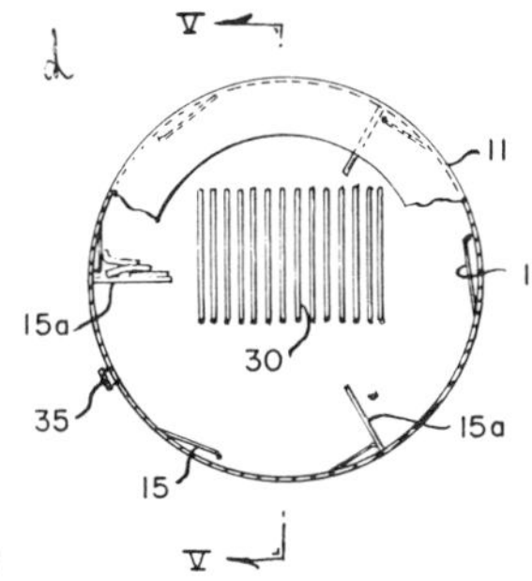

d.

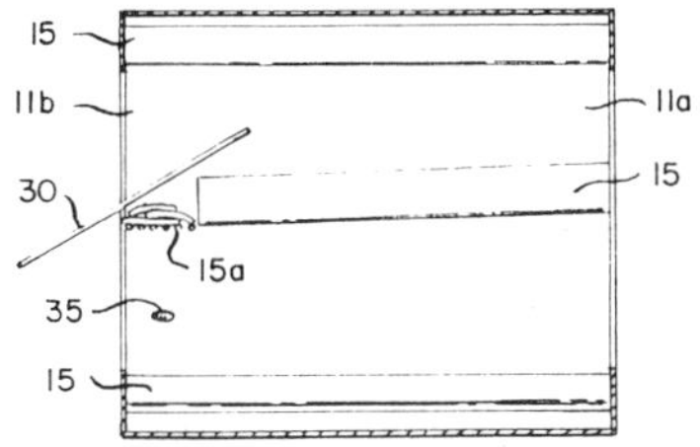

e.

Source: M.V. Satz, L.A. Edwards and T. Livsey; U.S. Patent 3,806,612; April 23, 1974

The operation of the machine is as follows. The product, such as fish fillets, is delivered on the conveyor to the infeed chute. When the product falls onto the chute it contacts sensing members which act on switch **20** opening valve **26** and actuating pump **25** which feeds the treating agent through mist sprayers against both sides of the product. The amount and rate of feed may be adjusted by valves so that the relative amounts of spray at each of the mist sprayers may be regulated. The product with any excess treating agent is then discharged from chute **16** into the interior of the drum which is rotating and which acts with slightly pitched baffles to move the product through the drum.

The product moves over the baffles which causes it to flex allowing deep penetration of the treating agent into the product as well as pickup of additional treating agent as may be needed from the excess which enters with the product and which acts as a surface lubricant for the interior of the drum. The product moves through the drum to the discharge scoops which collect the product and discharge it onto outfeed chute from which it goes to conveyor **31**. An excess solution at the discharge end of the drum is periodically discharged through opening **35** to a drain.

EDIBLE COATINGS

Proteinaceous Gel

H.J. Hestnes; U.S. Patent 2,933,398; April 19, 1960 has developed a process for preserving fish and fish fillets by impregnating and coating the fish fillets with a liquefaction product of a similar fish which is treated in such a manner that it is capable of forming a gel at temperatures of about 40°F. This proteinaceous liquid contains inherent gelling agents and, under certain conditions, other gelling agents prepared from protein or starch may be added. Prior to the formation of the gel, the concentration of the liquefaction product is increased and such concentration plus the gelling action has the effect of reducing the vapor pressure, i.e., increasing the surface tension of the preservative to the point where the surface energy exceeds the electrokinetic energy at a temperature of about 50°F.

As the temperature of the gel is progressively reduced below 40°F, the gelled mass contracts and exerts a squeezing action as in syneresis, and it fills the interstices in the flesh and forms a protective coherent envelope. In the gel the water is held as bound water, which cannot freeze at normal freezing temperatures, so the gel acts as an antifreeze, and even when the temperature is reduced below the freezing point of water no ice is formed within the mass of the gel-impregnated and coated flesh. The gel preservative eliminates air from contact with the fish thus inhibiting aerobic bacteria and the reduction of electrokinetic energy in the gel suppresses or retards activity of the catabolic enzymes which otherwise prepare the flesh for bacterial action.

The process may be described as follows. The edible waste of fish from the filleting table, including the heads and carcasses, are mechanically disintegrated, as by coarse crushing, and then hydrolyzed by applying heat and stirring. The temperature is maintained below the boiling point of the liquid. Hydrolytic action is accelerated by adding a protease, such as pepsin, papain, ficin or bromelin, and the pH is reduced by the addition of acetic or ascorbic acid, or the like. The process is preferably carried out in the presence of an edible salt such as sodium chloride, which accelerates the solution of some proteins in water.

The liquefaction process extracts and renders into what may broadly be called a solution substantially all the proteins and protein derivatives existing in the body fluids and in the cell plasma including the gel-forming proteins such as elastin, collagen, reticulin and myosin and also all the lipoids (fats) and other fluids including carbohydrates and the other easily hydrolyzed nonstructural proteins. The structural proteins (dead cell matter, bones, etc.) are not included in the liquefaction product. Once the proteins are broken down to their derivatives, the solution will not coagulate at boiling temperature and it may be thereafter raised to a boiling temperature to sterilize it.

After sterilization, the liquid is separated from the solids by suitable means and the liquid is drained off to form the preservative, leaving the solids in the form of a coherent cake which may be used as a by-product, as for instance, for animal feed or for fertilizer. The stabilization, as well as other important functions, may be accomplished by the introduction of a fermentable carbohydrate in the liquid, such as milk powder, which contains lactose, which may be added to the liquefaction product either before or after completion of the liquefaction process. Glucose, sucrose or any monosaccharide or polysaccharide may also be used.

The concentration of the fluid may be further increased by dehydration after the liquefaction process. It is desirable that by the addition of the carbohydrates or the dehydration or both, the vapor pressure of the liquid be reduced to about 75% of normal or lower. When a fish fillet is dipped into the solution at a temperature preferably above about 50°F, the fillet picks up a relatively large amount of the liquid depending upon the surface area exposed to the solution. By impregnation and surface coating the amount of solution picked up may vary from 5 to 20% of the weight of the fillet. Now when the temperature is reduced below 40°F the solution which impregnates and coats the flesh begins to form a gel and it is completely gelatinized and solidified at a temperature about 40°F.

The keeping qualities of the flesh dipped in the solution will depend upon the quality of the flesh, the amount of solution picked up, and the temperature at which the gelatinized flesh is kept. A reasonably fresh fish fillet with the skin on and which has picked up about 8% solution will keep for about one month's time in a cooler where the temperature varies between 33° and 40°F. At a temperature of about 25°F this fillet may be stored for about 3 months without noticeable reduction in freshness. A skinned fillet will pick up more solution, for instance, 20% by weight of the fillet and can stand a temperature down to about 20°F without ice formation. Such a fillet will remain fresh when stored for over 5 months.

When the fillets treated according to this process are deep frozen, for instance, at 10°F below zero, the fillets will have considerable advantage over ordinary deep frozen fillets because of the protective nature of the gel coating and impregnation. Particularly under optimum temperature conditions of storage between 32° and 40°F the flesh actually gains in freshness, richness of taste and nutritive value because of the greater concentration of the solution with which it is impregnated and coated.

Dispersion of Starch and Algin Gum

Large quantities of frozen seafood, such as shrimp, scallops, mackerel and red snapper, are marketed annually for human consumption. Frozen seafood, especially frozen shrimp, is regarded as decidedly inferior to the fresh product. When stored for periods of even a few weeks, frozen shrimp lose the characteristic pink and white color of the fresh product and, in addition, become corky and tough. It is believed this deterioration is caused principally by dehydration and by oxidation.

R.D. Earle and C.E. Snyder; U.S. Patent 3,255,021; June 7, 1966; assigned to A.E. Staley Manufacturing Company have developed an edible coating for frozen seafood which will minimize its deterioration during storage. The process consists of first inactivating the enzymes and bacteria in the seafood to be coated and frozen, and then forming a coating on the seafood by immersing it in a flowable aqueous dispersion containing an edible starch and a water-soluble algin gum. This coating is then firmly bound to the seafood by applying a water-soluble calcium-containing gelling agent or a water-soluble aluminum gelling agent to the coated seafood. The treated seafood can then be frozen or first breaded and then frozen.

Microorganisms and enzymes in the seafood can be inactivated by any suitable means which will not adversely affect the taste, color, texture or general quality of the seafood, such as heating the seafood in water at temperatures of from about 65° to about 100°C until inactivation is complete. At 70°C, inactivation is normally completed in about 30 minutes. Flavoring agents such as monosodium glutamate and/or corn syrup can be added to the water in which the shrimp are heated.

Another suitable means for inactivating microorganisms and enzymes in seafood involves soaking the seafood in an aqueous solution containing small amounts of available chlorine on the order of 25 to 100 ppm, obtained by bubbling chlorine gas into water or by dissolving calcium hypochlorite in water. Still another method of inactivating microorganisms and enzymes in seafood is by soaking the seafood in a weak solution of acetic acid.

After the inactivation step, the seafood is immersed in a flowable aqueous dispersion of starch and algin gum for a period of time sufficient to form a coating on the seafood. The mixture of starch and gum utilized should contain on a dry substance basis from about 98 to about 85% by weight edible starch and from about 2 to about 15% by weight algin gum, the weights based on the weight of the mixture. While the amount of starch-gum mixture in the dispersion can vary, it is most satisfactory to employ dispersions containing on a dry substance basis from about 10 to about 40% by wieght of the mixture based on the weight of the dispersion.

Any edible starch can be utilized in the process which will form a flowable aqueous dispersion with a water-soluble algin gum when combined in the given proportions. For example, edible native starches, edible hydrolyzed starches, edible oxidized starches, etc. may be utilized. Edible white dextrins are preferred principally because they form a light, clear film. Also since aqueous dispersions containing large amounts of white dextrins are relatively low in viscosity, such dispersions permit formation of a strong nongummy coating without excessive buildup of the coating in the grooves of the seafood.

Any water-soluble algin gum can be satisfactorily utilized in the process. Such gums include sodium alginate, potassium alginate, ammonium alginate, alginic acid, etc., and the like, i.e., salts of alginic acid and anions which form water-soluble salts. Principally because of its ready availability and low cost, sodium alginate is preferred.

In order to produce a coating which best protects seafood during long periods of storage, it is desirable to include in the aqueous coating dispersions an edible oil in amounts up to about 40% by weight based on the weight of the starch-gum mixture. Generally, it is preferable to utilize from about 5% to about 20% edible oil. Inclusion of an edible oil is desirable when the seafood is to be breaded for deep-frying since the oil-containing coating is especially resistant to breakage even when subjected to the high temperature of deep-frying. Edible oils which can be satisfactorily utilized include corn oil, safflower oil, cottonseed oil, soybean oil, etc.

After the coating step of this process, the treated seafood is immersed in an aqueous gelling solution containing a water-soluble source of calcium ions such as calcium chloride, calcium citrate, calcium acetate, calcium gluconate, etc. The seafood must remain in the gelling solution for a period of time sufficient to firmly gel the coating. For example, firm gel can be formed using a 0.2 molar solution of calcium chloride and a holding time as low as 1 minute.

After the gelling step, the treated seafood is preferably washed with water or with a calcium solution containing a much lower concentration of calcium than the original gelling solution. This wash is intended to remove any excess calcium. After washing, the seafood can be frozen and packaged. If desired, the seafood can be first breaded and then frozen. Any conventional breading mixture which is adaptable for use on frozen seafood can be utilized in conjunction with the process.

Example: 375 grams of a mixture containing 38 grams of sodium alginate and

337 grams of white dextrin were slowly added with accompanying agitation to a coating vessel containing 2,125 grams of distilled water. The resulting mixture was heated to 75°C and maintained at that temperature for 2 hours with continued agitation to obtain a uniform aqueous dispersion. 2 pounds of deveined, fresh shrimp were placed in an aqueous solution formed by dissolving 304 grams of monosodium glutamate in 2,183 grams of distilled water, the solution having a temperature of 70°C. The shrimp were held in the solution for 30 minutes during which time the solution temperature was maintained at 70°C. The shrimp were then totally immersed in the dextrin-alginate dispersion in the coating vessel.

After immersion in the contents of the coating vessel, the shrimp were placed in an aqueous solution prepared by dissolving 55 grams of anhydrous calcium chloride in 2,445 grams of distilled water. The shrimp were held in the calcium chloride solution for a period of 2.5minutes and then were removed and rinsed for 10 seconds in an aqueous solution consisting of 27.5 grams of anhydrous calcium chloride and 2,445 grams of distilled water. The shrimp were then drained and immediately frozen. The shrimp were stored in the frozen state for 3 months and then thawed. The thawed shrimp were observed to have retained their original flavor, texture and color.

IRRADIATION

Pasteurization of Packaged Frozen Fish

Early workers in the field of irradiation processing of foodstuffs subjected perishable foodstuffs to massive dosages of ionizing irradiation for the purpose of complete sterilization of the foodstuffs. Attainment of complete sterilization would permit indefinite prolongation of the storage capabilities of nonrefrigerated foodstuffs. Unfortunately, the massive dosages of ionizing radiation required to bring about complete sterilization cause undesirable alterations in the texture, color, and flavor of the treated foodstuff.

For these and other reasons, recent trends in this field have been away from sterilization and towards that of radiation pasteurization. In this connection, pasteurization means irradiation at a dosage sufficiently below that required for sterilization to avoid detectable alterations of texture, color, and flavor, but enough to prolong the shelf-life of perishable foodstuffs for a limited period of time in the presence of refrigeration. Irradiation pasteurization of fish, for example, would make this highly perishable food product available in substantially fresh form to regions presently beyond the areas of normal consumption.

As might be expected, irradiation dosages of foodstuffs for the purposes described above must be kept between limitations of sufficient irradiation to accomplish the pasteurization and excessive exposure to avoid the undesirable side effects. In addition, close control is required to assure a uniformity of dosage throughout each food package and successive units undergoing the irradiation. Furthermore, due to the expense of the sources required, it is necessary to obtain the most efficient use of a particular source in order to assure that the expense involved in irradiation pasteurization will not become prohibitive.

In the latter connection, proposed arrangements generally suffer the disability that they are very inefficient and occupy excessive floor area. Inefficient utilization of available irradiation energy when coupled with the need for large floor areas tends to yield costs which are high and thus inhibit widespread acceptance of irradiation pasteurization. Further, irradiation energy which is not attenuated and absorbed in the package foodstuffs, must be attenuated and absorbed in a shielding material surrounding the irradiation device, which adds additional costs to an installation for this purpose. Therefore, an inefficient arrangement of packages in relation to the source means greater capital and related expenses, including that due to the requirement of more sources which could perhaps be eliminated by the efficient utilization of few sources.

O.A. Kuhl and A.B. Oltmann; U.S. Patent 3,192,054; June 29, 1965; assigned to United States Atomic Energy Commission describe an irradiation pasteurization facility for packaged frozen fish. Figure 1.3a is a diagrammatic arrangement of fish packaging and freezing facility and the modification which would be required to introduce irradiation pasteurization; Figure 1.3b is a schematic layout of the irradiation facility, including the conveyor arrangement. Figure 1.3c is an isometric view of the facility shown in Figure 1.3b.

In Figure 1.3a a typical commercial arrangement for packaging and freezing fish for retail distribution is shown. In this arrangement eviscerated fish is scaled, filleted, trimmed and packaged. Then the packaged fish is frozen and stored or distributed in its frozen state. In order to incorporate this process, the only modification of the arrangement in Figure 1.3a required is as indicated in broken lines. Instead of the freezing and storage steps, the fish packages are irradiated and refrigerated for storage and distribution.

Referring to Figure 1.3b, an irradiation room **10** is formed by walls **12** of thick radiation shielding material such as high-density concrete and containing an elevator shaft **14** extending down from the floor. In addition, the floor of the room is provided with a pair of vertical downwardly extending tunnel shafts **16** and **18** which connect to a pair of substantially horizontal tunnels **22** and **24**, respectively. The latter extend out of the room and terminate in a pair of vertical shafts **26** and **28**, respectively, which open into a floor area used for feeding in and withdrawing the food packages **30**.

A continuous line or path **32** with arrow heads indicates the path of the packages through the various shafts and tunnels into the room for irradiation and then return. The packages are carried along their path by a conventional conveyor. Within the elevator shaft, an elevator **34** carries irradiation sources, plaques, or slabs **36**. The whole arrangement just described is located in a building **38** outlined in phantom.

Referring to Figure 1.3c for somewhat schematized details of the irradiation system there are shown the room in phantom, the source elevator shaft, and the conveyor passageways. The path of the packages is along line **32** which is a schematic representation of a monorail which may be used as described below from the load-unload area **A**, down through shaft **26**, tunnel **22**, up shaft **16**, and over elevator shaft **14**, where irradiation takes place. The packages **30** are returned to area **A** through shaft **18**, tunnel **24**, and shaft **28**.

Within the elevator shaft, the elevator supports the three irradiation plaques in spaced, upright positions as illustrated. The elevator is normally movable as shown by arrows **B** between the positions illustrated. In the upper position of the elevator, the plaques are located between adjacent packages for irradiating the latter.

FIGURE 1.3: FOOD IRRADIATOR AND METHOD

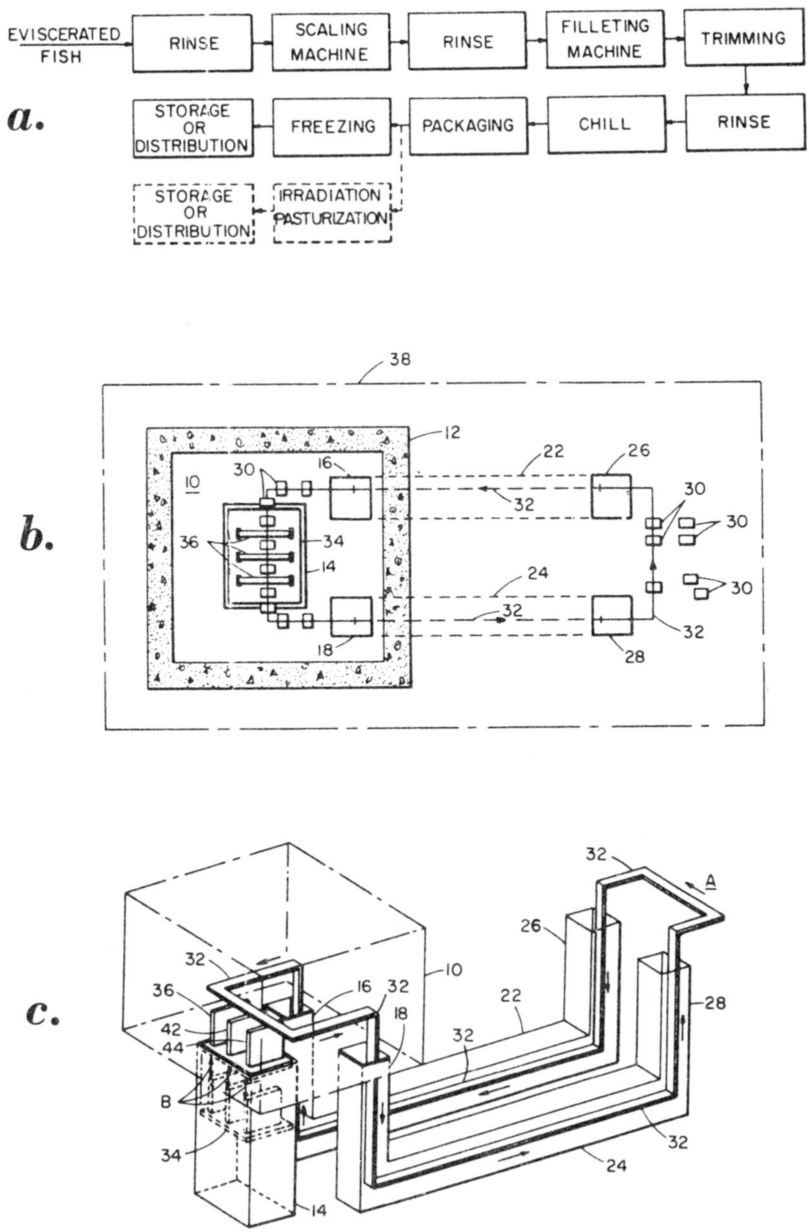

Source: O.A. Kuhl and A.B. Oltmann; U.S. Patent 3,192,054; June 29, 1965

When in the lower position, (as shown in phantom) the plaques are retracted to permit movement of the packages across the elevator shaft. The bottom of the elevator shaft is a storage pit filled with water into which the elevator is retracted when the irradiation cell is not being used. The bottom of the elevator shaft would contain sufficient water to cover the plaques.

Each of the plaques is a gamma radiation source which consists of cobalt 60 sandwiched between stainless steel sheets forming a cladding, as is understood in the art. Each source plaque is composed of a number of individual elements. For example, with an overall plaque size of 46" x 38", each source element is approximately 6.85" wide and 6.25" long. The cobalt metal when irradiated in the reactor would be nickel-plated and encapsulated in aluminum for activation. The individual cobalt 60 source elements then dejacketed from the aluminum irradiation containers are inserted into stainless steel envelopes, 3 to an envelope.

The open end of the envelope is remotely welded and the assembly tested for leaks. Four of these subassemblies, 0.85" wide and 19" long, are in turn inserted into a larger stainless steel envelope, the open end of which is crimped, welded, and tested. A total of 24 of these assemblies, each 1.90" wide and 38" long (exclusive of 1" extensions at each end which are used to mount the assembly in a rack supported by the elevator) make up each source plaque, having dimensions of 46" x 38".

The food packages are conveyed suspended from an overhead conveyor system along path **32** directly over the source elevator, where they are subject in turn to a carefully controlled dosage of gamma irradiation to effect the pasteurization of the contents.

It should be understood that this irradiation procedure can be carried out with one or two irradiation plaques and that the packages can be indexed two or even three or even more at a time when the elevator is down, in accordance with the amount of irradiation to be applied for the particular circumstance. It should also be noted that when the elevator is up none of the irradiation is wasted as all of it is being absorbed, not only by the directly facing packages but the next ones in line, by radiation energy passing through the packages.

Filleted Fish Placed in a Suspension Medium Before Irradiation

A great need, in the art of irradiation pasteurization, is for systems which occupy small floor areas and yet at the same time exhibit a high degree of radiation efficiency with a uniformity of dosage. A previous attempt to achieve these results is described in U.S. Patent 3,192,054. That patent describes a process which improved the state of the art by providing a method of controlled irradiation wherein radioactive source plates are moved up and down between individual food packages arranged in a continuously moving row.

Such a system offers reasonably good dose control operating at a maximum/minimum dose ratio of 1.3. Relative to prior systems, the parallel source plate arrangement also has a high production capacity as about 20 to 25% of the source energy is absorbed by the product. However, for irradiation pasteurization to become competitive with other methods of food processing it is necessary to design systems having even greater efficiency than heretofore obtainable.

The process developed by *J.B. Huff; U.S. Patent 3,434,850; March 25, 1969; assigned to the U.S. Secretary of the Interior* provides an increase in source utilization and uniformity of dosage while taking up much less floor space than conventional irradiators. This system is particularly applicable to the irradiation pasteurization of perishable foods such as fish. The apparatus will be described in more detail by a consideration of Figures 1.4a and 1.4b.

FIGURE 1.4: PROCESS FOR PRESERVING FISH BY IRRADIATION

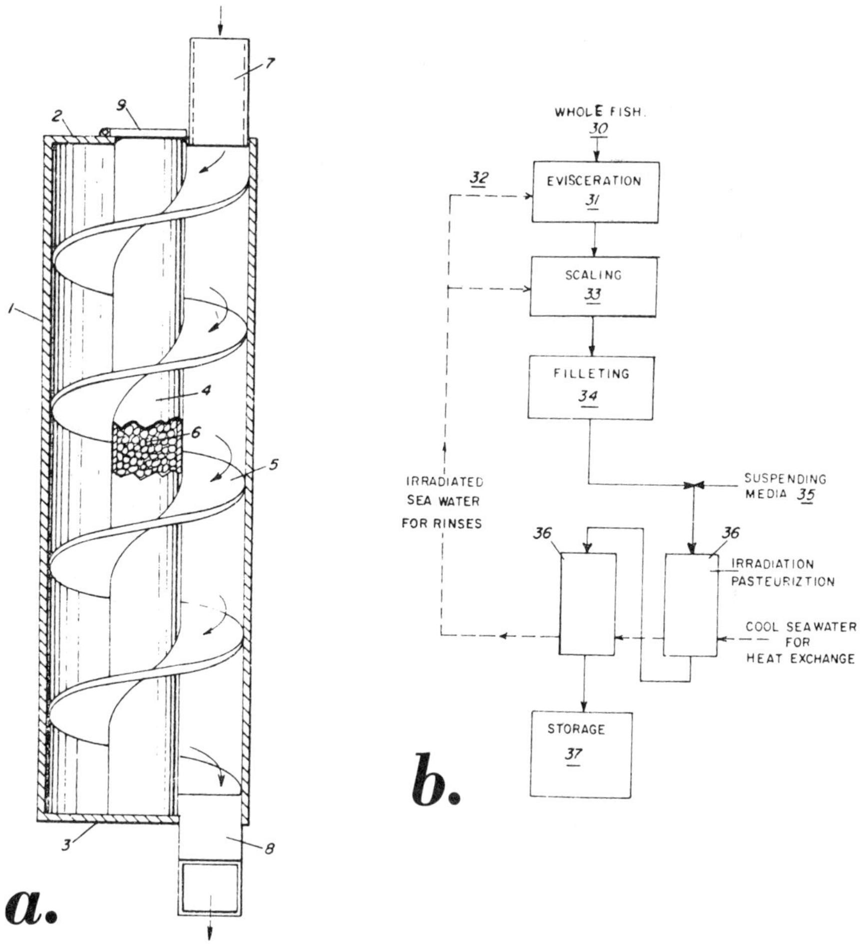

Source: J.B. Huff; U.S. Patent 3,434,850; March 25, 1969

Figure 1.4a represents an elevational view partly in cross section of an irradiator. In that figure, **1** represents an outer cylindrical casing having end portions **2** and **3**. Located concentrically with the outer cylinder is an inner hollow cylinder **4** which may be welded or brazed to the outer cylinder in such a manner that a helical path is formed by the intersection of the winding **5** and cylinders **1** and **4**. Chutes **7** and **8** are attached to the ends so that they communicate with the helical path. Within the inner cylinder there is positioned a radioactive source material **6**.

The source material fills the cylinder. Preferably, the source is pellets or balls of cobalt 60 or a cesium 137 salt, although a rod of cobalt 60 may also be used. Alternately, a can containing pellets of a cesium 137 salt can be used. The source assembly may be dropped into place through an opening in end plate **2**. A cover **9** may be used to cap the opening after the source material has been received.

Source **6** is preferably a cobalt 60 source, the strength of which is not a design limitation because sources from 1 to 1 million curies are obtainable. The radius of the helical path will be influenced by considerations such as percent utilization of the source and the maximum/minimum dosage ratio. As will be appreciated, a larger radius will enable a higher percent utilization of the source as more radiation will be absorbed in the product and less will pass through unabsorbed. At the same time, it will also be true that with a larger radius apparatus the ratio of maximum radiation to minimum radiation within the helical path will be greater than in a smaller radius apparatus.

Therefore when designing the radius of the apparatus, efficiency of source utilization must be balanced with uniformity of dosage. The helical winding **5** may vary in slope depending upon whether the material is being pumped through the apparatus, or whether the flow is dependent upon gravity. This basic unit can be fabricated a number of ways. For example, tube **4** can be first welded to helix **5**, which can then be machined to fit into outer tube **1**. Helix **5** may be soldered with tin or silver solder to outer tube **1**. End pieces **2**, **3**, **7**, and **8** may then be welded or soldered on. The important consideration is that the inside of the apparatus be smooth and free of crevices. The source may be dropped into place within the tube from a shipping cask by opening cap **9** which serves as a cover for the tube.

The apparatus and method of the process have a number of advantages over prior irradiators. Chief among them is the increase in source utilization that is possible. If, for example, a cobalt 60 helical irradiator of the type previously described is constructed so that the helical path is 15" in width the amount of radiation absorbed by the product can be calculated as 65%.

A 65% source utilization is roughly 2 to 3 times the amount obtained by prior art devices. It should be noted that this device takes up much less floor space than conventional irradiators using a series of vertical plates spaced in a horizontal row. It is expected that these advantages will make possible the incorporation of irradiation in environments which heretofore were unable to afford the cost and space required to use prior art equipment.

While the irradiation pasteurization of fish is known in the prior art as shown in U.S. Patent 3,192,054, the fish have always been delivered whole from the fishing boats to large land installations where they were pasteurized by irradiation. By use of this method it is possible to eliminate the need for fishing vessels to interrupt fishing activities to deliver their catch to a land installation. Using the described apparatus it is proposed to pasteurize the fish at sea. Briefly, a preferred method comprises eviscerating, scaling and filleting the fish aboard ship, adding to the fillets a suspending medium, as will be further described in greater detail, to create a suspension in which the fish are precluded from movement but which exhibits fluent properties enabling it to be pumped, passing the suspension through a helical irradiator and then to a refrigerated storage zone which may be aboard the same or a different vessel.

The method will be more fully understood when reference is made to Figure 1.3b where there is illustrated a flow diagram of a specific fish pasteurization process. In that diagram, whole fish **30** are passed to an evisceration zone **31** where the entrails, fins, slime, backbones and tails are removed. As an optional addition, the fish may be washed at this point with irradiated seawater **32**. After eviscerating, the fish are scaled **33** and filleted **34**. After they have been filleted, the fish are put into suspension by treatment with a suspending media **35**.

Solutions of starch, gelatin or agar may be used as the suspending media. It is desirable to use as little of the suspending media as possible, only using enough to cover the loosely packed fish and to keep them from moving within the suspension. Solutions of agar are the preferred suspending media and they have been found successful when used from about 20 to 40% by weight of the total mixture, depending upon the product being suspended. Satisfactory agar solutions may be prepared using as little as 0.2% agar in water. After the fish have been placed in suspension, the gel-like mixture is passed through a helical irradiation.

The dose rate required to pasteurize marine products varies from 75,000 to 350,000 rads. The fish may be given an appropriate dosage by varying the strength of the source, the length and radius of the helical path, and the flow rate. It is desirable although not essential to refrigerate the semisolid mixture to about 34° to 40°F before irradiation. After pasteurization the mixture is pumped through a tube or hose to a refrigerated storage vessel **37**.

TREATMENT WITH MICROWAVE ENERGY AND HOT GAS

The process developed by *M.R. Jeppson; U.S. Patent 3,409,447; November 5, 1968; assigned to Cryodry Corporation* provides for heat-treating a food product as a preliminary step in a preservative process, by simultaneously exposing the product to both microwave irradiation and a hot gas flow of controlled humidity. By varying the humidity of the gas flow in the course of the irradiation and by utilizing a specialized continuous processing chamber very pronounced economics and product improvements are realized.

The microwave energy interacts primarily with the water content of the product and thus a very rapid generation of heat occurs throughout the product volume. However microwave heating alone will not accomplish the desired blanching or cooking in the most advantageous manner. A first significant factor, which is in direct contrast to conventional heating, is that the surface regions of the product tend to remain cooler than the interior. This results since the heat is generated within the product driving water outwardly with the evaporation of the constituent water tending to occur primarily near the surface. Using microwave heating alone may result in under-heating of the product surface or over-heating of the interior with unnecessarily prolonged processing times.

A second highly significant factor is that microwave energy is an inherently expensive method of supplying heat to a product. Substantial economies are realized if some portion of the heat can be supplied by less costly means. These problems are met by the supplementary effect of the controlled gas flow. Additional relatively low-cost heat is supplied to the surface of the product so that adequate blanching or cooking of the superficial regions is assured.

In order to reduce the evaporative cooling at the product surface a very humid gas flow may be utilized during the initial portions of the microwave irradiation. When the desired degree of surface heating has been accomplished, the humidity of the gas flow may then be reduced to a very low value during the subsequent irradiation period to produce several advantageous results.

Most significantly, the drier hot gas flow continues to supply heat to the product while effecting a partial drying thereof. The result is exactly opposite to that of the conventional blanching or cooking techniques in which the water content of the product is increased. Owing to the reduction of moisture content, the product may then be dried, frozen, freeze dried, or dehydrofrozen, with substantially less energy requirements and with shortened processing time.

To utilize this method in the most efficient manner, the products are moved continuously through a tunnel, having electrically conductive walls, by a powered conveyor. Microwave energy is injected into the tunnel in a direction normal to the axis thereof and thus passes repeatedly through the product while undergoing reflection between opposite walls of the tunnel. To provide for more uniform atmosphere throughout the chamber, the controlled gas flow is preferably supplied through a conduit extending along the conveyor and having small gas outlets distributed along the length thereof. To provide for maximum exposure of the product surface to the gas flow, the conveyor is preferably an open-mesh belt or other similar porous element.

The temperatures and humidities of the gas flow are dependent upon the particular product and on whether blanching or precooking, and possibly surface coloration, of the product is desired but will in general range from about 170° to 212°F with very high humidity for the initial phase while lower temperatures and generally minimized humidity may be used for the subsequent irradiation period. Where surface crispness or coloration is desired, the temperature of the dry gas flow may be briefly raised, typically to values ranging up to 1000°F, after partial drying has occurred. An example will illustrate the adaptation of the method to a specific product.

Example: 24 shrimp, weighing 125 grams, were precooked by passage through a microwave tunnel of the class hereinbefore described which was operated at a power level of 23 kw at 915 megacycles. An initial 65-second passage through the tunnel in a steam atmosphere heated the shrimp to 200°F. Juice and moisture losses amounted to 22 grams. Appearance and color were excellent with no physical shrink observed.

Two additional passages through the tunnel were made with the shrimp exposed to a flow of 160°F dry hot air. The first additional pass reduced the weight from 103 to 93 grams without significant change in physical dimensions. The second additional pass reduced the weight to 83 grams and some shrinkage was observed. At this point the weight of the shrimp had been reduced by 34% and the amount of water therein had been reduced by approximately 45%.

The shrimp were subsequently frozen with approximately one-half the power requirements which would have been needed to freeze the original 125 grams. After storage in a freezer, the shrimp were reconstituted in water to a weight of approximately 120 grams and found to be of excellent quality. The patent gives a full description of the process, accompanied by drawings.

PREVENTION OF STRUVITE

By Phytic Acid

It is well-known that canned cooked fish such as tuna, mackerel, salmon, haddock, cod and the like, and canned cooked shellfish such as shrimp, lobsters and crab meat, develop transparent crystals known as struvite, upon standing. While neither chemically nor physically harmful, struvite crystals resemble glass fragments and, when present, render the fish or shellfish unfit to eat in the opinion of uninformed consumers.

Chemically struvite is hydrated magnesium ammonium phosphate having the theoretical chemical formula $MgNH_4PO_4 \cdot 6H_2O$, and it is crystals of this composition that slowly form in the canned fish and shellfish upon standing, usually on the inside surface of the can, and on the surface of the meat. The amount of magnesium ions, ammonium ions, and phosphate ions in the flesh of the fish and shellfish, and in the surrounding fluid or juice, varies considerably with the type of product and the method of canning it, but it is generally sufficient to form the objectionable struvite crystals.

It was found by *E. Geiger and S.H. Watkins; U.S. Patent 2,937,096; May 17, 1960; assigned to Van Camp Sea Food Company, Inc.* that the formation of struvite crystals in canned fish and shellfish is prevented by the addition to the fish, prior to canning, of a minor amount of a water-soluble compound or compounds containing or comprising a phosphoric acid ester of inositol, particularly the hexa- and tetraphosphoric acid esters. Especially suitable is phytic acid or its sodium and potassium salts. The empirical formula of phytic acid is $C_6H_{18}O_{24}P_6$ and the structural formula is believed to be:

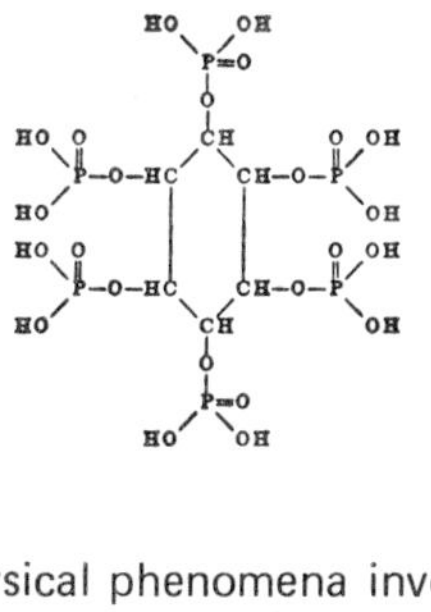

While the chemical and/or physical phenomena involved are not completely understood, it is believed that the ability of the inositol esters to inhibit the formation of struvite crystals is due in part at least to the production by these compounds of coordinative complexes with bivalent metals such as calcium and magnesium. In any event it has been found that the addition of a stoichiometric or excess amount of the inositol ester, based upon the amount of magnesium in the fish or shellfish, is sufficient to prevent the formation of struvite crystals for an indefinite period of time. Generally, from about 0.25 to about 1.5% of the additive, based upon the weight of the fish meat, is sufficeint, the specific amount required depending, as indicated above, upon the amount of magnesium in the material being treated.

Example: A 50% by weight aqueous solution (with respect to phytic acid) of phytic acid and sodium phytate (prepared by dissolving 25 grams of phytic acid and 35 grams of sodium phytate in sufficient water to make 100 grams of solution) was made up and 2.5 grams of this solution were added to 6 oz portions of cooked chunk style tuna. The resultant concentration of the phytic acid was 0.74%. Standard broth (0.35 oz) and vegetable oil (1.4oz soy bean oil) were then added and the canned and treated tuna was retorted in the conventional manner to produce the resultant struvite-free product.

While it is preferred to utilize a mixture of phytic acid and sodium phytate, either of these compounds, or the other compounds set forth above, may be used alone or in various admixtures. The manner and time of addition of the struvite-inhibiting material is not critical. Thus, it may be added to the fish prior to, during or after cooking, but preferably it is added after cooking and immediately prior to autoclaving. Most conveniently, the inhibitor is dissolved in the broth and added to the fish therewith.

The use of the inositol esters is not limited to the treatment of chunk style tuna packed in oil but applies also to grated and solid packs in both oil and brine, and to other fish and shellfish such as salmon, mackerel, haddock, shrimp, crab and lobster.

Salt Tablets Containing Ethylenediaminetetraacetic Acid

It has already been proposed to treat seafood prior to canning with an ethylenediaminetetraacetic acid (hereinafter referred to as EDTA) compound in order to preserve the natural color, flavor and appearance of the canned seafood and to prevent the formation of struvite crystals in the can; see for example U.S. Patents 2,669,520 and 2,680,076.

Since brine is commonly employed in the canning of seafoods, a tablet containing predetermined amounts of sodium chloride, EDTA compound, and, if desired, other canning and preserving assistants such as aluminum salts and the like would be desirable. One or more tablets could then be simply added to the seafood in canning or at a suitable stage prior thereto in order to provide in situ in the can the desired treating medium containing the sodium chloride, EDTA compound and the like.

C.R. Fellers; U.S. Patent 3,013,884; December 19, 1961; assigned to The Blue Channel Corporation provides a process which is based upon the discovery that satisfactory and improved tablets are obtained by thoroughly mixing sodium chloride and at least about 0.5% of a water-soluble EDTA compound by weight of the sodium chloride with sufficient water to yield a substantially homogeneous paste, drying the paste, grinding the dried mixture into particles at least about 85% of which have a size passing a Number 30 screen, thoroughly mixing about

0.05 to 2.0% (by weight of the mixture) of a cohesive agent into the ground mixture, and pressing the resulting mixture into tablets. This process has been found to eliminate fouling of the tablet punches and to enable production of firm, relatively hard, cohesive tablets having a uniform homogeneous composition throughout.

The sodium chloride employed should be composed of particles at least about 85%, and preferably at least 95%, of which have a size passing a Number 30 screen (U.S. sieve series), which has 0.59 mm openings. When the tablet is to be employed in the treatment of seafoods, it should be substantially free of iron, magnesium and copper, and should preferably contain less than about 1 ppm of copper or iron and less than about 10 ppm of magnesium.

The water-soluble EDTA compound preferably employed is the calcium disodium salt, but other compounds may be employed, such as EDTA and the mono-, di-, tri- or tetrasodium, potassium, lithium or ammonium salts of EDTA. Such compounds are generally available as fine powders of a more-or-less sticky nature. While preferably mixed in proportions of about 0.5 to 5% by weight of the sodium chloride, such proportions may be as high as about 50% in some instances.

The cohesive agent employed in the process may be selected from among the substances generally referred to as lubricants, mold lubricants and/or binders and the like. They must, of course, not be dangerous to health. Although calcium stearate is preferred, calcium and aluminum salts of higher fatty acids such as stearic, palmitic, oleic, linoleic, behenic and lauric acids and the like may generally be employed. Other cohesive agents useful include starch, degraded starches such as dextrin, glucose, sorbose, gelatin, casein, gum acacia, tragacanth, agar, methyl cellulsoe, carboxymethyl cellulose, polymerized N-vinyl lactams such as polyvinylpyrrolidone (PVP), lecithin, silica gel, and the like. The following example, in which parts are by weight in grams unless otherwise indicated, are illustrative of the process.

Example: Into a P-K twin shell blender (with intensifier bar) is charged 6,543 grams of sodium chloride and 226 ml of a 50% solution of the calcium disodium salt of EDTA, and the blender run until a thoroughly mixed, smooth stiff paste is obtained. This paste is dried in a tray-type hot-air oven at 110°C, and the dried mixture ground in a grinding mill into particles about 95% of which pass through a Number 30 stainless steel screen (U. S. sieve series). This ground mixture is then thoroughly mixed in the blender with 7.5 grams of calcium stearate, and 141 grams of iron-free aluminum sulfate screened to remove + Number 20 screen particles. The resulting mixture is pressed without fouling or caking of the punches into 75-grain tablets.

Electrical Process

All of the methods which have been advocated as being effective in preventing the formation of struvite depend on the addition into food products of so-called metal chelating agents, such as citric acid, phytic acid, and pyrophosphoric acid or their salts. These methods suppress struvite formation by fixing magnesium, which is one of the elements in the composition of struvite, as a chelation product.

During a long period of storage, however, the chelation effect seems to be lost because of changes or deterioration in the additive agent and other causes, and there is even the possibility of the agent becoming an accelerator.

S. Otsuka and H. Osada; U.S. Patent 3,547,657; December 15, 1970; assigned to Toyo Seikan K.K., Japan have solved this problem by providing a new technique for electrically processing marine products with the objects of removing undesirable salts from the products and improving the product flavor.

This is achieved by a process which comprises, generally, immersing a marine product such as fish meat in a suitable medium in a pretreatment stage prior to food processing and applying a DC potential to the product thus immersed, thereby to extract and remove electrically (electrophoretically) the salts in the seafood product, to lower fully the concentrations of the magnesium, the ammonium and the phosphate ions, and thereby to prevent crystallization of struvite.

In addition to prevention of struvite formation, the process has the further effects of effectively preventing discoloration by removing other inorganic ions and of improving the flavor of fish meats of relatively low degree of freshness by removing organic bases (amines), which are present as positive ions and cause deterioration of taste, and, at the same time, the process accomplishes precooking of the product due to the accompanying heat.

The process involves a kind of electrophoresis in which the seafood products such as a fish meat, for example, constitutes a stationary element and in which the substances (electrolytes) of either positive or negative charge or ions within the fish meat are caused to migrate (undergo phoresis) toward respective negative and positive electrodes provided at two ends of the stationary element and finally to be removed from it.

The velocity of the migration is affected by the voltage and current and by the strength (magnitude of ionization degree) of the electrolyte itself, the shapes and sizes of the ions, and the state of fish meat structure constituting the stationary element. Therefore, by adjusting the voltage and current applied from the outside, it is possible to extract and remove only the objectionable ions. The current per unit time can be easily controlled by adjusting the ionic strength of the electrolyte in the neighborhood of the electrodes.

Example 1: Four cuts of tuna meat, each of 100 gram weight, were used as samples, three of which were separately placed in a vessel and respectively processed for 1, 2, and 3 hours at 100 volts applied by an electrolyte (0.1% NaCl solution), partition walls, glass filters, or ion exchange membranes being placed between each of these three samples and the electrolyte to prevent direct infiltration of the electrolyte into the meat.

The essential process conditions were as follows: voltage, 100 volts; electrodes, 1 x 2 cm; potential gradient, 1.0 volt/mm; and current density, 24 ma/cm^2 (first 1 hour), 37 ma/cm^2 (second 1 hour), and 18 ma/cm^2 (last 1 hour). The resulting variations in the contents of VBN (volatile base nitrogen), NH_3, trimethylamine (TMA), P, and Mg were as indicated in the table on the following page.

[Unit: mg. percent]

	VBN	NH_3	TMA	P	Mg
Time (hr.):					
0 [1]	30.0	21.1	8.9	378.8	13.1
1	25.2	17.6	7.6	269.6	13.0
2	16.8	6.8	8.0	191.9	8.7
3	11.2	3.2	7.9	26.1	1.9

[1] Raw meat.

Example 2: A cut of tuna meat having a 3 x 4 cm cross section, 10 cm length, and 150 grams weight was immersed in a 0.1% solution of NaCl, and a direct current of 100 volts and 2 amperes was passed therethrough for 3 hours. The potential gradient was 10 volt/cm, and the current density in terms of the surface area of the meat was 83 ma/cm^2.

As a result, the contents of the phosphoric acid, Mg + Ca, and volatile bases (including ammonia) were reduced to values approximately 1/9, 1/10, and 1/3 respectively, of those before the electrical processing, as indicated in the table below:

	Phosphoric acid, (mg. percent)	Mg.+Ca. (mg. percent)	Volatile amines (incl. ammonia) (mg. percent)	Flavor
Before electrical processing	92.8	3.84	22.4	Odor of initial spoilage (amine odor).
After electrical processing	10.6	0.33	7.0	Fresh boil cured odor

Furthermore, since the amines, which are considered to be the cause of the unpleasant odor of fish meat in the initial stages of spoilage, were removed by the process, the flavor and other characterisitics of the fish meat were remarkably improved to an extent comparable to those of products processed from very fresh fish meat.

In addition, the temperature of the inner central part of the meat was raised to approximately 95°C by the process, whereby precooking was accomplished simultaneously. A further observable feature of the process was that the fish meat structure after the process was excellent and was maintained in a state of natural precooking. Furthermore, no evidence of deterioration or other abnormal state was observable. From the foregoing examples, it is possible to arrive at the following three conclusions.

(1) When electric current is passed through a cut of fish meat, the quantity of the Mg^{++}, NH_4^+, $PO_4^{\equiv}$, and volatile base nitrogen (VBN) contained in the meat is remarkably reduced and approaches zero with elapse of processing time.

(2) In electrical processing on a large scale, the removal of the Mg^{++} is incomplete, but in spite of this, no formation whatsoever of struvite in actual canning tests is observable.

(3) From the above finding (2) the known fact that struvite formation occurs to a great degree in canned tuna when the raw material meat of relatively low degree of freshness is used, it may be concluded that the primary cause of struvite formation in canned tuna is ammonia and not magnesium as commonly thought heretofore. In any case, it is possible by this electrical processing according to the present method to prevent fully the formation of struvite.

PREVENTION OF MELANOSIS IN SHELLFISH

The two substances usually used for the prevention of melanosis (blackening) of shellfish preserved by freezing are boric acid and sodium bisulfite or metabisulfite. With respect to the first, boric acid, its use has several disadvantages, such as affecting the taste of the shellfish so treated, the fact that in order to obtain positive results, it must be used in amounts that make it dangerous because of its toxicity, etc., for which reasons public health departments prohibit its use in most countries. Because of these marked characteristics of toxicity, it becomes mandatory to use it in small amounts, which leads to very uncertain results, and in most cases the appearance of melanosis is not prevented in shellfish preserved with it.

As for the use of sodium bisulfite, while it is authorized by many public health departments, it does not provide the results desired as regards the preservation and sale of the shellfish treated with it. Sodium bisulfite is a product that breaks down very easily once absorbed by the shellfish; it happens that its pH rises rapidly, becoming alkaline and thus prompting a more rapid decomposition of the shellfish itself. It also has the disadvantage of affecting the taste of the shellfish and discoloring it, the latter effect degrading its appearance and resulting in reduced sales.

It has been found that melanosis occurs in shellfish as a result of an enzymatic reaction and that the cause of this reaction, regardless of the enzymes that occur and of the catalytic agents within the shellfish itself, is an accumulation of oxygen in the enzymes which then produce melanosis. Thus, the fundamental means of preventing the appearance of melanosis is to prevent this oxygen accumulated in the shellfish at death from accelerating the enzymatic reaction.

After extensive biological studies, during which some shellfish were maintained in a perfect state of preservation, in some cases, up to approximately two years, *L.G. Alsina; U.S. Patent 3,859,450; January 7, 1975; assigned to Barreiros Hnos, SA Financiera Alcala, Spain* has found that suitable products for the treatment of shellfish before freezing for preservation are carbon dioxide and citric acid, the latter in an aqueous solution in specific proportions, used to saturate the shellfish through injection under pressure in appropriate devices in order to prevent oxygen accumulation.

It takes a longer or shorter time after death for evidence of melanosis to occur in shellfish, depending on their species, where they were caught, etc. If they are subjected to the treatment according to this process which is intended to eliminate the oxygen accumulated in them, within certain time limits, the results obtained are perfect and no melanosis appears regardless of how long the shellfish are kept preserved.

Referring to Figure 1.5, shellfish are initially placed in a suitable perforated basket **1**, mounted within a stainless steel container **2**, provided with a closing system or top capable of withstanding a given pressure, preferably at least 12 kg/cm^2. In its interior, the container is equipped with a paddle-type agitating mechanism **3** whose paddles are mounted in any convenient manner at their upper and lower ends so that the paddles can revolve around the basket. The lower ends of the paddles are attached at **4** to radial arms **5** which, in turn, are mounted to a central core **6**.

FIGURE 1.5: APPARATUS USED IN PREVENTION OF SHELLFISH MELANOSIS

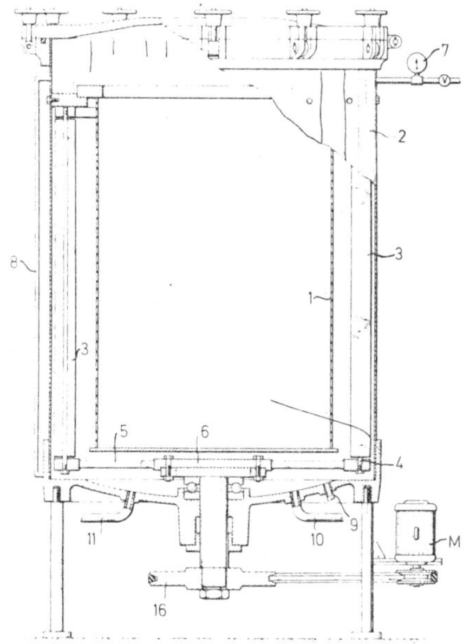

Source: L.G. Alsina; U.S. Patent 3,859,450; January 7, 1975

The container is provided with a number of conventional accessories for use in liquid pressure vessels, for example, a pressure manometer **7** and a liquid level indicator **8**, as well as a decompression and gas entry port, and a water inlet port **10**. The central disc or core of the agitator system is coupled, by any appropriate transmission arrangement, to a motor **M** which is adapted to drive it and the paddles at a predetermined speed, preferably at 140 rpm.

Once the shellfish are placed in the device, there is introduced into the container an aqueous solution of an innocuous acidic substance (as distinct from boric acid) that is, for example, citric acid, ascorbic acid, tartaric acid, or mixtures of the same, although preferably, citric acid. This solution is introduced into the container before or after the container is closed, from a tank or source thereof through the duct **10** and an associated valve (not shown).

The level of the aqueous solution supplied to the container, with the shellfish already within it, should in no case exceed 80% of the height of the container's upper opening, and the aqueous acidic solution should have a pH within the range of 2 to 4. This pH range has been found suitable for obtaining the most perfect results in the preservation of the shellfish, in accordance with this process.

After the container is filled with shellfish, it is hermetically sealed with its top.

As mentioned, the aqueous solution may be supplied to the container before or after it is sealed. In either case, sufficient fluid is supplied so that the shellfish are completely submerged in the aqueous acidic solution. Carbon dioxide is then injected under pressure, through port **9** and an associated valve (not shown) to a predetermined pressure, preferably between 3 and 6 kg/cm^2, depending on the nature, origin and other specific characteristics of the shellfish, and the agitator is put into operation. It is also possible to initially fill the container with carbon dioxide after the container is sealed and then supply the aqueous solution through duct **10**.

In either case the process achieves the same results with the selected conditions maintained for a predetermined length of time that varies as a function of the same shellfish variables mentioned above, after which time the pressure within the device is lowered slowly in any convenient manner. The possible breaking off of heads or loss of eyes of the shellfish are signs that the pressure to which they have been subjected is near the limit to which that specific kind of shellfish can be subjected, and they show that the pressure must be watched closely in order to avoid damaging the shellfish to be treated thereafter.

At the conclusion of the decompression operation, which should take from 1 to 2 minutes, the device is opened, the baskets of shellfish dried, and a new load of the same put in place, while more aqueous solution is fed to the container if necessary. The treatment is repeated in this way with successive loads of shellfish until it is noticed that the aqueous acidic solution is heavily soiled, at which time it is completely renewed.

During the treatment, when the carbon dioxide is injected into the device, it dissolves in the water, so that within the container there exist, at the same time, carbonic anhydride, citric acid and of course, carbonic acid. These products penetrate into the bodies of the shellfish owing to the pressure condition obtaining in the device, so that the cephalothorax, the gills, and other parts of the shellfish are completely impregnated with them.

It has been found that carbon dioxide is a very good antioxidant and a digester of oxidizable substances, so that at the pressure under which the process is carried out, two things occur: in the first place, the carbonic anhydride eliminates the excess of oxygen that there might still be in the shellfish, and in the second place, the carbonic acid, together with the citric acid, remain in the shellfish in an amount corresponding to its size.

When the shellfish treated in this manner, completely impregnated with the above-mentioned products, is frozen, the carbonic acid, owing to its stability at low temperatures, serves as a preserver against possible exudations caused by changes in temperature resulting from refrigerator changes during transport, etc., thus eliminating completely the danger of renewed oxygen accumulation within the shellfish. In the meanwhile, the purpose of the acid substance (citric acid, for example) present in the body of the animal, is to maintain a pH no higher than 6.5 during the period of preservation.

Experiments have shown that, preferably, the pH sould be kept between 5 and 6 during preservation. The amount of carbon dioxide normally used for every liter of water is of the order of 12 to 15 grams, while up to 3 kg of shellfish can be treated with 1 liter of water, with no inconvenience.

The results obtained with the method and the device according to this process are extremely interesting since the products used leave no toxic residue in the shellfish, nor do they change the color or the taste of the shellfish once they are thawed, while at the same time the flesh of the latter retains its original texture for a much longer time than in the case of preservation treatments used up to now; finally, the appearance of melanosis is prevented completely, even in shellfish with a large head, like the lobster, that can be frozen whole and undergo no change for a very long time.

Subsequently, and for the purpose of increasing the keeping qualities of shellfish treated according to the process after thawing, it has been found that the addition of sodium or potassium metabisulfite in amounts as small as 1 to 4 grams of metabisulfite per liter of aqueous solution is helpful.

Example: A container, similar to that illustrated in the drawing was filled with approximately 75 kg of shrimp from the Caribbean Sea and the container then hermetically sealed. Next, carbon dioxide was supplied to the container until a pressure of 3 kg/cm^2 was attained, in approximately 10 minutes. Thereafter, an aqueous solution of citric acid and metabisulfite was supplied to the container until the shrimp were covered and a pressure of 5 kg/cm^2 was attained.

The paddles were then agitated for 5 minutes during which period pressure in the chamber dropped to about 4 kg/cm^2. However, the pressure is preferably maintained at 5 kg/cm^2, by adding more solution or carbon dioxide to the container. After the 5 minute period, the pressure was allowed to decrease to zero in about 3 minutes.

In this example of the process the aqueous solution contained 6 grams of citric acid per liter of water and 4 grams of metabisulfite per liter of water. The shrimp thus treated were frozen for 6 months without any organoleptic change and no melanosis symptoms. Moreover, upon thawing, the shrimp are safe for consumption for 48 hours if maintained at 5°C.

The solution used in the above-described example may be used for treatment of four separate batches of shrimp before being disposed of. Where the process is performed on board ship, after four uses, the solution is stored in a settling tank for later reuse.

TUNA AND TUNA-LIKE FISH

Tuna fish comprise one of the world's leading fishery resources and are widely distributed in temperate, semitropical and tropical waters. Insofar as United States landings are concerned, the Pacific coast catch conventionally includes albacore *(Germo alalunga)*, bluefin *(Thunnus thynnus)*, skipjack *(Katsuwonus pelamis)*, and yellowfin *(Neothunnus macropterus)*. The Atlantic coast catch, which is quite small as compared to the Pacific coast catch, conventionally includes bluefin and little tuna *(Ethynnus alletteratus)*.

The above recognized species of tuna vary widely in size with albacore (white meat tuna) having a maximum weight of about 80 lb and with the usual commercial catch weighing less than 40 lb and averaging from about 12 to 25 lb; Pacific bluefin having a maximum weight of about 250 lb and with the usual commercial catch averaging from about 20 to 40 lb; skipjack having a maximum weight of about 40 lb and with the usual commercial catch averages 4 to 20 lb; and yellowfin with a maximum weight of about 400 lb and with the commercial catch averaging about 30 to 40 lb.

In contradistinction therewith, the Atlantic bluefin can reach a maximum weight of about 1,500 lb or more and the commercial catches average either between 8 to 65 lb or from 65 up to 600 lb, and the little tuna has a maximum weight of about 20 lb with a commercial catch averaging between 5 and 10 lb. In addition to the above five recognized species of tuna, two additional species of tuna-like fish are processed commercially (canned tuna style) although, when canned, they cannot be domestically labeled as tuna.

These additional species include both the Atlantic and Pacific species of bonito *(Sarda sarda* and *Sarda chiliensis*, respectively) and the yellowtail *(Seriola dorsalis)*. The Pacific bonito has a maximum weight of about 25 lb, about twice that of the Atlantic species, and the commercial catch thereof usually averages from 6 to 8 lb, while the yellowtail has a maximum size of about 80 lb with the commercial catch averaging somewhat less than 15 lb. In this chaper the term tuna will, because of the conventional commonality within the industry, be considered to specifically include only the above described seven species of fish.

The following is a description of the nature of the essential butchering and cleaning operations that have been conventionally employed by the tuna canning industry which presently butchers and cleans the preponderance of the frozen tuna landed, written by J.M. Lapeyre.

The majority of the tuna catch conventionally arrives at a cannery in a frozen or partly thawed-partly frozen condition and is initially subjected to a thawing operation preparatory to butchering. Immersion in water, water spraying or air-current exposure are commonly employed techniques to hasten thawing. Since, however, the delivered tuna are not only quite disparate and nonuniform as to size, but also may vary in condition from hard frozen to (when thawing is begun aboard ship during the return voyage) completely thawed.

Requisite thawing times under current practices vary widely and introduce problems both with respect to the size of thawing areas and facilities that are required and to the efficient scheduling of operations subsequent thereto. Thawing times of up to 12 hr or more are not unusual. Thawing of the fish is progressive and smaller tuna and exterior parts of larger tuna reach the desired state of thaw while the inner parts of the larger tuna are still frozen.

Deterioration of unfrozen tuna, especially whole uneviscerated tuna, is rapid at temperatures sufficient to hasten thawing. The quality of the unfrozen tuna diminishes with time, and the first thawed whole tuna and outer portions of other tuna begin to diminish in quality before the last thawed portions have become unfrozen.

Once thawed, the tuna are tabled and individually hand manipulated and butchered by longitudinally slitting the belly to permit removal of the visceral organs, after which the visceral cavity is thoroughly washed out with water and the tuna inspected for spoilage. In some instances, tuna are eviscerated aboard ship prior to freezing, in which cases, of course, the in-plant evisceration procedure is unnecessary. It is important to note that the above slitting and evisceration procedure is the only butchering operation performed on the tuna while it is in the raw condition. As will be explained, all other cleaning is performed after the tuna has been cooked.

Following evisceration operation, the tuna is placed in precooking baskets or racks. While basketed or racked, the tuna are subjected to a batch-type precooking operation which effects a sufficient denaturing of the tissues holding the loins to the bony and cartilaginous material, that generally comprises the skeletal structure, to permit the loins to be readily separated therefrom by subsequent manual operations. Such precooking involves exposure of batches of eviscerated whole tuna to temperatures in the order of 212° to 220°F for a sufficient period of time to effect the requisite tissue denaturing in the vicinity of the spinal formations adjacent the longitudinal center of the tuna.

Although a size grading of the tuna is oftentimes effected in forming the individual batches in an effort to achieve optimum cooking times therefore, a variation in size of tuna is usually still present with a concomitant unavoidable variation in the degree of cook within the tuna. In addition, the time of the precook for individual batches varies widely due to the inevitable disparate sizes of tuna involved. For example, the cooking time may vary from 1½ hr for very small tuna up to 8 to 10 hr or more for larger tuna.

Following the precook, the precooked tuna must, because of its friable and fragile nature while warm, be cooled thoroughly to firm up the flesh before the manual cleaning operation can be performed. Cooling also reduces the temperature to a level which is comfortable to the touch so as to permit manual cleaning with a minimum of wasted motion. The cooking time required will vary widely both in accord with the nature of the cooling facilities employed and the size of the tuna. Cooling periods can vary in amounts from a few hours up to 12 or 16 hr or even more and during which time the tuna are subject to the deleterious effects of exposure to air and enzyme action.

Even more important than the above, from an economic standpoint, are the weight losses attributable to uncontrollable shrinkage that inherently results from the above described conventionally employed precooking and cooling operations. According to a report of the Fish and Wildlife Service of the U.S. Department of the Interior, shrinkage in amounts of up to 30% of the weight of the tuna takes place during the precooking and cooling operations, with two-thirds to three-fourths thereof taking place during the cooking as distinguished from the cooling operation.

Such shrinkage is effectively caused by loss of oils, solids, dissolved protein and other tissue constituents and moisture during cooking and by loss of moisture due to evaporation and drip during cooling. Some reduction in weight of the raw tuna is inherent and necessary, but uncontrolled loss is wasteful and costly. Since the cost of the basic raw tuna represents about 70% of the processor's cost, it has been long recognized that any savings that could be effected in the shrinkage that inherently accompanies the precooking and cooling operations would be of the greatest importance.

Subsequent to the precooking and cooling operations, the whole cooked and cooled tuna are routed to tables where they are individually manually manipulated and cleaned. Conventionally, in the cleaning of the cooked tuna, the head and tail are removed by breaking off of the same, and the skin and fins are manually scraped off, after which the tuna is manually split lengthwise and the backbone removed. Each half is split again longitudinally to form a pair of discrete dorsal loins, i.e., the right and left epaxial muscles, and a pair of discrete ventral loins, i.e., the right and left hypaxial muscles.

The blood and associated dark meat portions are then scraped away and the loins, edible flakes and waste products are selectively routed to separate processing stations for further operations thereon. It should be noted that, in conventional cleaning practices, the precooking to which the tuna is subjected is a processing step designed to ease the manual cleaning operations since after being precooked the tuna comes apart very easily.

As noted above, manual labor is conventionally employed for the above described cleaning step and the rate at which tuna can be cleaned varies with both the skill of the worker and with the size of the tuna being handled. The costs involved and the essentially tedious and disorderly nature of the cleaning operation have created a recognized and long-standing need for mechanization of the cleaning step. However, the disparate sizes of the whole tuna being processed and the inherent longitudinal dissymmetry of anatomy of same, coupled with the friable nature of the cooked tuna have been material factors contributing to the problems involved.

Following the cleaning operation, the separated longitudinal loins, which resemble the shape of french bread loaves, are subjected to a guillotine-type shearing operation across the grain of the cooked loin to form can depth segments thereof for a solid pack or to form smaller segments for a chunk pack thereof, both of which are attended by undesired fragmenting of the loin meat at the locus of the cut. After such cross grain cutting, the segments of tuna are either hand or machine packed into cans. The fragments are separately packed and demand a much lower price on the market.

It should be noted from the foregoing that the in-plant processing time, from the commencement of the thawing operation to the final can-filling operation, can be as much as 38 hr, and even this period is frequently exceeded by some processors. The bulk of this time period is devoted to thawing the tuna prior to cleaning to facilitate manual cleaning of the fish.

In spite of the fact that an upgrading of the canned product can be attained when tuna meat is uniformly cooked after cleaning, the industry has continued the above outlined processing procedure for approximately 50 yr. Processes reviewed in this chapter are arranged in the order in which they would be carried out in tuna processing.

THAWING FROZEN FISH FROM SHIPS

Temperature-Controlled Thawing

R.T. Erickson; U.S. Patent 3,050,403; August 21, 1962; assigned to Star-Kist Foods, Inc. has developed a process for canning tuna-like fish which permits more economical operation of a fish canning plant, eliminates the necessity of either providing a continuous flow of steam retort cooling water or alternatively of re-cooling such water, and insures precise temperature control of the thaw water.

Additionally, because it is not necessary to employ seawater for thawing purposes, it eliminates the danger of water pollution. It moreover permits the solids extracted by the thaw water to be utilized. The cleaning and packing schedules of the fish canning plant may be more accurately coordinated inasmuch as the temperature of the thaw water can be carefully controlled. By virtue of such careful temperature control it is possible to effect uniform cooling of the fish, which permits considerable savings in the fish butchering operation. Finally, the method reduces can spotting.

The operation of the process is illustrated in Figure 2.1. The fish to be thawed are deposited in the bins **10** and **12** of the thawing tank structure **T**. At or prior to this time, the steam retort structure **S** will have been operating. The condensate received within the can retort chambers **40**, as well as the cooling water forced thereinto through the line system **44**, will have been directed into the confines of the hot water sump **48**.

The temperature of the water within this sump may vary between 170° to 190°F. This hot water will be forced by centrifugal pump **50** through lines **52, 54** and filter **56** to the hot water line **58** which empties into the hot water tank **60**, where the temperature of the water may vary between 170° to 190°F.

FIGURE 2.1: THAWING INITIALLY FROZEN FISH FOR CANNING

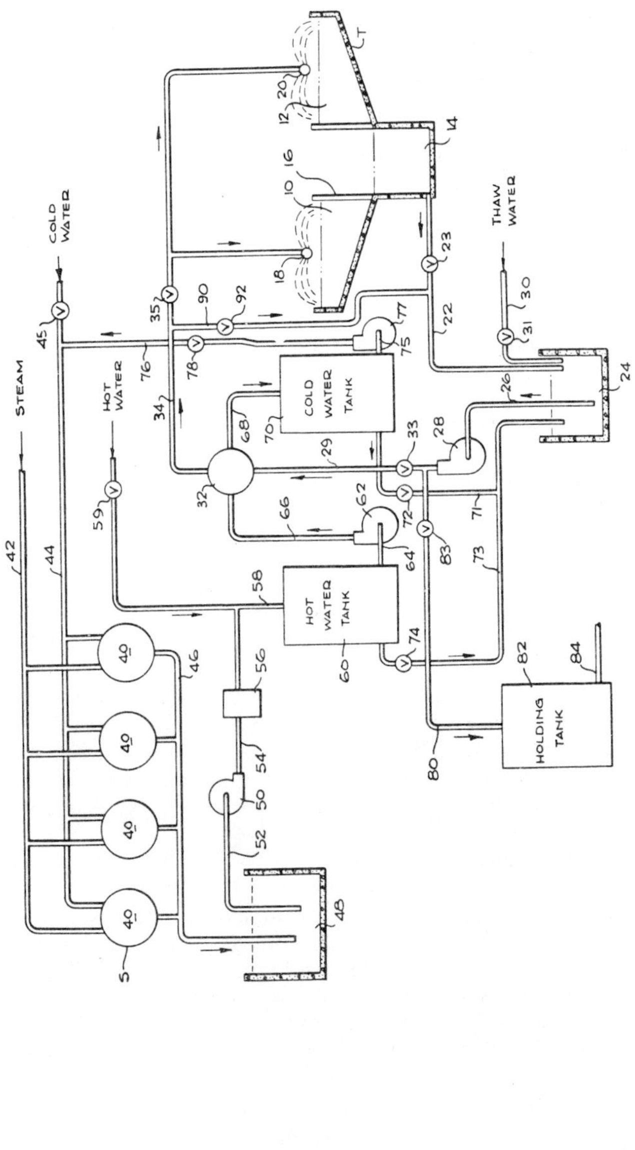

Source: R.T. Erickson; U.S. Patent 3,050,403; August 21, 1962

Water from this hot water tank is forced by pump **62** through lines **64** and **66** to the heat exchanger **32**. This water received by the heat exchanger will be discharged from the heat exchanger through line **68** into the cold water tank **70**. The heat of the water flowing into the heat exchanger will be transferred to the water entering the heat exchanger through line **29** although these bodies of water are maintained physically isolated. The water entering the heat exchanger through the line **29** is discharged from the heat exchanger through line **34**.

The temperature of such water entering the heat exchanger through line **29** will be raised by heat transfer to approximately 45° to 55°F. This heated water will pass by means of line **34** to the nozzle means **18** and **20** and such water constitutes the thaw water directed onto the fish within the bins **10** and **12**. It will be apparent that by proper valving the temperature of the thaw water may be carefully controlled.

The water used for thawing is directed into the concentrate sump **24**, by means of line **30**. Such thaw water may either be fresh or may be seawater. Generally, however, it will be seawater. After the thaw water from the nozzles has passed over the fish to be thawed so as to reduce the temperature thereof it will flow into the trough **14** and thereafter through line **22** into the confines of the concentrate sump. By contact with the frozen or partially frozen fish its temperature will be reduced to approximately 34°F. This cooled water will be forced upwardly through lines **26** and **29** by the centrifugal pump **28** back to the heat exchanger **32**.

The lines **72** and **73** permit any desired transfer of water from the hot water tank or the cold water tank to the concentrate sump. The line **80** permits the thaw water entering the concentrate sump to be transferred to the holding tank **82**. The water contained within this holding tank may be redirected to the concentrate sump for successive thawing operations. The line **84** leading from the bottom of the holding tank permits the contents thereof to be removed. Where these contents include a substantial proportion of fish solids extracted by the thaw water, it may be desirable to process such contents so as to obtain a nutritional product, such as an animal feeding concentrate. Alternatively, the contents of the holding tank **82** may be directed to sewage.

Although the system may be operated on a batch basis, it also lends itself to a continuous process. A very important advantage of the process is that it reduces or eliminates spotting of the cans during the retorting operation. This spotting occurs where the retort cooling water contains contaminants, such as various mineral salts, that produce spots on the cans when the latter dry after retorting. In time, portions of these spots may rust so as to produce can damage.

With this method the steam condensate from the retorts may be accumulated in the hot water tank during successive retorting operations until such condensate forms the major portion of the retort cooling water. This condensate is pure of spot producing elements and, accordingly, spotting of the cans will be avoided. This antispotting characteristic will prove particularly important in areas having limited water supplies or areas having especially high percentages of spot forming elements.

Thawing Blocks of Frozen Fish Without Cooking

Since fresh caught fish are often frozen aboard a factory ship in blocks which facilitate storage during the voyage, a process for quick thawing of such blocks before canning is necessary. Such thawing must take place without any cooking of the fish.

A.C. Jason and H.R. Sanders; U.S. Patent 3,383,218; May 14, 1968; assigned to The Minister of Technology, England have developed a process which relates to the thawing of frozen blocks of fish of generally uniform thickness. It has previously been proposed to effect the thawing of such blocks by passing therethrough an alternating current the potential of which is regulated so that the mean rate of dissipation of energy per unit volume is insufficient to give rise in the block to progressive localized overheating but is preferably maintained at or substantially at a rate consistent with this requirement.

The current is supplied by the aid of electrodes of corrosion resistant metal, preferably stainless steel, applied over the whole area of the two opposite largest surfaces of the block. Below a certain temperature (about -5°C in the case of frozen fish) the resistance of the block is so high that it is preferable to heat the blocks up to this temperature by absorption of external heat before subjecting the blocks to resistance heating, and it was proposed to effect such heating by passing them through a preliminary heating zone such as a water spray, a radio frequency oven or a tunnel through which air at room temperature is circulated.

According to this process the electrodes themselves are heated so that they will supply heat by conduction in addition to serving to convey current for resistance thawing. This thermal conductive heating is discontinued towards the end of the thawing process because as thawing proceeds the rate of thermal conduction is reduced progressively while the electrical conductivity rises and if heating of the electrodes is sustained too long the surfaces of the block become overheated. As before the current is regulated to avoid progressive localized overheating.

The electrodes may be heated in any convenient way but electrical heating is preferred because it is convenient and then only a single supply of electricity is needed for both forms of heating. A very suitable arrangement comprises a heating element of insulated cable form sandwiched between a pair of corrosion resistant metal, preferably stainless steel plates for each electrode, the element being set in a sinuous or similar pattern to effect substantially uniform heating at the surface of the block.

With a uniform heating of the block by means of uniformly heated electrodes a planar heat front travels into the block from each electrode with the result that the variation in resistance of the material of the block as its temperature rises is also uniform so that the current flowing between the electrodes is uniformly distributed and does not tend to concentrate in localized regions to produce local overheating.

The efficiency with which thawing is effected by the current passing through the block is also affected by the frequency of the alternating current flowing between the electrodes and its magnitude and that for different materials to be thawed different frequencies and magnitudes may produce optimum efficiency in the different cases.

It is therefore preferable to arrange for both the frequency and the magnitude of the accompanying alternating current supplies to the electrodes to be controllable to enable optimum efficiency to be obtained by adjustment of one or both these parameters. The following table summarizes the procedure in two typical cases of application of the process to the thawing of frozen fish.

Fish	Boned herring	White fish fillets
Block size (inches)	15 x 10 x 1⅝	15 x 8 x 1⅝
Block weight (lb.)	7.5	7.2
Total plate power (W)	300	300
Potential applied to block (volts r.m.s.)	42	26
Duration of simultaneous resistance and thermal conduction heating (minutes)	25	30
Duration of additional resistance heating (minutes)	5	12
Energy supplied to heater plates (kW minutes)	7.5	9.0
Energy for resistance heating (kW minutes)	2.4	3.1

TREATMENTS USED BEFORE COOKING

Improving the Color with Cysteine

In the processing of tuna-type fish which includes yellowfin, skipjack, albacore, bluefin, big eye, etc., there is a phenomena generally recognized in the tuna processing art as greening of the fish. This greening of the fish flesh occurs during the initial cooking of the fish, such as the precooking of the fish prior to the canning thereof. Upon the precooking of such fish a very visible green discoloration occurs rendering the fish unfit for consumer acceptance and commercial packing. The precise reason why this greening occurs is not known and it is impossible to determine which fish in a particular batch will be subject to the greening effect.

A method for controlling this color is described by *A. Khayat, G.W. Courtney and H.J. Dunn; U.S. Patent 3,851,078; November 26, 1974; assigned to Ralston Purina Company.* Whole fish selected from the tuna family such as albacore, skipjack, yellowtail, bluefin and big eye, hereinafter all to be referred to as tuna, are obtained in the normal manner and subjected to a standard eviscerating operation.

The whole fish while still in an uncooked or raw state is then passed to a position wherein injection needles may be forced into the tuna fish in order to inject a predetermined amount of a reagent selected from the group of cysteine or homocysteine into the tuna flesh. The precise configurations of the injection needles and the location of the injections into the tuna fish are not critical.

However, it is necessary that the injection of the cysteine or homocysteine be performed at an effective level and in such a manner that during the precooking of the tuna flesh good distribution of the material is achieved throughout the tuna fish such that the greening or discoloration does not occur. It has been found that the injections of quantities of cysteine or homocysteine of less than about 0.3% by weight of the tuna fish are effective in improving the color and, in the case of homocysteine, of reducing or eliminating the greening or discoloration of the fish.

As an alternative to treating the whole fish with cysteine or homocysteine, it is also possible to treat the tuna flesh by means of forming a slurry of the tuna fish to be treated, after the tuna is deboned. Following the deboning operation the tuna meat is then subjected to a grinding operation which is sufficient to place the flesh into a condition suitable for forming a slurry. The slurry may be of any desired solids content such that it is suitable for handling.

A suitable solids content would be approximately 20%. If desired, following the formation of this slurry an antioxidant may be added to the slurry at a rate of about 0.02% of the fat content of the slurry. This is merely to prevent oxidation of the fat during the subsequent processing thereof. Following the homogenizing of the slurry, cysteine or homocysteine may be added to the slurry at a concentration up to about 0.3% by weight of the tuna fish.

If desired, the slurry following homogenization and prior to the addition of the cysteine or homocysteine may be subjected to a centrifuging operation in order to separate the soluble and insoluble portions of the slurry. The insoluble portion of the slurry would essentially consist of the myoglobin parts and the nonheme proteins. The soluble or hemoglobin portion of the slurry would then be treated by adding the cysteine or homocysteine at the desired level.

Additionally, it is desirable although not necessary to adjust the pH of the soluble fraction to the range of about 7.8 to 8.3 and preferably about 8. This pH adjustment may be carried out with any food grade alkaline material. The soluble fraction portion is then subjected to a heating sufficient to coagulate the protein therein. The heating would be in the range of 72° to 77°C. Following the heating to coagulate the protein, the slurry is then subjected to a centrifuge or filtering step to recover the coagulated protein. Approximately 70% of the protein is recovered.

The recovered coagulated protein is then placed into another slurry which may then be homogenized in a manner such that all of the material will pass through a filter having 1/16" holes therein. The water content added to the coagulated protein in order to form this slurry will be sufficient to get the desired final moisture of the end product. The insoluble protein fraction which was previously separated is then blended or mixed back with the homogenized coagulated protein in obtaining the final product. The final product should have a moisture content in the range of about 60 to 65% to be in a condition suitable for use in forming restructure protein products.

The final product upon cooking exhibits good color characteristics and does not green or discolor with the cooked final product exhibiting a lighter color than cooked tuna flesh which has not been treated with cysteine or homocysteine. If further color improvement is desired, it may be desirable to add 0.1 to 0.2% by weight of cysteine or homocysteine to the final product mixture because of the insoluble protein added to the mixture and the possibility that some of the cysteine was discarded in the liquid during the centrifuging following the heat treatment of the soluble portion.

Example: Albacore which had previously been eviscerated and deboned was ground to obtain 350 g of whole fish (approximately 68 g of protein). Three volumes of water were added to the ground fish to form a slurry. BHA was added to the slurry at a level of 0.02% of the fat content.

The slurry was homogenized at 1°C for 6 min by an ultrasonic device at 90 to 100 watts and passed through a screen having 1/16" openings therein. The slurry was then centrifuged at 8,700*g* for 10 min to separate the insoluble protein fraction from the soluble protein fraction. Approximately 49 g of protein were obtained in the insoluble fraction and approximately 23 g of protein in the soluble fraction. 0.3% cysteine by weight of the tuna was added to the soluble fraction and the pH adjusted to 8.3 with food grade alkaline material.

The soluble fraction slurry was then heated to 76°C for 5 min while continuously stirring. The coagulated proteins (approximately 16 g) were separated by filtration and formed into a second slurry. The insoluble protein fraction was then blended with the second slurry to obtain the final tuna product. The final tuna product was blended with soy curd with the final product exhibiting improved or lighter color properties than products prepared with tuna flesh which had not been treated with cysteine.

Pumping with Phosphates Before Cooking

In general, the white and light meat tuna are canned and sold for human consumption, while the dark portion of the tuna is sold as cat food or otherwise used in animal food mixtures. Obviously, it is highly desirable to be able to increase the yield of white and light meat because of the high price it commands as human food.

W.E. Swartz; U.S. Patent 3,493,392; February 3, 1970; assigned to Calgon Corp. has found that if tuna is pumped with a solution of molecularly dehydrated phosphates or an orthophosphate alone or in admixture with salt, the yield of white and light meat can be increased as much as 15% or more as compared to the yield of untreated tuna where the cooking is carried out in steam alone. In addition to the increased yield of white and light meat, the flavor and odor are more delicate, being less fishy. Moreover, the cooked meat is more tender and less dry in the treated as compared to untreated tuna.

Preferably, a solution of salt and molecularly dehydrated phosphate, such as sodium tripolyphosphate or mixtures of sodium tripolyphosphate and sodium hexametaphosphate, or sodium pyrophosphate or a solution of sodium orthophosphate is pumped into the thawed flesh of the tuna before cooking to provide a level of about 0.5% added phosphate in the flesh and thereafter tuna is cooked as usual and the meat removed in the usual manner.

While a solution of salt and phosphate is preferred, the phosphate alone will also provide marked improvement in yield of white and light meat as well as in flavor and odor, although not to the degree generally achieved by the salt-phosphate mixture. The level of phosphate addition should preferably be between 0.1 and 1.0%. This process is peculiarly advantageous in its effect on white and light meat yield.

Example: Two whole albacore, weighing approximately 17 and 19 lb respectively, were used. After thawing, they were eviscerated and the gills and head were removed. One received no treatment. The other was injected through the skin with a 5% solution of sodium tripolyphosphate using a large hypodermic needle to provide a level of about 0.4% phosphate in the meat. After holding for 1 hr at room temperature, they were cooked together in the autoclave.

A temperature of 220°F and pressure corresponding to such temperature in the autoclave was reached in 30 min and this temperature was maintained for 1½ hr. The internal temperature of the meat at the end of the cook period was 168° to 170°F. After cooling in air at room temperature for 2 hr, they were covered and refrigerated overnight. The following day the cooked weights were taken and the white and light meat was separated and weighed. Yield data are as follows.

Treatment	Cooked Yield Based on Initial Weight (percent)	White or Light Meat Based on Initial Weight (percent)	Added Phosphate as Tri-Poly (percent)
Control, none	71.0	47.7	–
Treated pumped, 8% of 5% tripoly	77.1	50.8	0.42

The yields are calculated on the basis of the thawed uneviscerated weights. The added phosphate is calculated from an analysis of the cooked white or light meat. The treated meat was definitely superior to the untreated in odor, flavor and texture. Portions of the cooked meat were packed tightly into 2 oz screw capped jars. These were cooked by placing in a preheated autoclave and raising the temperature to 240°F as if the cooked meat were canned. This required 15 min. After cooling to approximately room temperature, no cooked out juice could be obtained from either sample. If anything, the cooked meat was drier than it was before cooking.

In a similar process, *W.E. Swartz; U.S. Patent 3,620,767; November 16, 1971; assigned to Calgon Corporation* describes the effect of the phosphate treatment on bonito, to improve its yield, taste and the texture of its light meat. The process is illustrated by the following example.

Example: Four frozen bonito were thawed in ice water overnight to an internal temperature of approximately 33°F. The fish were then weighed and pumped with a manual stitch pumper at 40 lb pressure using a three needle manifold. Injection was at 1" intervals first on one side of the bonito and then on the other. The fish were then drained, weighed and cooked at the same time at 220°F for 1 hr and 15 min. They were then cooled at room temperature and weighed. Phosphate and salt analyses were run on the cooked meat. The data appears in the table below.

		-----	-----	Percent	-----	-----
Sample	Treatment	Pump	Cook* Yield	P_2O_5	Added Phosphate	Salt
1	5% tripoly	10.1	91.1	0.92	0.59	0.20
2	5% tripoly and 10% salt	5.7	87.3	0.84	0.45	1.42
3	None	–	85.0	0.57	–	0.21
4	5% meta	5.1	88.2	0.91	0.48	0.17

*Based on initial raw weight.

PRECOOKING TUNA

Cleaning and Precooking Method

The excessive and uneven cooking necessitated by precooking the whole fish, as described in the beginning of this chapter, has the following distinct and very serious disadvantages: The excessive cooking time and subsequent cooling time required before the fish can be further processed; the excessive time during which the fish are exposed to the air and consequent enzyme action; the losses resulting from shrinkage due to the loss of natural oils, juices and moisture; the lack of uniform cooking, the exterior portions of the fish being usually overcooked while the interior portions thereof are undercooked; the undercooking of large fish and the overcooking of small fish cooked in the same batch; running of the color of the dark undesirable portions of the fish into the light edible portions of the fish during cooking and subsequent cooling; and the excessive cost resulting from increased labor, fuel and cooking facilities.

L.K. Anderson, P.E. Jann and E.H. Carruthers; U.S. Patent 2,954,298; Sept. 27, 1960; assigned to Fish Processes Corporation describe a process for preparing tuna for canning which effects a material reduction in precooking time and which provides an improved method for the removal of tuna loins from the skeletal structure of the fish. This process may be more clearly explained by reference to Figures 2.2a through 2.2g.

The skeletal structure of a tuna fish comprises a backbone or spinal formation indicated by **11** which is approximately centered with respect to the fish and extends from the head into the tail. The bones of the spinal formation are connected by the intervertebral pod **12** (see Figure 2.2b). The intervertebral pod is of somewhat cartilaginous nature and allows the spinal formation to bend or flex.

Extending vertically upward and integral with the bones of the spinal formation are what are known as neural spines **13** and extending vertically downward and integrally secured to the bones of the spinal formation are what are known as haemal spines **14**. The neural and haemal spines are shown in Figure 2.2b. The spines and the bones of the spinal formation are sometimes termed the backbone.

Referring to Figure 2.2c, attached to the bones of the spinal formation by ligaments at **15** and extending outwardly on each side of the spinal formation are epicentral ribs **16**. These epicentral ribs, unlike the neural and haemal spines, are not integral with the bones of the spinal formation. They are more readily loosened from the spinal formation than are the spines, when heat or other tissue or ligament weakening media are applied to them. Between adjacent spines and adjacent ribs and extending longitudinally of the fish and outwardly from the spinal formation is a tough membrane **17** (Figure 2.2d) which is known as the myoseptum.

The membrane between neural spines is called the dorsal skeletogenous septum and the membrane between the haemal spines is called the ventral skeletogenous septum. The myosepta between the epicentral ribs while less pronounced than the myosepta between the spines, at least in tuna, are tough and strong. The tuna is thus divided into four sections distinctly separated from each other by continuous walls which define natural planes of cleavage.

These natural planes of a cleavage are made up of the bones which extend essentially radially with respect to the spinal formation and the membranes extending between the bones. These partitions divide the meaty or muscle portions of the fish into two dorsal loins **18** sometimes termed the right and left epaxial muscles and two ventral loins **19** sometimes termed the right and left hypaxial muscles. These fleshy or muscle portions of the fish are called loins. This method thus takes advantage of these natural planes of separation to accomplish the separation of the fish into four loins in a minimum of time, the cleaning of the fish in an expeditious manner, the facilitation of the removal of the skin, and the application of better methods of precooking and cooling the fish prior to canning.

FIGURE 2.2: PREPARATION OF TUNA FISH FOR CANNING

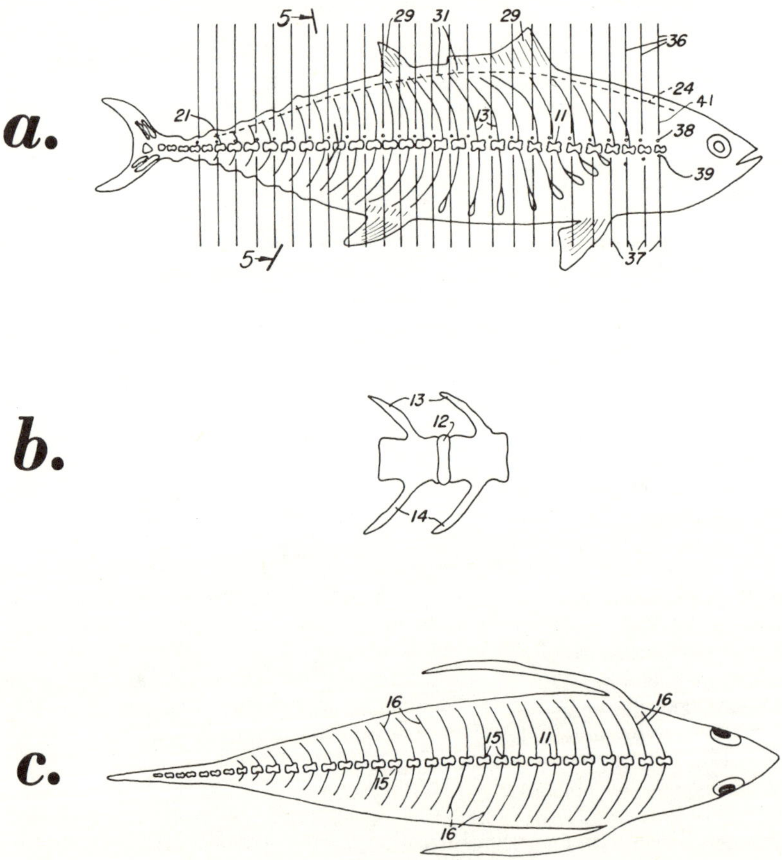

(a) Side View of Skeletal Structure
(b) Enlarged View of Spinal Column
(c) Top View

(continued)

FIGURE 2.2: (continued)

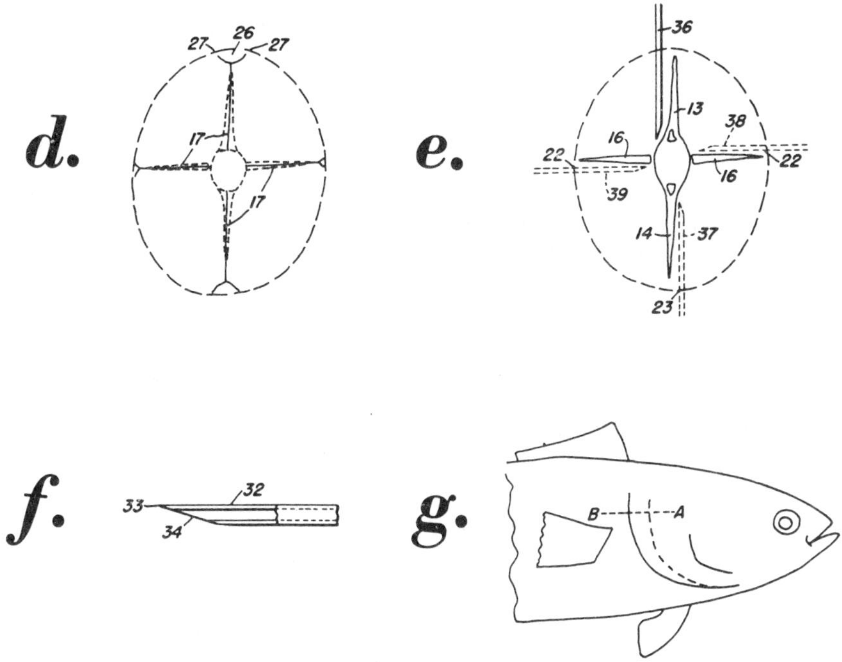

(d) Cross Section Showing Loins
(e) Sectional View Along Line **5-5** of Figure 2.2a
(f) Steam Needle
(g) Collar

Source: L.K. Anderson, P.E. Jann and E.H. Carruthers; U.S. Patent 2,954,298; September 27, 1960

After the fish has been eviscerated, and the head removed in the usual way, two cuts are made through the collar found adjacent the gills, one cut on each side of the body on the line **A–B** shown in Figure 2.2g. The tail is then removed by severance in the region of the caudal peduncle, **21** (Figure 2.2a). The reason for removal of the tail is that it has strong tendons which if not removed make parting of the loins very difficult.

The skin is then slit by means of a knife longitudinally of the fish on both sides thereof substantially in the plane of the epicentral ribs along the lines indicated by **22** in Figure 2.2e. The skin is also preferably slit along a line longitudinally of the fish approximating extensions of the belly split previously made, that is, along the line indicated by **23**. Across the back of the fish as indicated by the dotted lines **24** in Figure 2.2a, a relatively tough strip of cartilage and fine bones exist known as the supracarinales muscle. This strip, indicated at **26** in Figure 2.2d, may be torn out longitudinally by grasping the loose end formed by the cut at **21**. This strip is roughly V-shaped in cross section and carries the first and second dorsal fins **29** and the interneural bones **31**. After the longitudinal incisions **27** and the cut **21** have been made the stripping is easily accomplished and not only are the dorsal fins removed but also substantially all the interneural bones are carried away with strip. A similar strip occurs on the periphery of the fish at the junction of the two ventral loins and may be similarly removed if desired.

The tuna is now in condition to have applied thereto means for denaturing the proteins along the planes of cleavage. Denaturing is preferably accomplished by heating with steam. The preferred method of applying the steam to the fish is by means of hollow tubular members or needles **32** shown in Figure 2.2f. These tubular needles have a sharpened end **33** to facilitate piercing of the flesh and a discharge opening **34**. A conventional 3½" No. 13 stainless steel needle is satisfactory.

Employing dry steam at about 20 lb pressure and 260°F which gave a temperature nominally somewhat greater than 212°F as the steam emerged from the hypodermic needles gave very satisfactory results. The layers of flesh adjacent the cleavage planes were denatured to an extent sufficient that the four loins of the fish could be readily separated or parted intact as whole loins from the skeletal structure. The heating time required with steam at this temperature is of short duration.

The next step is to skin the fish. In this process, the skin is attached to four long loins, each of which is of a nature such that the fish may be placed skin down substantially flat on a planar surface. This is of material advantage in facilitating the skinning of the fish and cannot be accomplished when the fish is in the round.

In the preferred method of skinning the separated loins are placed skin down for approximately 15 sec on a perforated plate through the perforations of which steam is discharged and impinges on the skin. Preferably the steam is superheated to about 200°C and emerges through the perforations of the plate at approximately atmospheric pressure. With superheated steam at about 200°C with a very short interval of cooking, the flesh underlying the skin is not cooked to any appreciable extent but the layer of material beneath the skin sometimes termed the tela subcutanea is preferably denatured.

The skin remains moist for relatively easy removal, and can be removed by holding the fish in contact with a rotating brush with relatively stiff bristles. To secure best results in skinning the loins should not be desiccated as drying of the fish tends to cause the skin to adhere to the underlying flesh. To avoid desiccation of the skin, if the fish are to be placed in cold storage for any considerable length of time they should preferably be glazed.

The next step in the fish cleaning process is to remove the dark meat. Some of this dark meat is easily removed as it has an integument or covering which serves as a parting so that it can be easily pulled away from the loins. Other portions of the dark meat must be scraped away with a knife. Adjacent the dark meat there is usually white or light meat that has been discolored by running of color from the dark meat into the light meat. This light but slightly discolored meat is preferably scraped or cut away also. It may be packed separately from the light meat and sold at a discount or it may be sold as scrap. After the loins have been completely cleaned they are ready to be precooked. If the loins are from a large fish they may be split to convenient size for precooking. In any precooking batch it is desirable that the loins or cut up loins from large fish be of substantially uniform size.

After the loins have been precooked they are allowed to cool sufficiently to enable them to be packed. Here again a distinct advantage results from the application of this method because of two factors: the loins of fish being smaller than the whole fish, cool more rapidly; and since the loins need not be manually handled after cooling they need only be cooled sufficiently to enable them to pass through the can packing machine. After the salt and vegetable oil have been added, the filled can is sealed and retorted in the usual manner.

Cooking Tuna Inside and Outside Simultaneously

In an improvement on the previous process, *E.H. Carruthers, L.K. Anderson and P.E. Jann; U.S. Patent 3,152,912; October 13, 1964; assigned to Fish Processes Corporation* offer a process for precooking the tuna both from the inside and from the outside. A suitable apparatus is shown diagrammatically in Figure 2.3a.

The fish are placed as shown in dotted lines on to a belt conveyor indicated by **51**. As the belt travels in the direction indicated by the arrow **C**, it passes through a steam chest or steam tunnel indicated by the numeral **52**. The steam tunnel comprises a walled member having an outer wall **53** and an inner wall **54** which are supported by members indicated at **56**. The inner wall is provided with a multiplicity of perforations **57** preferably throughout its extent both vertically and horizontally.

Steam is applied to the external portions of the fish through these perforations as the fish travel along the conveyor. A duplicate steam chest on the opposite side of the tunnel has been indicated by **58** and **59**. Steam is brought to the steam chests by a manifold **61** from which pipes **62** bring the steam into the steam chest. A steam return line has been indicated at **63**. A similar manifold and steam pipes are provided on the other side of the tunnel.

The conveyor may be driven continuously in which case an elongated steam tunnel would be required and the relationship of the length of the tunnel and the speed of travel of the conveyor would be regulated so that the fish would be cooked by the time they reach the discharge end of the tunnel. If, on the other hand, a steam chest is employed, it would be desirable to drive the fish belt intermittently. The fish would then be allowed to remain in the steam chest during the pause for a sufficient length of time to effect the cooking.

When the fish is cooked both internally and externally, the time of the cook may be substantially reduced. It will be appreciated that the temperature and time can not be particularly set forth for the reason that fish vary in size from small skipjack of 8 or 10 lb to large tuna of the yellowfin and bluefin varieties which may weigh up to 300 or more pounds. While it is desirable that the fish be needled or cooked internally while they are passing through the tunnel or steam chest, it is also contemplated that the needles may be inserted in banks extending around the tuna and longitudinally thereof from substantially the head to the tail either prior to or subsequent to the external cooking.

FIGURE 2.3: COOKING TUNA INSIDE AND OUTSIDE SIMULTANEOUSLY

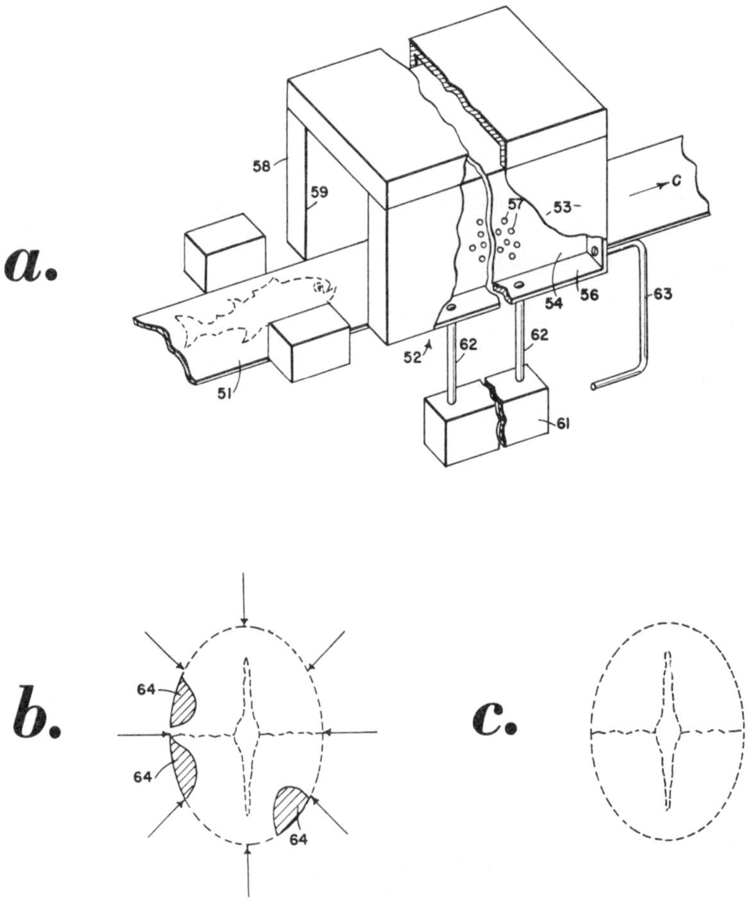

Source: E.H. Carruthers, L.K. Anderson and P.E. Jann; U.S. Patent 3,152,912; October 13, 1964

Preferably, however, the banks of needles should be applied while the fish is in the steam chest or while the tuna is moving along the belt. It will be understood if the tuna is moved continuously that the bank of needles would be carried by the conveyor while if an intermittent operation is employed, the bank of steam needles would be supported from the interior side walls of the steam chest. When the needles are applied internally, occasional raw spots are encountered, particularly when the tuna is not cooked externally. As indicated by the arrows, it is also possible to increase the amount of steam applied so as to reduce raw spots even when external cooking is not used by increasing the number of angles at which the needles are applied as indicated by the arrowed lines in Figure 2.3b.

As indicated in that view, the bank of needles in each cross section of the fish comprises eight needles. Any desired number of needles may be employed in the bank both radially and longitudinally of the fish. When the multiplicity of needles cooking the fish internally, as shown in Figure 2.3b, are employed, raw spots are minimized. This is particularly true if the fish is cooked externally simultaneously. In Figure 2.3c, a fish completely cooked is indicated diagrammatically by the application of a multiplicity of needles, illustrated in Figure 2.3b, and by the use of a steam chest or steam tunnel as illustrated diagrammatically in Figure 2.3a.

Precooking at Lowered Pressures

The conventional method of precooking fish gives rise to several marked disadvantages. In order to cook the inner portions of the fish at the desired temperature the external portions thereof are often subjected to the action of the steam to such an extent that they are scorched. This scorching results in a darkening or browning of the meat. Since it is important that the meat to be canned is of a white or light color, such brown discoloration must be removed by skillful hand scraping. This scraping not only reduces the yield of usable meat but it additionally requires an added expenditure of labor.

It has also been determined that because of the high cooking temperatures the rate of heat exchange between the steam and the meat of the fish causes the meat's cell structure to expand so rapidly that the fish juices undergo excessive evaporation. Such excessive evaporation causes a decrease in the ultimate yield realized from the fish.

Since the cost of the raw fish generally represents approximately 70% of the total cost of the canning operation, this is a serious problem. An additional disadvantage of the conventional precooking operation results from the considerable number of fish loins that are split because of exposure to the high temperature steam. Such splitting not only renders the cleaning operation more time-consuming but it also lowers the percentage of meat available for solid pack canning.

R.T. Erickson and M. Loewe; U.S. Patent 2,919,988; January 5, 1960; assigned to Star-Kist Foods, Inc. describe a method to eliminate each of the aforementioned disadvantages of the existing conventional method of precooking fish for canning. The process may be conducted with existing equipment after effecting slight and comparatively inexpensive modification of such equipment, and appreciably raises the yield of usable meat.

Figures 2.4a through 2.4d inclusive serve to illustrate the process. Precooking oven **O** is of generally rectangular configuration and includes a loading door **10** at its left hand portion and an unloading door **12** at its right hand portion. The interior of this oven is supplied with steam by conventional steam nozzles **14**. These nozzles are in communication with a suitable source of steam (not shown).

The interior of the oven is also formed with one or more openings **18** having communication with a source of vacuum, such as a vacuum pump or steam jet (not shown). A suitable water drain (not shown) is provided for the floor of the oven. Referring to Figure 2.4b, the fish to be prepared for canning are disposed within wire baskets that are loaded upon wheeled racks **22**. As indicated, the loaded racks **22** are moved into the precooking oven through the loading door, the unloading door being closed. Thereafter, the loading door **10** is closed and with both doors shut the interior of the oven is substantially isolated with respect to the atmosphere.

FIGURE 2.4: PRECOOKING OF TUNA TO RAISE YIELD OF HIGH QUALITY MEAT

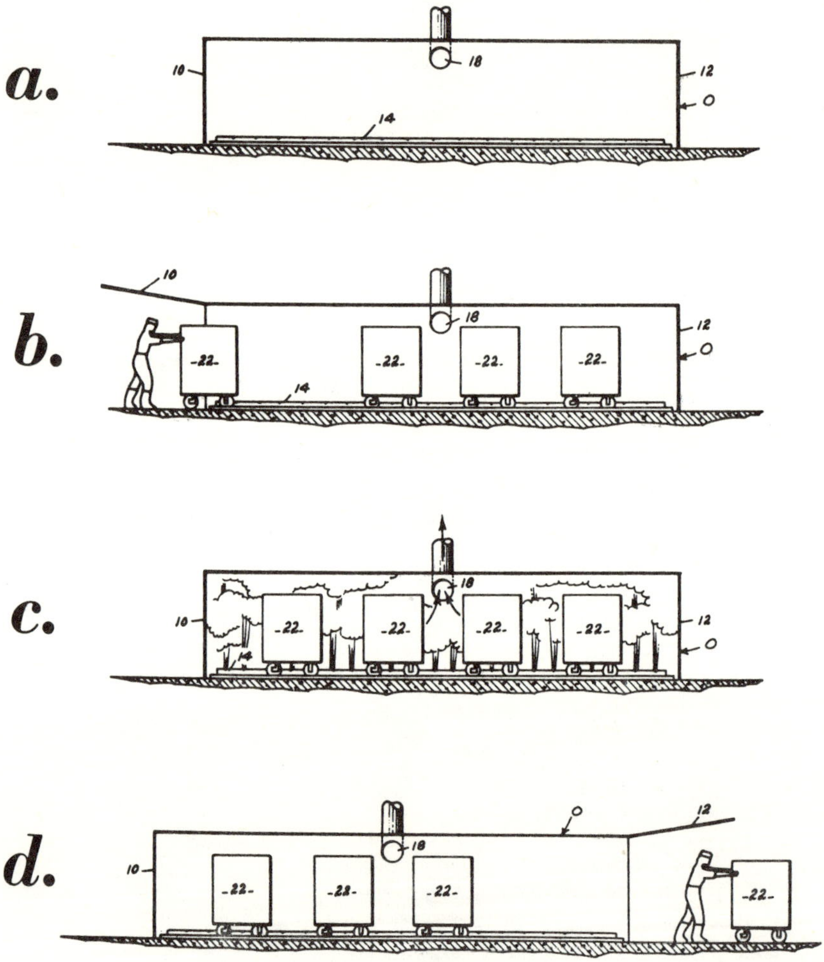

(a) Side View of Oven
(b) Loading of Oven
(c) Side View of Cooking
(d) Unloading of Oven

Source: R.T. Erickson and M. Loewe; U.S. Patent 2,919,988; January 5, 1960

Referring to Figure 2.4c, with the loading and unloading doors shut, steam is admitted to the oven through the nozzles **14**. As this steam is admitted the air originally contained within the oven will be replaced either through the opening **18** or through conventional one-way vents (not shown). When the air has been substantially displaced from the oven, the steam will be admitted at a slower rate. This rate of steam admission should be less than that required to replace the loss of steam caused by condensation as the steam heats the fish, by radiation from the oven walls and any other normal losses.

Accordingly, such condensation will effect a pressure drop within the oven below atmospheric pressure. Since the oven interior then exists at subatmospheric pressure the fish may be cooked at a temperature considerably lower than the conventional temperature of 212° to 220°F, the exact temperature being proportional to the pressure and the time of cooking. Preferably, any air entering the oven by leakage will be drawn off through the opening **18**.

In practice, a pressure range of 10 to 14 psia has been found satisfactory, with a pressure of approximately 11.5 psia being preferable. The temperature within the oven should be in the order of 200°F. Preferably, the oven will be maintained free of air with such air being completely replaced by steam, i.e., water vapor. The fish should be cooked under these conditions until the temperature at the densest portion of the fish meat reaches 160° to 180°F. This generally requires from 1 hr and 15 min to 7 hr, depending upon the size of the fish being cooked.

It has been determined that as a result of the utilization of the lower cooking temperature of the described method, the external portions of the meat will not be scorched. Additionally, since the cell structure of the fish meat will expand at a comparatively low rate, excessive evaporative losses of the fish juices are avoided. Moreover, the usual splitting of the fish loins is likewise avoided.

Fish Wrapped in Semipermeable Film

T.P. McConville; U.S. Patent 3,024,114; March 6, 1962; assigned to South Coast Fisheries, Inc. improves on previous processes by wrapping the fish in a semipermeable film prior to the precooking operation with such film restricting evaporation of the natural fish juices during the precooking operation, while restraining direct contact of oxygen or steam with the fish meat, such film further permitting the escape of the gases naturally present within the meat, the escape of such gases being necessary in order to prevent the canned meat from having an undesirable taste and odor.

According to this procedure tuna-type fish are precooked in conventional open topped wire precooking baskets which are each provided with a sheet of semipermeable film having greater dimensions than the basket so that the sides and ends of the sheet will overlap the upper edges of the basket when such sheets are placed therein.

It has been found that various synthetic plastic materials may be employed for such sheets. Polyethylene has proven to be especially satisfactory while polystyrene, cellophane, Pliofilm and Saran have also been utilized. It is essential that such film be substantially moisture-proof, yet it is desirable that certain gases be able to freely pass through the pores of the film. After the film has been placed within the basket one or more fish are deposited within the confines of the film, and portions of the film extending above the upper edges of the basket are folded inwardly over the top of the fish. Preferably, these overlapping flaps are not sealed but merely abut one another. The baskets are then moved into a precooking oven, of generally conventional construction, provided with steam nozzles.

When all of the baskets have been moved into the oven, the doors are closed and steam is admitted through the nozzles raising the temperature within the oven to between 212° to 220°F. The time required for the precooking step will vary in accordance with the size and quantity of the fish being processed. Generally, however, this time will vary between 3 to 10 hr.

During this precooking operation the semipermeable film will restrain direct contact of the atmosphere (and hence oxygen) or a mixture of the atmosphere and steam within the oven with the fish meat. Accordingly, the discoloration which normally results from such contact during the conventional precooking operation will be prevented.

Additionally, the film will restrain the unrestricted evaporation of the natural fish juices which takes place in a conventional precooking operation. It is essential, however, that the film allows the gases naturally present within the fish and released during the cooking to escape. Unless such gases are permitted to escape the meat will take on a disagreeable odor and taste which would render it undesirable for human consumption.

At the conclusion of the precooking step, water is admitted to the interior of the precooking oven by nozzles to effect rapid cooling of the precooked fish. During this cooling step, the film will restrain evaporation of the moisture present within the fish meat while continuing to restrict direct contact of oxygen with the fish meat.

Additionally, the film will prevent the meat from soaking up any of the water introduced into the precooking oven. When the temperature of the fish has been reduced to the point that they may undergo manual handling, the baskets are removed from the oven, and the fish removed from the film. The utilization of the film will eliminate any sticking of the meat to the wire baskets. Accordingly, none of the meat will be lost as would be the case where such film was not utilized. Additionally, the job of cleaning the baskets for subsequent use is made far easier than would be the case if such film were not employed. The fish meat separates readily from the film.

SKINNING

By Caustic

The precooking of tuna fish generally takes place in steam ovens and lasts for a period of time between two and eight hours depending upon the size of the fish. After precooking the fish are allowed to cool for up to twenty-four hours at which time they are taken from the baskets and placed on trimming tables where workers remove the skin, bones, and blood spots. From both a time and a labor standpoint, the skin removal step is a major one in the overall fish processing operation.

The workers must strip the skin with their hands using a suitable knife utensil as an assist, which procedure makes the skinning operation both tedious and time consuming. In fact, when fish are skinned in this manner, forty workers are required to skin enough tuna to keep one conventional tuna canning machine continuously supplied with tuna. Accordingly, a more expedient and labor saving means for skinning tuna, as well as other species of fish useful for commercial processing, is desirable.

S.A. Mencacci, J.H. Strasser and L.B. Toran; U.S. Patent 3,806,616; April 23, 1974; assigned to International Machinery Corporation SA, Belgium have found that fish with scales, such as tuna, may be treated with caustic solution to make their skin easily removable. The caustic solution will not readily affect the scales themselves to any significant degree but will only decompose and at least partially dissolve the skin in which the scales are embedded to make the skin as well as the scales easily removable.

Tuna fish which have been brought to a tuna plant from the fishing vessels are first precooked with hot water or steam to prepare the fish for the caustic treatment. The precooking step softens the connective tissue in the skin to make the tissue more receptive to the caustic solution and also firms the underlying meat or muscle to prevent the caustic from having a detrimental effect thereon. The skin is defined as including the epidermis, dermis, and underlying subcutis.

After precooking, the fish are exposed to a caustic solution which saponifies the fats in the skin and causes at least a partial dissolving of the skin. Any skin which is not decomposed is loosened from the meat by the particularly effective action of the caustic solution on the fats and on the collagen in the dermis and subcutis. Once the skin has been decomposed by the caustic solution the fish are subjected to skin removal means which removes the remaining skin from the meat along with substantially all of the caustic solution that remains on the fish after the caustic treatment.

The tuna are then beheaded, eviscerated, and directed into a cooker which will remove any remaining residue of caustic solution. Cooking the fish makes the meat suitable for subsequent handling and for easier bone and blood spot removal. The term caustic solution defines a solution of a strong alkali material, such as sodium hydroxide or potassium hydroxide, in sufficient strength to attack and decompose animal matter.

Referring to Figure 2.5, a diagrammatic representation of an apparatus for carrying out the process is shown. It will be seen that a fish **F** having scales, such as a tuna, which is treated in accordance with the process encounters six separate operations. The operations are specifically: precooking, pricking, caustic application, washing, evisceration and cooking.

It is to be understood that the temperature and exposure time ranges for the above operations, as set forth here, have been determined on the basis of the treatment of tuna which were not completely frozen but which had been partially thawed to approximate the condition at which such fish normally arrive at a commercial processing plant. During the precooking operation the tuna fish are confined between a pair of opposed endless screen conveyors **20** which are driven (by means not shown) to advance the fish in the direction of the arrow. The screen conveyors define a trough open at both the top and bottom which will support fish of varying sizes within a predetermined range. Mounted above and below the conveyors are water headers **22** and **23**, respectively, having spray nozzles **24** and **25**, respectively, evenly distributed along their length. The headers are connected to pressurized water supplies which are not shown. The water supplied to the headers is preheated to have a temperature in the range of 90° to 100°C, and with the water in this temperature range, the conveyors are driven at a speed such that the fish are held in the precooking portion of the apparatus for 5 to 15 min depending on the thickness of the skin of the fish. 100°C steam could also be used to precook the tuna fish rather than the heated water.

Fish leaving the precooking operation are passed first between upper and lower prickers **26** and **28**, respectively. The prickers are rotatable members each having an annular fish engaging surface **29** which is concave in longitudinal cross section. Each pricker is provided with a plurality of randomly oriented needles **30** which extend normally away from the surface in a position to prick the skin of the fish as it passes adjacent to the prickers. The needles are long enough to extend deep into the skin. Downstream from and immediately adjacent to the upper and lower prickers are side prickers **44** and **46**. These are similar in shape and construction to the prickers **26** and **28** but are larger to conform to the large side surfaces of a fish.

The prickers distribute a plurality of perforations **60** in the skin of the fish as the fish are advanced through the pricking operation. The pricking of the fish enables a more efficient and effective skin removal in the case of tuna fish. Upon leaving the pricking operation, the fish are passed between screen conveyors **62** which define a trough in which the fish are confined as they are advanced through both the caustic application and the wash operations.

Immediately upon being passed to these conveyors a caustic solution is sprayed onto the fish. A caustic solution which has been found suitable for saponifying the fats and decomposing the collagen in the skin of a tuna is a sodium hydroxide solution with about a 5 to 10% concentration by weight. The caustic solution is sprayed on the fish through an upper header **64** and a lower header **66** each having a plurality of spray nozzles **68** distributed along the length thereof whereby the solution will be distributed about the entire outer surface of the fish.

FIGURE 2.5: SKINNING TUNA FISH

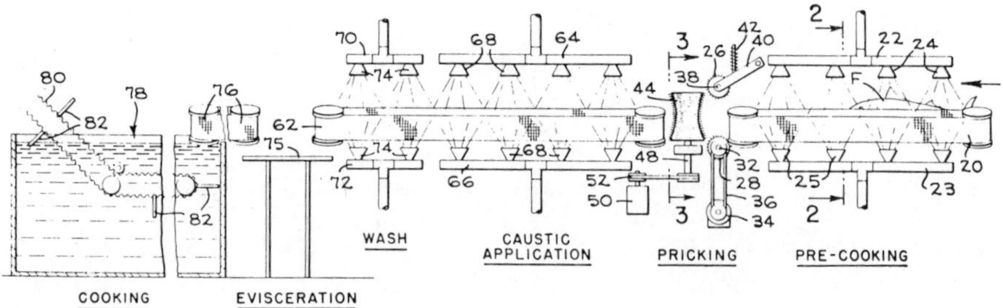

Source: S.A. Mencacci, J.H. Strasser and L.B. Toran; U.S. Patent 3,806,616; April 23, 1974

For best results, the caustic solution should be heated to a temperature in the range between about 90° and 100°C, and it should be sprayed on each individual fish for a period of time between about 5 and 15 min. The speed of the conveyors is adjusted so that the desired caustic exposure time is achieved. The required exposure time to the spray as well as the temperature of the solution is dependent upon the thickness of the skin of the fish, however, for a typical tuna fish of the albacore variety the aforementioned ranges were found to be the most efficient.

In some instances, where the fish being processed have an abundance of scales, it is desirable to introduce a wetting agent, such as Tergitol, into the caustic solution to reduce its surface tension whereby the caustic can more easily pass beneath the scales to penetrate the epidermis and attack the lower layers of the skin.

While still being transported by the conveyors the fish are passed into the wash operation where hot water is used to wash the remaining decomposed skin as well as the excess caustic solution from the underlying meat of the fish. Upper and lower water headers **70** and **72**, respectively, are positioned about the conveyor and are each provided with jet spray nozzles **74** distributed along their length for spraying high velocity hot water on the fish.

The water preferably has a temperature in the range of from about 90° to 100°C, and it is sprayed on each fish for a period of time of from about 1 to 5 min after which time the skin will have been completely washed from the remainder of the fish and collected in a suitable receptacle (not shown). After washing, the fish are deposited on an eviscerating table where a minimal number of workers are used to eviscerate each fish. The fish are then placed by the workers between a pair of transfer conveyors **20** and **62** so that the fish will be conveyed to a cooking bath.

Example: A section of tuna fish was skinned successfully by first steam precooking the fish in a pressure cooker at 100°C for 10 min. The skin of the fish was then perforated with a fork to allow the caustic to more quickly penetrate the skin. A 5% concentration solution of NaOH at a temperature of 95°C was then brushed on the skin of the fish for a 5 min period. At the end of that time, the decomposed skin was washed away from the meat of the fish with 95°C water. The water was sprayed across the fish for a 1 min period after which it was noted that all of the skin of the fish had been thoroughly removed with no noticeable damage to the meat.

By Enzymes

A process developed by *G.B. Fehmerling; U.S. Patent 3,729,324; April 24, 1973; assigned to Ocean Research Corporation* will be fully described in Chapter 4, in the section on bivalves, entitled Enzymatic Separation of Undesirable Tissue from Clams. This process is also applicable for the removal of skin from various fish, including tuna, as illustrated by the following examples.

Example 1: Ten whole skipjack tuna, previously steamed to backbone temperature of 140°F in steam cabinets, cooled to room temperature, were used in this demonstration. The fish weighed from 6 to 10 lb each and the solution consisted of 40 g of a mixture of enzymes and 35 gal of water.

The enzyme mixture comprised 50% proteolytic, 5% glycoside hydrolase, 2% pectolytic, 35% lipolytic, 3% pentosanase, and 5% cellulolytic. Five fish selected at random were placed in a stainless steel wire basket suspended in a stainless steel jacketed kettle in which the enzyme solution was maintained at 120°F by circulating water at 120°F through the jacket of the kettle. The solution was agitated by bubbling compressed air through the solution surrounding the fish. Tuna were immersed in the enzyme solution for a period of 10 min.

Five fish selected at random were treated in exactly the same manner as described above except that the water contained no enzymes, these fish were used for controls. To compare the ease of skin removal, the enzyme treated fish were placed individually in wire baskets and subjected to sprays of water from a hose with a nozzle providing water pressure of about 40 psi. Approximately 85% of the skin was readily removed from the tuna by water spray with very slight attendant flesh loss accompanying the skin removal.

The control fish, however, resisted efforts of skin removal by water spray except with extremely vigorous spraying which caused a heavy flesh loss and was considered by tuna canning experts as being entirely prohibitive. It is believed that the fish skin is rendered more readily removable due, at least in part, to the softening of the fatty connective tissue between the skin and the flesh.

Example 2: Frozen skipjack tuna were heated in conventional steam cabinets on racks to an interior temperature at backbone of 140°F. Twelve tuna were cooled on racks to room temperature. Six tuna were placed on a stainless steel mesh rack and immersed in an aqueous solution containing 0.026% enzyme mixture buffered to a pH of 6.00 with monopotassium phosphate. The enzyme mixture contained 58% proteolytic, 5% glycoside hydrolase, 2% pectolytic, 35% lipolytic.

Temperature of the solution was maintained at 125°F in the jacketed kettle containing the solution and fish by circulating water at 125°F through the jacket of the kettle. The tuna remained in the solution for 35 min, the solution continuously agitated by blowing compressed air through it. Tuna were removed and sprayed with water at 40 psi on all sides to remove the undesirable edible skin from the desirable edible flesh. Skin removal was judged to be 80 to 90% effective without removing any edible flesh from the tuna.

The remaining six tuna were immersed in a water bath under conditions identical to those described herein but without addition of the enzymes. Upon removal from the bath they were subjected to the same type of spray washing with a result of about 10 to 15% removal of skin and severe loss of edible tuna flesh.

Example 3: Six frozen eviscerated, whole albacore tuna were selected for this experiment weighing from 30 to 40 lb each. Three tuna were placed in an aqueous solution containing 0.045% enzyme mixture containing 45% proteolytic, 10% glycoside hydrolase, 42% lipolytic, and 3% pentosanase.

The mixture was maintained at 125°F by circulating water at 135°F through the jacket of the kettle containing the fish and solution. A higher temperature of heating water was necessary to compensate for cooling of the enzyme solution by the frozen fish. Enzyme solution was agitated around the fish by bubbling action of compressed air introduced into the kettle.

Tuna were subjected to this treatment for 30 min. Three tuna were treated in an identical manner in a water bath without the enzyme solution, as controls. The six tuna, appropriately tagged for identification, were placed in a steam cabinet for 4 hr until backbone temperature of 140°F was reached. The tuna were removed from the cabinet and allowed to cool overnight at room temperature. Skin from all of the tuna was accomplished by hand skinning.

Skin on the enzyme treated fish could be grasped at the tail of the fish and pulled free of edible flesh almost in one whole piece with very little force necessary. No hand scraping with knives was required except where breaks in the skin occurred during catching and freezing operations. These small areas required the normal scraping where steaming had caused dehydration and discoloration of the tuna flesh.

Skin on the control fish was still tightly adhering to edible tuna flesh. Only small patches near the tail and along the back fin of the fish could be removed by pulling with the fingers. In some instances, desirable edible flesh adhered to the skin. Approximately 80% of the skin needed vigorous hand scraping with knives to separate skin from edible flesh.

Experienced tuna processors estimated that skin removal from albacore tuna was improved in efficiency by about 75% using enzyme treatment over conventional methods. Incidentally, it should be noted here that the albacore tuna is considered by tuna processors to be the most difficult of all large species of fish to skin because of the extreme adherence of the connective tissue between skin and flesh.

From the foregoing examples, it should be evident that the process enables those processors of marine creatures to more rapidly and economically separate the desirable edible flesh from the undesirable edible tissue, and particularly to remove skin from various fish or to remove the digestive, reproductive, or other tissue, from desirable edible flesh that has been previously removed from various shellfish.

IMPROVED COOLING OF PRECOOKED TUNA

The conventional method of cooling tuna after its precooking is to move it, still in the racks holding the filled wire baskets, into a cooling room, a large room usually held at ambient temperature and humidity. There the fish are left to cool to a temperature where they can be handled by hand, and are firm enough to be cleaned. This conventional cooling process usually takes about 12 to 24 hr, but has often been followed by a 1 to 3 day wait.

The conventional method of cooling the precooked tuna has many drawbacks. For one thing, the surface of the meat turns brown during cooling, resulting in a dark pellicle which has to be removed for it is important that the canned fish meat be white or light colored. However skillfully the brown meat is scraped away, the operation not only reduces the yield of usable meat but also adds to the labor cost. Another problem is the dehydration of the meat, for much of the water in the natural fish juices, possibly 5 to 12% of the total water content of the meat, is lost. A third problem is that, even under optimum conditions, the cooling room could not be maintained truly sanitary.

Mold and bacteria can get on the fish and grow there, causing off-color and off-taste effects. Partly because the temperature and humidity within the cooling room changes under varying atmospheric conditions, it is extremely difficult to schedule the cooling of the fish so that the fish can be sent to the cleaning and packing operations as soon as the precooked fish is cool. Hence, considerable deterioration in meat quality takes place during cooling.

S.H. Lassen; U.S. Patent 3,180,738; April 27, 1965; assigned to Ralston Purina Company has overcome these difficulties and eliminated the cooling room entirely by a process which includes cooling the precooked fish in the cooker while holding the pressure therein at approximately atmospheric and while preventing the access thereto of oxygen, by admitting to the cooker inert gas in quantity sufficient to hold the pressure in the cooker above atmospheric.

The process rests partly on the discovery that the discoloration and degradation result from oxidation, partly on the discovery that the growth of mold and bacteria can be prevented by the substantial exclusion of oxygen from the cooker, and partly on the fact that carbon dioxide gas, used as the inert gas or in the inert gas may enhance the quality of the fish meat.

The process may be practiced with an autoclave or cooker that provides a combination precooking and cooling chamber. The chamber may be supplied with cooking steam by conventional steam nozzles and is also provided with one or more gas inlets that are used during the cooling and are connected by a conduit to a supply of inert gas (e.g., nitrogen or argon) under pressure. Carbon dioxide may be used or may be mixed with an inert gas such as nitrogen or argon.

For cooling, a heat exchanger and a circulating fan are helpful. During the cooling step, but not during the cooking step, cold fluid goes to the heat exchanger returning, if desired, to an evaporator and condenser or other type of cooling apparatus. Fish, in wire baskets on wheeled racks, are cooked by steam at 212° to 220°F in the chamber. When precooking is completed, the flow of steam into the chamber is cut off.

Then the chamber is cooled, either by simply waiting while the heat is dissipated through the walls of the cooker into the atmosphere, or preferably, by use of the cooling system with a heat exchanger. The cooling of the chamber results in reducing the pressure therein, and there is a tendency for vacuum conditions to develop. However, before the chamber drops below atmospheric, a valve admits carbon dioxide-containing gas or inert gas without carbon dioxide, such as nitrogen, into the chamber. Since the relative humidity remains 100, there is little evaporation, and most of the moisture present within the fish is retained therein.

When carbon dioxide is the inert gas or is in the gas, it tends to lower the pH of the fish slightly and thereby helps to prevent the formation of struvite later on after the fish has been canned. If desired, the carbon dioxide gas or the inert gas may be blended with other suitable gases, such as a suitable reducing gas (e.g., hydrogen, sulfur dioxide, or nitrous oxide), in addition to nitrogen or other inert gas to enhance the color of the meat. In practice, assuming the pressure within the cooker approximates 16 psia at the end of precooking and the beginning of the cooling period, that pressure will drop only slightly, if any, below atmospheric before the gas is admitted, so it remains at substantially atmospheric or slightly higher at the end of such period.

The temperature during this time drops from approximately 218° to about 80°F. The time required for this cooling generally lies between 1 to 4 hr, and during this time the cooker remains sterile and anaerobic. The time at which the fish are to be removed from the precooking and cooling chamber can be adjusted to the schedule of the cleaning and packing personnel, thereby avoiding costly overtime. They can be stored there over the weekend, if desired. At the proper time, the wheeled racks are removed from the chamber and transferred directly to the cleaning room.

This method results in a considerable reduction in discoloration of the fish during the cooling period, primarily because of the prevention of oxidation. Also less water is lost. Since the discoloration of the meat of the fish is at a minimum and since less water is lost, the percentage of yield is appreciably increased. Additionally, the costly hand scraping ordinarily required to remove such discolored pellicle is eliminated and the skin does not have to be scraped off, but can be peeled off.

Since the humidity within the precooking and cooling chamber is maintained at the desired 100 relative humidity, the water in natural fish juices will not be subject to evaporative losses during the cooling period. It has also been determined that the fish loins are considerably firmed as compared to the existing methods, and accordingly, the cleaning and packing operations are facilitated.

BUTCHERING

Skinning and Cleaning Tuna in Frozen State

J.M. Lapeyre, to whom we are indebted for the description of the tuna processing industry which appears at the beginning of this chapter, has developed an entirely different method of providing cleaned tuna meat suitable for subsequent processing and packaging. This is described in *J.M. Lapeyre; U.S. Patent 3,593,370; July 20, 1971.*

Unlike the previous practice of skinning and cleaning the fish after cooking, this method performs the operations of skinning and cleaning of the edible portions of the tuna while the latter remain in the original frozen or substantially frozen condition which was effected upon the fresh tuna at sea. The occasional practice of beginning the thawing operation on board the returning fishing vessel must be discontinued if the benefits of this process are to be fully realized. Such discontinuance will also serve to simplify shipboard operations.

With a view to clarity, it should be noted that conventional butchering and cleaning practices ignore the physical properties of frozen tuna. In contradistinction to conventional practices, this process takes advantage of certain physical properties of frozen tuna and/or parts thereof which properties exist only while the tuna is in a frozen state.

Frozen tuna flesh has good mechanical strength in compression. Whereas it does not become as hard as ice formed from pure water at conventional freezer temperatures, it loses its natural supple and deformable gelatinous quality, and becomes hard enough to readily withstand, without damage, those compressive loads as may necessarily be imposed upon it during mechanical handling and/or working of the tuna or parts thereof. Consequently, frozen whole tuna or parts thereof may be handled, for mechanical purposes, much like soft wood. The natural flesh grain or directional alignment of the edible muscle fibers is generally parallel to the longitudinal axis and backbone of the fish. Frozen tuna flesh, somewhat like wood, has relatively weak resistance to cleavage parallel to its natural grain and along certain planes or junctures. This is especially true at the vertical junctures of the right and left epaxial and hypaxial muscles with the vertical skeletal structure, and the horizontal junctures of the epaxial and hypaxial muscles with the horizontal skeletal structure.

The condition of being frozen makes it possible to perform with ease mechanical operations on tuna which would be difficult or impossible if attempted on unfrozen or cooked tuna. Frozen whole tuna and/or parts thereof may be sawed, drilled, routed, carved, die cut, broached, scraped, cleaved and longitudinally split, without diminishing the quality of unworked portions of frozen fish.

After the conventionally practiced cooking of the unskinned fish, the skin of the tuna becomes soft and weak in strength, and is easily torn or fragmented. It is readily rubbed and/or scraped free of the cooked fish by hand with the aid of simple knives or hand scrapers. The conventional practice in the industry of hand skinning after cooking is therefore facilitated by the standard precooking operation. However, damage and wastage of the delicate and friable flesh underlying the skin of the cooked fish is, due to the manual scraping off of the skin, severe, and much of the otherwise perfect loin meat is degraded to fragments or lost to scrap during the conventional skin scraping operation.

In contradistinction, this process includes the removal of the skin while the edible flesh is in a substantially frozen condition. In its natural state, unlike the cooked state, tuna skin is tough and pliable and has good resistance to tearing due to a network of strong fibers with which it is reinforced. Freezing of the skin does not substantially weaken or diminish the toughness or tear strength of the skin but while frozen it is substantially rigid and hard, and it may be removed while in the frozen condition by sanding or rasping.

It thaws quickly, however, when exposed to tap water or ambient air at room temperature, and after thawing to the point of pliability it can be pulled or peeled away from the underlying flesh which if maintained at least in a partially frozen state will remain substantially intact. The skin therefore may be removed in either the frozen condition by sanding or rasping and the like, or it may be removed after at least a partial thawing operation. The latter is preferred since the partial thawing is simply effected and since the apparatus required for pulling or peeling off of the skin from a frozen section, is of the simplest kind such as a pair of hand held pliers.

During the shipboard freezing of the whole tuna, the contents of the abdominal or visceral cavity become substantially solidified. Rather than thaw the fish as is the conventional practice prior to opening the visceral cavity by longitudinally slitting it, it is far more expeditious to crosscut the tuna into sections and remove the viscera from one end of each of the cut sections. In this manner, the possibility of spoilage of the fish, due to thawing, is eliminated. The viscera, if kept frozen, may be stored for separate processing or disposal without the possibility of it becoming unpleasantly odorous. When viewed in cross section, it will be noted that the interior parts of the fish (except for the contents of the visceral cavity) are symmetrical about a vertical plane hypothetically passed through the backbone which is at, approximately, the geometrical center of the fish. In addition to taking advantage of the physical properties of the tuna while it is frozen, this process, as will be shown, makes use of this fact of bilateral symmetry within the fish.

In its broader aspects, this process includes subdividing the principal loin-bearing portions of the tuna generally transverse to the longitudinal axis and the natural flesh grain thereof into a plurality of discrete bilaterally symmetrical frozen sections having a predetermined thickness and thereafter skinning the sections and longitudinally cleaving and/or cutting the sections to effect separation of the scrap parts of the sections from the loin parts while the latter are in a substantially frozen condition.

The process also includes removal of the skin from the frozen sections of tuna by subjecting the skin to at least a partial thaw so as to render it pliable and thereafter pulling or peeling off the skin from the section while the latter remains in a substantially frozen condition. An important advantage of the process is the permitted maintenance of the sea frozen condition, throughout the butchering and cleaning operations, so as to maintain the highest possible quality in the fish until the same is eventually cooked or packaged frozen.

Another important advantage is a permitted flexibility in commercial tuna product production since it permits the packer to produce either or both canned tuna and frozen raw tuna products, such as frozen steaks and the like, from the same supply of whole fish. Other advantages include a substantial reduction in the amount of hand labor required to butcher tuna and a substantial reduction in processing time by the elimination of the need to thaw whole fish.

In Figure 2.6, where symbols refer to like or corresponding parts throughout the several views, the same designation has been used. Figure 2.6a is a perspective view of a tuna which has been subdivided into a plurality of cross sections. Figure 2.6b is a perspective view of an exemplary section taken at midlength of the fish. Figure 2.6c is a perspective view of the major portions of waste material being partially removed from the section.

Tuna, frozen at sea, is received whole at processing plants in frozen, or substantially frozen condition. In accordance with the principles of this process, the received frozen tuna, after desirably being subjected to a preliminary grading or sorting as to size and species, is initially subdivided into a plurality of discrete lateral sections, as for examples, sections numbered **1** through **9** of Figure 2.6a, by cutting the loin bearing portions thereof generally transverse to the longidutinal spine or backbone, as represented in **146** of Figure 2.6b.

Preferably, such lateral sections are of uniform thickness, at least over the extent of the main body of the tuna, and are longitudinally dimensioned so as to be ultimately complemental to the type of pack for which the tuna is intended. Such operation can readily effected by a variety of manual, semiautomatic, or automatic sawing or cutting equipment. The subdivision step not only permits immediate removal of head portions **112** and **112a** and tail portions **114** for separate processing, either immediate or future, but also serves to convert the whole tuna into discrete ready handleable, manipulatable and sortable segments that are of convenient size and shape and, due to their frozen condition, having physical properties, as outlined above, to facilitate subsequent processing steps and operations. It is apparent from Figure 2.6a, that while all of the sections are generally oval shape and all of approximately uniform thickness, the diametric size and internal anatomic character or configuration of the individual sections for a given tuna will vary in accord with the location from which they are derived. In addition, the diametric size of the lateral sections derived from a given tuna will vary in accordance with the size of that particular tuna.

FIGURE 2.6: BUTCHERING FROZEN TUNA

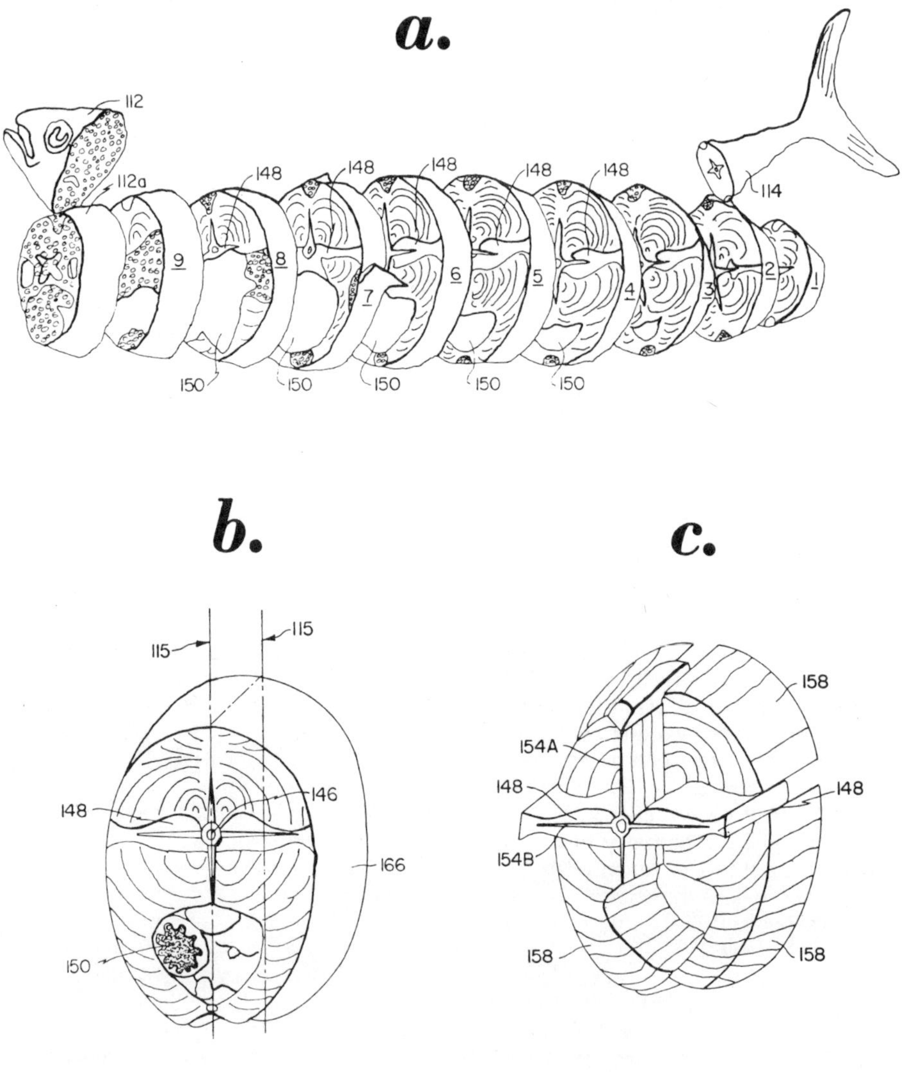

Source: J.M. Lapeyre; U.S. Patent 3,593,370; July 20, 1971

Figure 2.6a is intended to be a schematic illustration of the general internal character or anatomic configuration of a tuna when it is divided into sections. It should be noted that the visceral cavity **150** varies in size, depending upon the location within the fish from which the section is derived. Likewise, the blood meat portions **148** vary in cross sectional configuration depending upon the location from which the section is derived.

The above described subdivision of the whole tuna into discrete lateral sections provides segments of tuna, each of which is laterally symmetrical about a dorsal to ventral plane, as represented by lines **115** in Figure 2.6b, and makes possible the sorting of individual sections obtained from pluralities of tuna into predetermined groupings according to the general internal anatomical characteristics or configurations thereof to markedly facilitate subsequent operations such as evisceration, when required, skinning, removal of the blood meat, skeletal structure, cartilaginous materials and the like.

Moreover, when the whole tuna are sorted as to size and species, as mentioned earlier, and the resulting multiples of the whole tuna of the same size and species are then subdivided into transverse sections, the sorting of the sections in accordance with their anatomical configurations automatically effects a dimensional sorting of the sections as well, which markedly facilitates those subsequent processing operations wherein size may be a critical parameter.

After the subdivision and sorting operations, the sections are subjected to a thermal treatment wherein the peripheral band of skin (**166** in Figure 2.6b) is caused to at least partially thaw. This thawing returns the skin to its natural pliable condition, which condition, coupled with its inherent toughness in the raw condition, permits it to be grasped and pulled free of the section.

Subsequent to the skinning operation, or simultaneously therewith, the frozen viscera may be removed by forcing same out through the larger end of the visceral cavity. The removal of the viscera in this manner may be accomplished with the aid of a simple plunger-type punching device, which is configured to conform to the general cross-sectional outline of the visceral cavity at its smaller end.

In those sections where the cross-sectional areas of the visceral cavity approximates that area which can be covered by the human thumb, removal of the frozen viscera may be accomplished simply by applying thumb pressure to the small end of the visceral cavity, thereby manually forcing out the frozen viscera through the larger end of the visceral cavity.

With a view toward clarity, it should be pointed out that, due to the inherent longitudinal dissymmetry of the visceral cavity, the contents thereof will usually have a tapered and roughly truncated conical shape, one end of which, depending upon where the section of fish has been taken, is usually larger than the opposing end.

Subsequent to the above outlined evisceration procedure, the skeletal structure as represented by **154a** and **154b** in Figure 2.6c, along with the blood meat portions, as represented by **148**, may be removed from the section by coring out. Such an operation is facilitated through the use of cutting dies of cookie cutter principle, suitably configured to cleave the section of tuna along pattern lines which generally follow the junctures of the more valuable loin meat, represented by **158** and the waste material.

Sorting Tuna for Canning

In another patent, *J.M. Lapeyre; U.S. Patent 3,594,191; July 20, 1971* describes an improved method for the commercial processing of canned pack tuna which includes subdividing the principal loin bearing portions of uncooked tuna generally transverse to the longitudinal axis and the natural flesh grain thereof into a plurality of discrete lateral sections having a predetermined thickness with preferential operation including the subdividing of the tuna while in substantially frozen condition into lateral sections of effectively uniform thickness which, after cooking, will complement, either singly or in multiples, the depth of the can to be filled.

In its more narrow aspects, the process includes the additional steps of grading as by sorting and routing, similarly sized and/or anatomically configured laterally cut sections into predetermined groupings to permit both uniformity and automation of further processing, preliminary cleaning of the visceral portions of the lateral sections while in frozen or partly frozen condition, separation of the edible loin portions of these sections from the remainder, and accumulation of the loin portions into a mass of substantially uniform thickness from which selectively sized portions may be removed for introduction into cans or other suitable containers.

The process includes also the separation of the edible loin portions and their accumulation while they are in a raw condition, into a composite mass of suitable geometric configuration having a substantially uniform thickness in cross section so that the natural flesh grain of the loin portions is maintained in generally parallel relationship. The composite mass of edible material may, if desired, be subdivided prior to cooking so that the resultant divisions thereof will, after cooking, be sized to closely complement not only the depth of the can but also the volume thereof.

Among the advantages of the process is a reduction in wastage and upgrading of quality with attendant economic savings including a permitted minimization and standardization of thawing, precooking and cooling times, a substantial reduction in overall processing time from receipt of the frozen tuna to delivery of the canned product, a substantial reduction in the amount of hand labor required to process a given tonnage of tuna, a marked increase in the tonnage of tuna processable per worker and a marked reduction in required physical effort by the workers due to the reduction in size of the workpieces being handled.

Still other advantages include a readily permitted mechanization and automation of operations leading, in substantial degree, to continuous flow operations which may include compact continuous cooking, automatic or semiautomatic eviscerating and skinning of the tuna, and selectively separating the thin belly meat portions from the remainder thereof.

Further advantages include a permitted expeditious removal of the blood meat prior to cooking, if desired, and simplified removal of the loin meats from the skeletal structure with concurrent provision of loin meats cut to desired length and in proper grain orientation with a minimum of attendant flaking or grating.

The patent offers figures and a complete description of the process.

Separating Loin Portions from Waste Portions

In yet another process, *J.M. Lapeyre; U.S. Patent 3,800,363; April 2, 1974; assigned to The Laitrom Corporation* describes a system which provides means for separation of the edible loin portions of tuna slices from associated waste portions along the actual boundary of the edible loins of slices being butchered, rather than along a boundary presumed for particular slices, thus making an improvement in his previously patented butchering system.

In accordance with the process, an automated butchering method and system is provided for determining the actual boundary between the edible loin portions of a frozen transverse tuna slice and the associated waste meat portions, and for providing electrical signals representative of the detected boundary, in response to which cutting apparatus can be controlled for precise separation of the edible loin portions. Substantially all of the edible loin portions are thus separated for subsequent cooking and canning without contamination by a layer of blood meat or skin which can remain if tuna is butchered by conventional techniques.

In practicing the process, a frozen whole tuna is subdivided by cutting the loin-bearing portion into a plurality of discrete transverse slices of predetermined thicknesses. The boundary of the edible loin portions of each transverse frozen tuna slice is determined by scanning a surface of the slice to provide a signal indicative of the reflectivity thereof, which differs for the loin portions, and the waste meat portions which include the blood meat, skin and viscera.

Many different scanning instrumentalities can be employed to provide the signals indicative of the edible loin boundaries. Scanning of the transverse tuna slice can be accomplished, for example, by mechanical means whereby a slice is moved relative to a sensor to provide the delineating signals. Usually, the sensor would be stationary while the tuna slice is caused to move past the sensing station.

Another form of scanner employs electronic or optical scanning wherein a beam of light or other suitable radiant energy is caused to traverse the tuna slice along a predetermined path to provide the intended output signals. The resulting signals can be subsequently employed to control the movement of cutting apparatus for separation of the edible loins from the remainder of the slice.

The scanning of both surfaces of a frozen tuna slice can also be done to provide signal indications of the loin boundaries of the respective surfaces for use in adjusting the cutting angle of the cutting apparatus to follow the loin boundaries which can be tapered through the slice by reason of anatomical configuration.

The patent describes the use of various types of scanners, including photosensors, video detectors, and an operator to move a stylus manually along the boundary of the loin portions of the slice. The cutting apparatus for cutting the edible loin portions along the predetermined boundaries is preferably a high pressure water jet cutter. Such water jet cutters typically operate at a pressure of about 30,000 to 75,000 psi and employ a jet of about 0.005 to 0.015 inch diameter to cut through the frozen tuna slice. A pressure of 40,000 psi and a 0.007 inch nozzle have been found satisfactory as an example. Such cutters are typically operative at a rate of 300 inches/minute and offer the major advantage of providing precision cutting without contact by a cutting tool. The water jet cutter is typically mounted in a movable head which is capable of movement along mutually orthogonal axes such that the cutting head can be controlled by rectangular coordinate driving signals to position the cutting head and cause it to follow a path along the predetermined boundary of the edible loin portions of the tuna slice. Other cutting apparatus can also be employed as desired, such as a knife blade or ultrasonic cutter.

The viscera exhibits a lower reflectivity than the edible loin portions and can also be delineated during the scanning operation. However, it is not usually necessary to remove the viscera by cutting, as in the frozen state the viscera can be simply removed by forcing it out through the larger end of the visceral cavity. A suitably configured plunger, conforming to the general cross-sectional area of the smaller end of the cavity, can typically be employed for this purpose. Such a plunger can be positioned automatically in response to the electrical signals provided by the scanner.

The skeletal structure of the tuna slice is not readily discernible by optical scanning techniques, however, the position of this structure can be readily determined by reason of the bilateral symmetry of the anatomical configuration of the transverse slice and the central disposition of the skeletal structure. The neural spines which generally separate the right and left loin portions are disposed along an axis which substantially bisects the slice.

The neural spine position can thus be ascertained by determining the center line of the blood meat portions which are delineated by scanning of the slice, as described. Removal of the edible loin portions from the associated bone structure can be accomplished by cutting the transverse slice along planes parallel to the neutal spines. Such a cut can be accomplished by a water jet cutter or knife blade, for example, to remove the neural spines and a slight portion of contiguous meat.

This cut can be taken either before or after scanning of the slice. If the cut is provided after scanning of the entire slice, the boundaries of the edible loin portions and the contiguous blood meat portions complete the separation of the edible loins. In the event that the bifurcating cut is taken prior to the scanning operation, the right and left halves of the slice are then individually scanned to delineate the edible loin portions which are then separated.

CANNING

Fancy Solid Pack

The bulk of tuna packed in this country is derived from fishing areas off the coast of Central and South America. Most of the tuna fish caught in these areas are known by the species names yellowfin, skipjack and bluefin. Because of the distances involved from the fishing areas to the packing plants, most of which are located in San Diego, or San Pedro, California, the fish are frozen in shipboard as they are caught and arrive at the packing plants in a frozen condition. Such fish, if they are cleaned and canned immediately after thawing, are usually hard and firm and a good so-called solid pack of tuna may be accomplished with the method and machine of *E.H. Carruthers; U.S. Patents 2,601,093; June 17, 1952;*

and 2,669,378; February 16, 1954. These methods are useful for packing most tuna, whose flesh is firm and hard, but when the flesh of the fish is soft and flabby more difficulty is encountered in packing. The container when opened does not present to the housewife a solid pack of tuna which is entirely pleasing in appearance. Soft tuna may arise from a number of different conditions. For example, the character of the fish may present packing problems or the tuna may be held too long in a warm atmosphere prior to canning. However, the main source of soft fish is albacore tuna.

Albacore tuna is caught along the coast of lower California, primarily along the coast of California, but does occur to and beyond the Canadian Pacific border. Albacore tuna due to the fact that it is caught not too far from the packing plants, is seldom frozen prior to being brought to the packing plants. Usually, the canners make an effort to can such unfrozen tuna promptly, but since most of the albacore tuna is caught during the months of July, August, and September, the air temperature is relatively warm and the fish are relatively soft due to exposure after being caught to warm air. Under such conditions, a satisfactory pack of tuna is difficult to accomplish.

E.H. Carruthers; U.S. Patent 3,124,469; March 10, 1964 has developed a process and apparatus for packing albacore tuna into each of a series of containers, in which the tuna is first guillotined into pieces of the desired height to be placed in the containers, and then continuously fed to a forming station or section where the tuna is molded into the desired shape to be fitted into the containers to be filled, the action being one of forcing the material at the forming station into forming elements to fill any possible voids encompassed by the forming elements.

The preguillotined tuna is fed to a forming area where a pair of forming jaws gradually close and form a section or slug of tuna corresponding in cross-sectional area to the cross-sectional opening in the can or other container, the forming action being opposed by the feeding of the tuna to the end that voids encompassed by the forming elements are filled and the excess material is forced in a direction opposite to the direction to which the material is fed to thereby provide a slug or chunk of tuna which is substantially voidless and of constant volume and therefore of substantially constant weight. In this way, the juices extracted from the fish, to a large extent wasted by current methods, are returned to the containers.

In general, the apparatus comprises a support, a rotatable turret assembly, a driving mechanism for driving the turret assembly, a can feeding and driving unit, a filled can discharge chute, a fish feeding mechanism, a fish forming section and mechanism for operating the elements thereof, a mechanism for filling the containers with fish, and cutoff means. As shown in Figure 2.7a, the support **11** comprises a table **22** carried by suitable legs (not shown), a pedestal **23** bolted, as shown at **24**, to the table and a fixed cam plate **26** bolted to flanges formed on the top of the pedestal.

The rotatable turret assembly, generally indicated by **12**, comprises a turret **27** axially located with respect to the center of the machine seated upon and rotatable with respect to an annular flange **28** formed on the pedestal. Suitable means (not shown) are provided to supply lubricant to an oil groove **29** between the rotatable turret and the stationary pedestal. The turret assembly further includes a turntable **31** which supports the mechanism described below.

FIGURE 2.7: SOLID PACK TUNA

a.

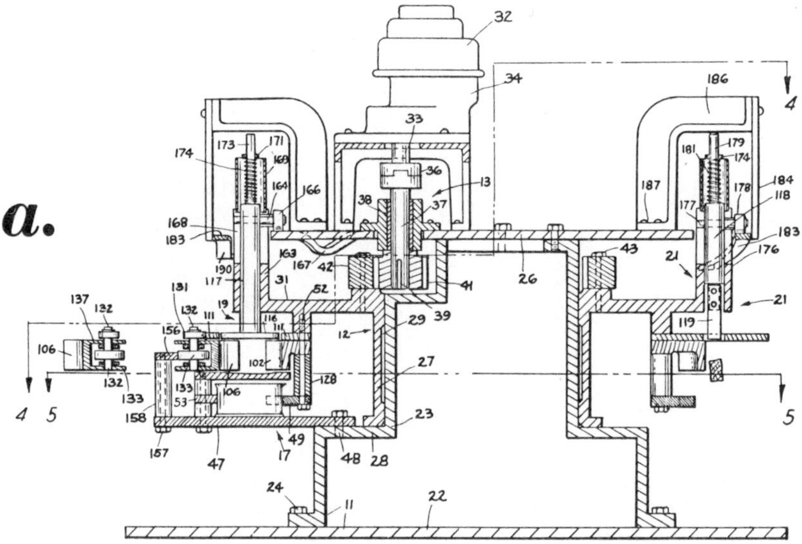

b.

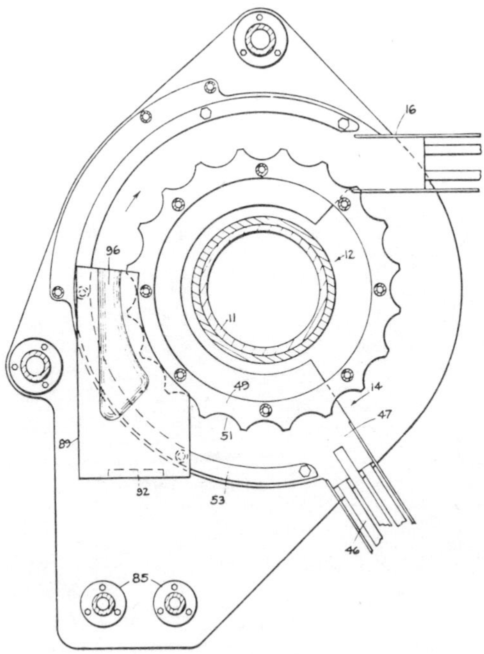

(continued)

FIGURE 2.7: (continued)

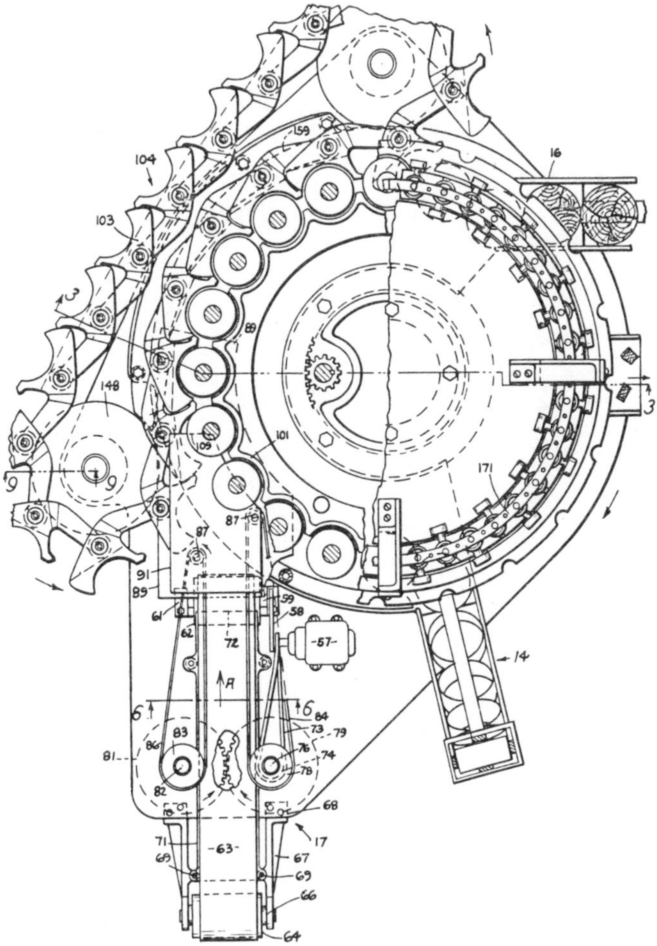

c.

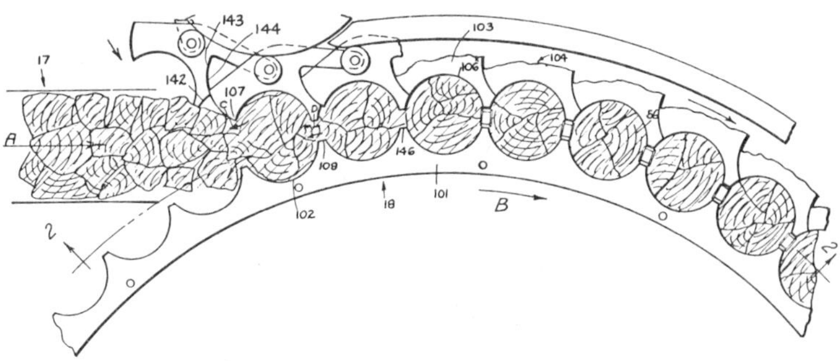

d.

(continued)

FIGURE 2.7: (continued)

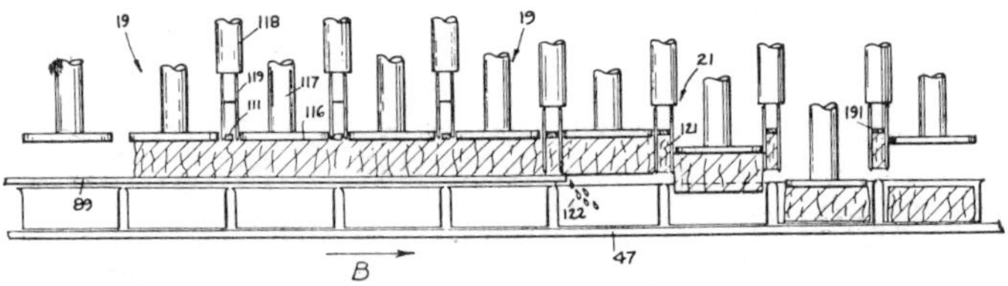

Source: E.H. Carruthers; U.S. Patent 3,124,469; March 10, 1964

The driving means for the turret assembly, generally indicated by **13**, comprises an electric motor **32** preferably of the variable speed type. The variable speed motor is connected to a drive shaft **33** by suitable reduction gearing enclosed within a casing **34**. The shaft is connected by a flexible coupling **36** to a driven shaft **37** mounted in a suitable bearing assembly **38**.

Keyed to the driven shaft, as indicated at **39**, is a spur gear **41** which mates with and rotates an internal ring gear **42** axially located with respect to the machine. The ring gear is bolted to the turntable assembly by bolts **43** whereby upon rotation of the motor the ring gear and rotatable turret assembly are rotated about the spur gear.

Referring to Figures 2.7a and 2.7b, the can feeding and driving unit comprises a can chute **46** into which cans are fed usually from a can loft located above the fish filling floor. The cans are deposited from the can chute onto a stationary can plate **47** which extends partway around the machine from the empty can entrance to the filled can discharge.

The can plate is bolted by a series of bolts to the annular flange **28** of the pedestal **23** in spaced relation to the table **22**. A can drive ring **49** having can embracing cutouts **51** is bolted by bolts **52** to an annular boss formed on the underside of the turntable **31**. A can guide ring **53** is mounted on the can plate and forms a partial circle extending around the can plate from the can entrance chute to the can discharge chute **16**. The margins of the cutouts or can pockets and the can guide ring force the cans to rotate over the can plate in a fixed circle path about the vertical axis of the machine.

The cans enter the intake chute, travel over the can plate as driven by the can drive ring in accordance with the rotation of the turret and are discharged off the can plate at the can discharge chute. During travel through this arcuate path, the cans are held rigidly and register with the forming chucks and filling mechanism.

Preferably, the cans move over the can plate in substantially lip-to-lip relation. The fish is first guillotined by cutting the four loins of each fish transversely along their length. Machines for guillotining fish are well-known in the art and need not be described. It is sufficient to state that the fish are guillotined into slices or pieces of a height substantially to be placed in the containers. In tuna fish, as in most fish, the loins extend longitudinally of the fish and are separated from each other into four quadrants by bones and membranes. The flesh of the fish extend in annular rings outward from the central spinal column.

Tuna fish will withstand pressure exerted substantially normal to these layers but will not withstand any substantial pressure exerted longitudinally of the layers. If any substantial pressure is exerted longitudinally of the loins the fish breaks up into a mass suitable, for the most part, only for packing as grated tuna. The method and machine has been designed to exert only such pressure as will not break up the fish, the purpose being to produce what is known in the industry as fancy solid pack tuna.

Referring to Figure 2.7c a motor **57**, preferably of the variable speed type, has two pulleys mounted on the extended shaft thereof, one of which drives a V-belt chain or other driving means **58**. The belt drives a shaft **59** supported in suitable bearings mounted in a bracket **61**. The bracket is bolted to the table **22**. The shaft is rigid with a roller **62** which drives a fish feed belt **63** in the direction indicated by the arrow **A**.

The fish feed belt extends over a second idler roller **64** mounted in trunnions **66**. The trunnions are carried in brackets **67** fixed to the underside of the table as shown at **68**. Carried by the brackets, as shown at **69**, is a feed belt support plate **71** which is spaced from the table by spacers and which supports the fish feed belt and terminates at **72**.

The second pulley mounted on the motor shaft drives a belt **73** which in turn drives a pulley **74**. The pulley is mounted on and rigid with a vertically extending shaft **76** which also has mounted thereon a pulley **78** and a gear **79**. The gear drives a second gear **81** rigid with a shaft **82** upon which is mounted a pulley **83**.

The shafts are journaled in bearings **85** mounted upon the table. The pulleys drive side fish feed belts **84** and **86**. The fish drive belts confine the tuna on the fish feed belt and are approximately the same height as the height of the pieces of tuna. The side feed belts pass over idler pulleys **87**. The side feed belt **84** is longer than the feed belt **86**.

The side fish feeding and confining belts pass between a lower transition plate **89** and an upper transition plate **91**. These transition plates are spaced apart a distance substantially corresponding to the height of the pieces of tuna being packed. At its forward end the lower transition plate is relieved, as indicated at **92**, to allow room for the driving roller and the fish feed belt to turn downward and smoothly transfer the pieces of fish to the lower transition plate **89**.

The upper transition plate is thin in cross section. The upper transition plate is smaller in area than the lower transition plate. The lower transition plate has a shallow trough **96** formed in its upper face, the purpose of which will later appear. The process is most clearly illustrated in diagrammatic form in Figures 2.7d and 2.7e. As shown in Figure 2.7d, the fish feeding section, generally indicated by **17**, feeds the fish in the direction of the arrow **A** and crowds or forces it in that direction toward the fish forming section **18**. As previously described, the fish adjacent the end of the fish feeding section and during part of the slug forming operation, is confined both above and below by the transition plates. The fish forming section includes a fish forming ring **101** having fish forming sections **102** and driven in the direction of the arrow **B** from the main turret. These fish forming sections are segmental cylinders and are of approximately a height to fit between the transition plates and are approximately of the height of the pieces of tuna to be placed in the cans.

The other part of the fish forming sections comprises a multiplicity of links **103** mounted on a chain assembly, generally indicated by **104**. These fish forming links have segmental cylindrical cutouts **106** forming fish forming sections of cylindrical shape also of approximately a height such as to fit between transition plates **89** and **91** and also of approximately the height of the pieces of tuna to be placed in the containers.

As the chain is driven the forming sections or chucks **102** and **106** gradually come together and compress the fish into a cake or slug of approximately the diameter, although preferably slightly smaller, than the diameter of the cans into which the fish is to be transferred. It should be understood, with the pieces of fish being crowded by the feeding action in the direction of the arrow **A** and the space between the fish forming sections being relatively wide adjacent **107**, that an excess volume of tuna is crowded into the space between the fish forming sections. This crowding action fills voids and insures that an excess of tuna will be encompassed by the fish forming sections.

It will be noted from Figure 2.7d, that the fish forming sections do not quite come together at any point. Adjacent the numeral **107**, the fish forming sections are gradually coming together and partially squeezing out the excess tuna in the direction indicated by the arrow **C**. This force is directed against the stream of fish being fed in the direction of arrow **A**. Adjacent **108**, the fish forming sections are more nearly closed and the excess fish embraced by the fish forming sections or chucks is being crowded back against the stream of fish as indicated by the arrows **D**.

The result of the above operations is to fill the volume enclosed by the upper and lower transition plates **89** and **91** and the fish forming sections **102** and **106** with a predetermined volume of tuna substantially without voids. Since the fish is usually of substantial constant density, the weight of enclosed fish is substantially a constant. Should the weight of the fish in the cans as they are discharged from the discharge chute be other than the required weight of tuna to be placed in the cans, either the motor **57** or the motor **32** may be varied in speed to effect a variation in the volumetric fill encompassed by the fish forming sections to vary the weight of tuna placed in the containers.

As previously mentioned and as appears from Figure 2.7e, the cans are driven by the can drive ring **49** (Figures 2.7a and 2.7b) which is driven by the turret. The turret also drives the fish forming ring **101**, and drives the chain, generally indicated by **104**.

Thereby, the cans and tuna cakes or slugs are driven in synchronism with each other. As appears in Figure 2.7e, the cans move in substantially lip-to-lip relation although some space may be allowed if desired. However, a space between the lips of the cans increases the small wafers of tuna which must be cut from between adjacent cakes as appears from Figures 2.7d and 2.7e.

For this reason, the expression substantially lip-to-lip relation should be interpreted as meaning an actual lip-to-lip relation or a reasonable spacing between the cans. As will be observed from the examination if Figure 2.7c, the upper transition plate **91** terminates at **109**. Upper transition plate is extremely thin in cross section and a drive plate **111** rides over a portion of it. The drive plate has a multiplicity of cutouts **112** approximately circular in shape. These cutouts have notches **113** to accommodate knives. The cutouts or openings in the drive plate accommodate plungers **116** carried by plunger rods **117**.

The cutoff between adjacent slugs of tuna is accomplished by a multiplicity of plungers **118** each of which carries two knives **119**. The cutoff knives are flat but can be slightly curved to conform to the curvature of the cans and as can be observed from Figure 2.7e are forced downward through the tuna and cut out a wafer of tuna **121**. This wafer of tuna is of a thickness corresponding to twice the width of the lip of a can plus the spacing between cans, if any. After the wafer has been cut, the plungers are forced downward and the plunger faces **116** force the cakes or slugs of tuna into the cans which are moving in synchronism with the slugs as will be observed from an examination of Figure 2.7e.

One of the important aspects of this process is that in previous machine methods of packing tuna fish, the juices extracted from the tuna, due to the compression of the tuna, were wasted to a large extent. These juices are high in nutrient value and by this process, these juices may be recovered.

As previously mentioned, the lower transition plate **89** has a trough **96**. As the tuna is squeezed at the forming stations, the juices extracted from the tuna drop into this trough which is slightly deeper at its forward end so that the juices drip into the cans of tuna as they pass the forward edge of the lower transition plate. Since the cakes of tuna are uniformly squeezed and the extraction of juices is substantially uniform, each can receives approximately the same amount of returned juices.

Use of Tray-Type Cookers

Tray-type fish cookers are well-known wherein cans of fish are inverted and are carried through the entire cooker while in the inverted position. This process has the following disadvantages. First, the unchanged position of the fish in the cans travelling through the cooker cause certain parts of the fish, i.e., those parts of the fish which are in firm contact with other fish and those parts of the fish which are isolated from the heating mediums because of pockets formed in the packed fish, to become improperly cooked.

Second, vapors become trapped in the cans which cause condensate to form in the cans when the cans are eventually returned to an upright position at the end of the cooking operation.

Third, the fish tend to stick to the grid of the supporting tray upon discharge from the cooker thereby adversely affecting the appearance of the fish. Another undesirable feature of many well-known tray-type fish cookers is that they include a cooking chamber and a dehydrating chamber which are in linear alignment with each other thereby rendering the apparatus undesirably long. There is also a high heat loss from the walls of linear cookers because the ratio of their wall area to their volume is large.

S.A. Mencàcci; U.S. Patent 3,161,526; December 15, 1964; assigned to International Machinery Corporation SA, Belgium has developed a tray-type fish cooker in which the containers of fish are moved from air inverted to an open-end-up position while in the dehydrating chamber of a cooker. This apparatus has high thermal efficiency.

In this method of cooking fish the cans of fish are inverted during approximately the first half of the dehydrating operation to allow juices to gravitate from the cans, and then the cans are returned to an upright position during the second half of the dehydrating operation to dissipate vapors from the cans and to dislodge the fish from the carrier grill prior to discharge from the dehydrating chamber.

The apparatus also provides an improved closed hot air recirculation system for the dehydrating chamber of the fish cooker and an improved mechanism for feeding inverted carriers to conveyors passing through the cooker. The patent contains a number of drawings and a detailed description of the apparatus.

Chunk-Style Tuna

U.S. Patent 2,602,578 describes a machine for packing tuna fish in the so-called chunk-style pack. This machine is in successful operation in a large number of tuna canning plants in this and other countries and its operation has been satisfactory. However, some canneries desire what may be termed a fancier pack of chunk-style tuna. In utilizing the machine from this process, the tuna is first cut into chunks by a guillotine or other means and then conveyed to a rotatable basin which has scraper blades cooperating therewith to direct the chunks of tuna into a continuously presented series of measuring pockets for subsequent discharge into cans.

The scraper blades have a tendency to break up the chunks of tuna into sizes smaller than desired by some canners and, in fact some of the tuna is broken up into flakes. Moreover, the machine is provided with a cutoff knife for the purpose of obtaining volumetric control, and hence, weight control. This cutoff knife trims the upper face of the tuna so that the upper face of the cake or slug of tuna, when the can is opened, appears somewhat smooth instead of having the distinct chunky appearance which some canners desire.

In addition, with this machine, it is much more difficult to control the put-in weight than, for example, with the machine as shown in U.S. Patent 2,669,378. Since tuna is an expensive fish, it is desirable to control the weight within relatively exact limits.

E.H. Carruthers; U.S. Patent 3,181,957; May 4, 1965 describes a process for overcoming the disadvantages of the previous apparatus, in which flaking of the tuna and its appearance in the can when opened, is minimized. The chunky characteristics of the can of tuna is more clearly defined and more accurate weight control can be obtained.

In this method, a machine is used like the one described in U.S. Patent 2,669,378. The machine produces cakes of solid pack tuna and then breaks them up into chunks and packs them into cans. The process comprises a system for forming loins of fish into a roll of constant weight per unit of length, advancing the roll until two slices or discs of tuna are cut off (controlling the combined weight very accurately by variable speed drive), dropping the discs of tuna so that when they strike a surface they are broken into chunks of substantial size, packing these chunks into a can, and tamping them down after they are in the can.

In two other processes, *E.H. Carruthers; U.S. Patents 3,321,317; May 23, 1967; and 3,346,403; October 10, 1967* has made further changes and improvements on his apparatus of U.S. Patent 2,602,278. These processes are primarily concerned with insuring that the can will be packed with its correctly labeled weight despite variations in texture and moisture content of the tuna being packed. The method comprises the following steps:

> Flowing a volume of fish to a rotatable basin having filling chambers, the basin always carrying an excess of fish to insure that the chambers will be filled with more fish than is to be packed in each of the containers;
>
> Guiding the pieces of fish into the chambers;
>
> Precompressing the fish in each of these chambers and then moving the fish upward so that some excess fish lies above the tops of the chambers;
>
> Trimming off the excess;
>
> Compressing the fish in each of the chambers again and moving the fish upward so that at least some excess fish lies above the tops of each of the chambers;
>
> Trimming off the excess to obtain a substantially constant volume of fish in each of the chambers of substantially constant density; and
>
> Transferring the fish in each of the chambers to one of a series of containers presented at a filling station.

The patents contain a detailed description of the apparatus for effecting these steps, and numerous diagrams of the apparatus parts.

Preventing Fish from Adhering to Can Filler

S.A. Mencacci; U.S. Patent 3,700,386; October 24, 1972; assigned to International Machinery Corporation SA, Belgium has developed an apparatus for filling cylindrical containers with chunk tuna or the like and for minimizing the tendency of the fish to adhere to the surfaces of the filling apparatus. U.S. Patent 2,926,095 describes a method and apparatus for canning fish in which a formed charge of fish is displaced from a forming chamber into a container by means of a reciprocating piston.

It has been experienced with apparatus of this type that, due to the sticky nature of the raw fish as well as to the formation of a vacuum between the fish charge and the piston, flakes or chips of fish are often pulled loose from the free surface of the canned fish charge upon withdrawal of the piston. This imparts an unappealing appearance to the contents of the can when subsequently opened by the consumer. Although the quality of the fish might be excellent, it will psychologically be downgraded by the consumer due to its appearance. It will easily be understood that this is a serious drawback for the fish canning industry.

It has also been experienced that at times the fish charge adhered to the piston and was partially pulled out of the can upon withdrawal of the piston causing loss of the charge and possible jamming of the apparatus.

The fish canning apparatus described here includes a conveyor that moves fish fillets into a rotary knife that cuts chunks of fish off of the fillet so that the chunks may be moved into a compression chamber. An extrusion piston in the compression chamber then moves the chunks laterally against a curved forming surface which with the aid of cooperating shaping knives forms the fish into can size charges.

Each charge is then pushed into the associated can by a transfer piston, and the piston is oscillated about its own axis when approaching and leaving the transfer end of its stroke to minimize adhesion of the fish to the transfer piston due to suction and due to the sticky nature of the fish. The abovementioned disadvantages are eliminated by oscillating the transfer piston about its own axis as it is transferring a charge of fish into a container and immediately after the charge has been transferred.

Figure 2.8a is a schematic perspective showing the working principle of a fish canning apparatus to which the process may be applied. Figure 2.8b is a schematic side elevation, partly in section, of part of an apparatus embodying this process and showing the drive means for the transfer piston. Figure 2.8c is an enlarged side elevation of a portion of the apparatus of Figure 2.8b. Figure 2.8d is a horizontal section taken along lines **4-4** of Figure 2.8c. Figure 2.8e is a vertical section taken along lines **5-5** of Figure 2.8d.

The apparatus works as follows. Fish fillets **1** (Figure 2.8a) are fed by an infeed mechanism **2** into a compression chamber **3**, the excess of fish being cut off by rotary knife **4** provided with fish receiving relieved portion **4a** and a large diameter cutting portion **4b** which cuts the fish and temporarily shuts the chamber and the infeed side. The charge of fish **5** thus brought into the chamber is then pushed by an extrusion piston **6** against a forming surface **7** corresponding to a portion of a side wall of a container or can **8**.

Subsequently, shaping knives **9** and **10** are brought toward each other, the knives each having profiled surfaces **11** and **12** respectively to form the remainder or a portion of the fish charge into a shape corresponding to the complementary side wall of the can. The thus obtained fish cylinder or charge is pushed into the can by a transfer piston **13**, the filled can being discharged at **14** and an empty can being brought into alignment with the pusher **13** by a turret **15**, the same cycle of operations as described being repeated for each can.

The to and fro or reciprocatory linear motion of the piston **13** is imparted by a cam **16** (Figures 2.8b, 2.8c and 2.8d) mounted on a continuously driven cam shaft **17**, via an oscillating lever **18** provided with a cam follower **19** and an adjustable rod **20**. One extremity of the rod is pivoted on a slide **21** carried by the lever **18** and the other extremity of a piston rod **22** of the piston **13**. The piston is mounted for rotation on and about the axis of the rod **22** (Figure 2.8c). An oscillating motion is imparted to the piston about the rod **22** by a cam track **23** (Figures 2.8d and 2.8e) provided in a disc **24** carried by a shaft **25** driven from the cam shaft by conical gears **26** and **27**.

FIGURE 2.8: APPARATUS FOR CANNING TUNA

a.

b.

c.

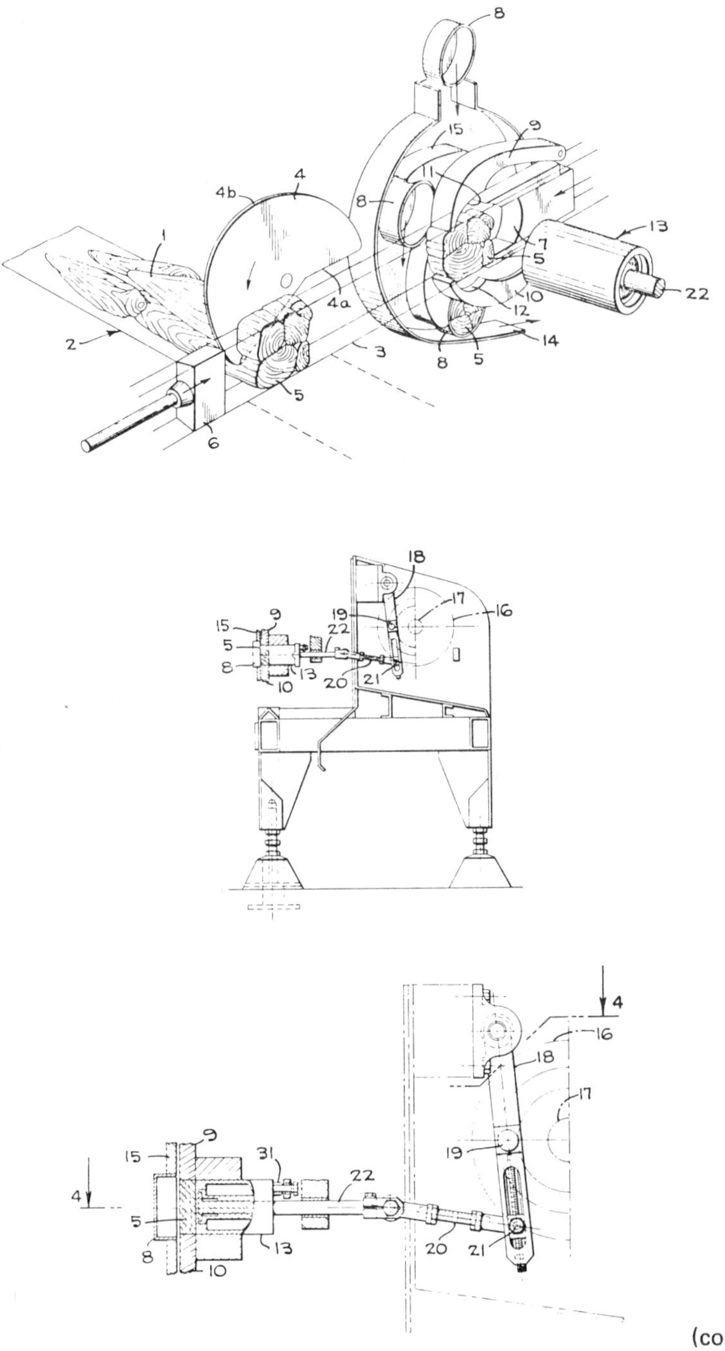

(continued)

FIGURE 2.8: (continued)

d.

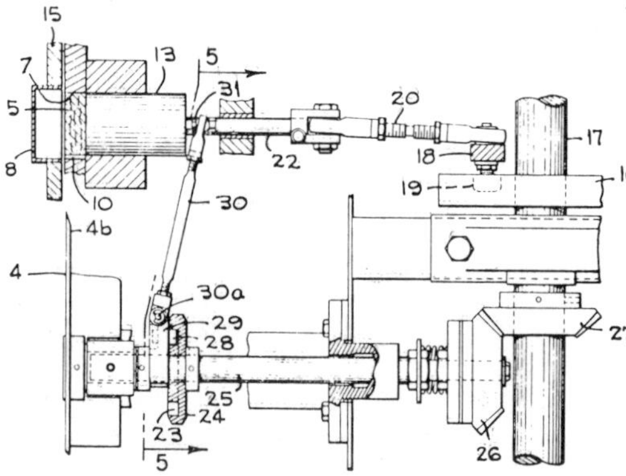

e.

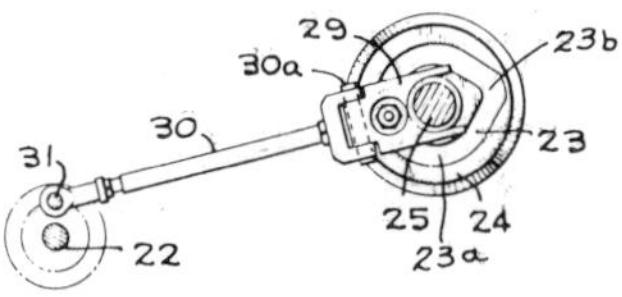

Source: S.A. Mencacci; U.S. Patent 3,700,386; October 24, 1972

The cam track is followed by a roller **28** that is pivotally mounted on a rider **29** to which rod **30** is hinged at **30a**, the other end being connected by a swivel to a projection **31** of the back of piston **13** for pivotal motion about an axis parallel to the piston axis and about another axis that is perpendicular to the piston axis. This projection is eccentrically mounted with respect to the rod **22**.

The shape of cam track **23**, as seen in Figure 2.8e, is shown as including a large circular portion **23a** and a small eccentric portion **23b** such that an oscillating motion is communicated to piston **13** during a relatively small portion of each revolution of shafts **25** and **17**. The relative timing of cam track **23** and cam **16** is such that the piston **13** oscillates during the end of the transfer and the beginning of the subsequent withdrawal stroke. A relative sliding motion is thus imparted between the piston surface and the fish charge.

The above described structure for operating the transfer piston causes the piston to more readily pull free from the charge of fish and provides a flat appearing surface on the fish which is devoid of loose particles thereby greatly improving the appearance of the fish when the can is subsequently opened by the consumer.

FISH OTHER THAN TUNA

SARDINES

In the prior art commercial preparation of sardines for consumption, the following general method is utilized. The sardines, after having been salted and dressed, are packed by hand in open cans. The cans are placed in baskets or racks and then moved to a steam chamber where the sardines are cooked by the steam while in the cans.

After having been cooked in the steam chamber, the baskets of cans are then removed from the chamber, and the baskets are tilted to drain the condensed steam, brine, and fish oil from the can. After draining the sardines in this fashion, the cans, while still in their baskets, are passed through a drying chamber through which air is blown to properly reduce the moisture content of the cooked sardines. The cans are then subsequently closed and sealed by conventional canning procedures.

Cooking by this method has a number of disadvantages. First of all, the undesirable natural oil of the sardines is cooked, but nevertheless remains in the cans and subsequently is retained on the fish. This oil not only imparts a brown color to the sardines, but also impairs their flavor. The steam also tends to dilute the brine in the can by condensation.

J.D. Toft; U.S. Patent 3,137,581; June 16, 1964 has obviated the shortcomings of the conventional method of canning sardines by a process, which has the following main steps. The sardines are salted and dressed and placed in individual cans ultimately to be sealed after the preparation of the sardines is completed. The open-top cans of sardines are then placed in trays which are mounted one upon another in a suitable rack. The racks of sardines are then cooked in a brine tank.

Sardines contain a considerable amount of oil or fat, located between the skin and the flesh. The hot brine separates the oil from the sardines, the oil rising to the surface of the brine where it forms a layer of oil film on the top of the brine. (By separating the oil from the sardines during this brine cooking process, and by maintaining the fish out of further contact with the oil due to the immersion of the racks within the brine, an enhanced product is obtained.) The film of oil at the surface of the brine is then skimmed off and the oil is separated from any brine collected therewith for subsequent use, as desired. It is important that the sardines, after being cooked, have the right salt content. In actual practice the cooking brine must be maintained within a very narrow range. Consequently, the salinity must be checked periodically, and, if incorrect, brought back to the proper reading either by dilution with hot water or by dumping in additional salt.

After the sardines have been cooked in the brine, they are removed, while in the racks, to a closed, insulated cooking chamber. Because of the heat already retained in the sardines, cans, and the like, the sardines continue to cook while in the cooking chamber. Due to this arrangement, it will be seen that a considerable quantity of fuel is saved by conserving the cooking heat within the sardines and the cans.

After passing through this stage, the brine remaining in the cans is drained off. The racks are then passed through a drying chamber where the sardines are cooled still further; after which they are ready for ultimate sealing, and may be run through a conventional closing and sealing machine.

Figure 3.1 is a diagrammatic, perspective view of the apparatus used in the process. As previously mentioned, the sardines are prepared for treatment by removing the heads, salting the fish, and placing the sardines in individual cans, which open-top cans will ultimately be sealed and otherwise used for marketing. After the cans have been so filled, they are placed in wire trays, the cans at this time not having tops. The sardines within the cans are thus visible along the upper surface of each of the trays. A number of trays are mounted one above the other in a rack, sufficient space being provided between each of the trays to permit the brine to come in contact therewith, as will be apparent.

The racks, indicated generally at **10**, are moved into the brine tank **12** by suitable conveyor means **14**. In this regard, the racks of sardines are preferably immersed wholly below the level of the brine, the temperature of which is preferably kept, as closely as possible, at about 206°F. The racks of sardines are cooled in the brine tank for about 5 to 7 minutes, depending on the size of the individual sardines. During this cooking stage, the brine separates the oil from the sardines, the oil rising to the surface of the brine.

While the racks of sardines are cooking in the tank, a curtain of air is blown over the surface of the brine through manifold **16**, moving the film of oil which has risen to the surface of the brine to the opposite end of the tank to the region of a separator **18**. The oil is removed from the surface of the brine by skimming from the surface; the brine passes into the separator, and the oil and any brine removed therewith are separated, the oil being dumped into a collecting tank **20**. Suitable valve means **22** are disposed at the bottom of the collecting tank to drain the fish oil therefrom, and water take-off means are provided at **24**.

A make-up brine feed tank **26** is provided, which is connected to the cooking tank by suitable piping **28**. A brine heat exchanger **30** is provided to allow for the proper maintenance of brine make-up feed temperature, which is important to the proper cooking of the sardines. The brine within the tank is maintained at its desired temperature by heating coil **32**, connected to a water heating means.

FIGURE 3.1: METHOD FOR CANNING SARDINES

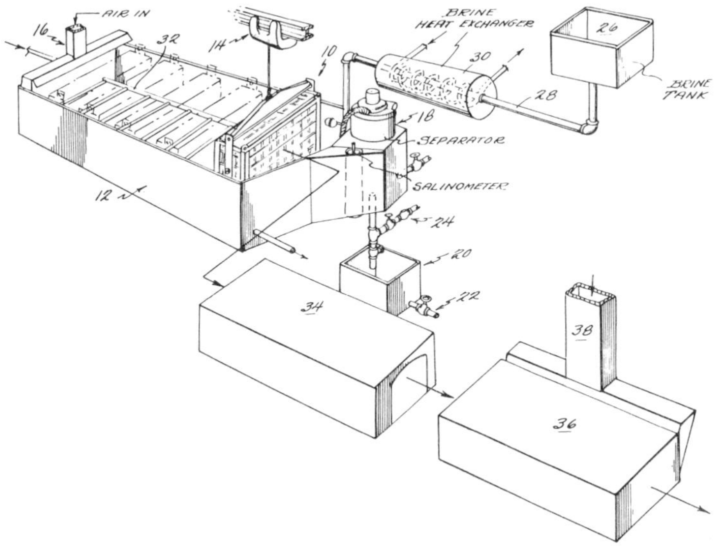

Source: J.D. Toft; U.S. Patent 3,137,581; June 16, 1964

After the racks of sardines have been properly cooked, they are removed from the cooking tank by the conveyor and thence moved on a cart to a closed cooking chamber **34**, which is preferably insulated. Because the racks, the trays, the cans and brine and sardines therein at the time of being removed from the cooking tank are at a temperature of about 200°F, the sardines continue to cook in the cans while in the cooking chamber.

After about twenty minutes continued cooking, and after the sardines have cooled down to about 145° to 150°F, the chamber is opened and the carts and racks are removed. The brine remaining in the individual cans is then drained off, as by tipping the enclosed rack. At this time, the racks and trays may be sprayed with cold water from a hose to both cool the sardines somewhat further, and also to wash any brine from the exterior of the can.

The racks are left in this position for about one hour while the carts are pulled through a drying chamber **36**. The drying chamber may be in the nature of a tunnel about 50 feet long through which air at a temperature of about 110°F is blown at a rate of about 7,000 cubic feet per minute. After about an hour in the drying chamber, the racks are tilted back to their upright positions so that there will be no possibility of sardines sticking to the underside of the tray next above. The drying and cooking process is then continued for about another half hour in the drying chamber, and at the end of this stage, the sardines have been adequately dried and cooled to about 100°F. The cans are then ready for ultimate sealing, and may be run through a conventional closing and sealing machine.

REMOVAL OF FISH HEADS

In the dressing of fish, the head is usually removed by cutting through the fish at the gills and the gall bladder removed manually prior to other cleaning. It has been found that the gall bladder is slightly toward the head from the pectoral fins of the fish and, when the fish is cut transversely through the body immediately rearward of the pectoral fins, the gall bladder will be in the severed head portion. Such a transverse cut will include only a negligible amount of edible flesh.

A process developed by *H.F. Heck; U.S. Patent 3,588,949; June 29, 1971;* provides an apparatus for removing fish heads by cutting transversely through the fish at a location to include the gall bladder within the severed head portion. The apparatus has a movable member for supporting a head portion of the fish and pectoral fin engaging members on the movable member to position the head portion relative to a cutter whereby, when the fish and the movable member are moved toward and beyond the cutter, the head portion is severed.

This method comprises pulling a fish by the tail to move rear sides of pectoral fins into firm engagement with spaced upright rods thereby holding a head portion in a fixed position while moving the fish into engagement with cutting means to sever the head portion from a body portion of the fish at a location substantially on a plane adjacent the pectoral fins so that the gall bladder is in the head portion removed from the fish body.

Figure 3.2a is a perspective view of an apparatus for removing fish heads; Figure 3.2b is a rear elevational view of the fish head removing apparatus; and Figure 3.2c is a plan view of the fish head removing apparatus. Referring more in detail to the drawings: The numeral **1** generally designates an apparatus for removing fish heads in production dressing of fish by cutting transversely through a fish **2** at a location to include the gall bladder (not shown) within a severed head portion **3**. The head portion to be severed is held in a selected position relative to cutting means and moved to engage this means at a location to sever the head portion from the fish so that the gall bladder is kept within the severed head portion. The preferred location of the cut is immediately rearward of the pectoral fins of the fish.

The fish head removing apparatus is mounted on a frame **4** which has legs **5** supported on a floor surface **6**. The frame supports an upper stationary platform **7** for supporting a body portion **8** of the fish, as later described. The cutting means is illustrated as a band saw **9** supported on the frame and driven by a suitable motor **10** mounted within an enclosure **11** positioned below the stationary platform.

In the illustrated structure, the band saw has a drive member **12** rotatably mounted within the enclosure **11** and operatively connected to the drive motor for driving a saw blade **13** around a driven member (not shown) rotatably mounted in a housing **15** positioned above the stationary platform. The saw blade is positioned adjacent one side edge **16** of the stationary platform. A fixed partition

17 is mounted on the stationary platform at the side edge and extends upwardly from the stationary platform thereby being positioned to maintain the body portion on the stationary platform and separate it from the severed head portions.

FIGURE 3.2: APPARATUS FOR REMOVING FISH HEADS

a.

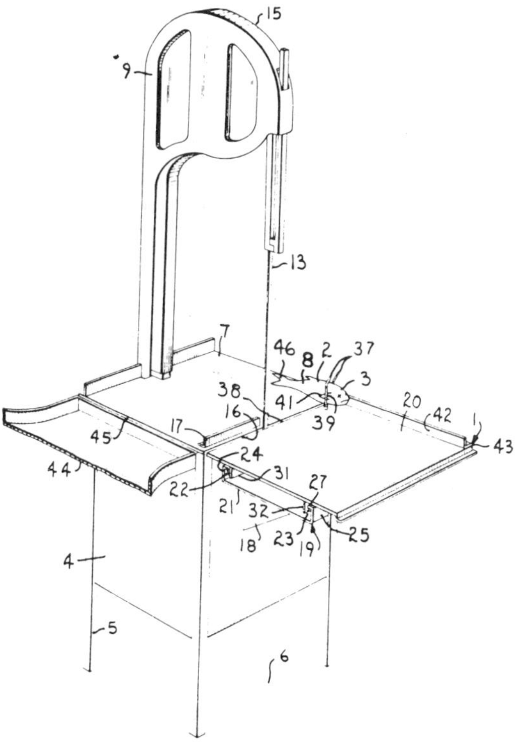

b.

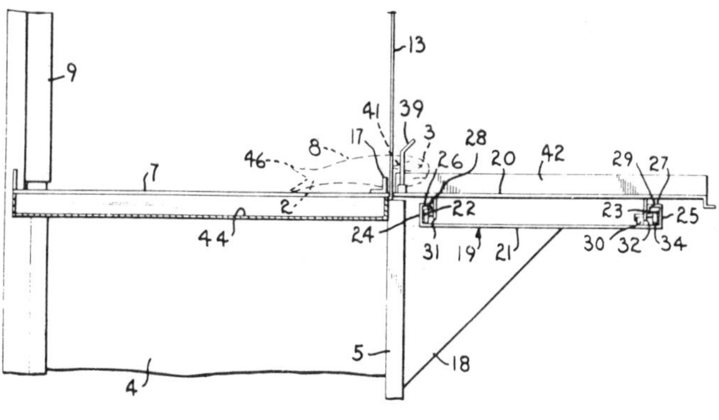

(continued)

FIGURE 3.2: (continued)

c.

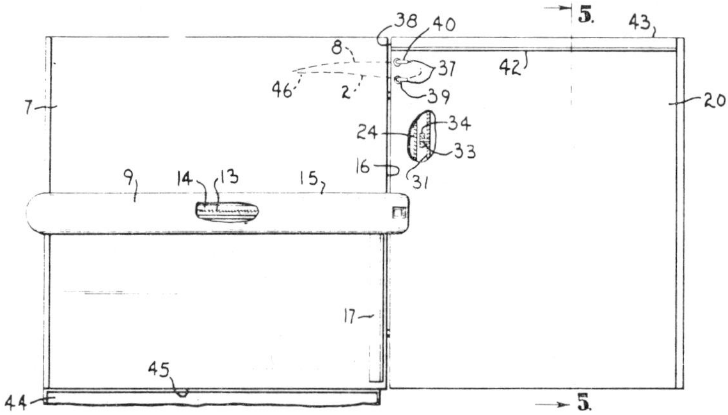

Source: H.F. Heck; U.S. Patent 3,588,949; June 29, 1971

It is desirable to support the head portion of the fish while cutting it with the band saw, therefore, brackets **18** are supported on the frame **4** and extend outwardly therefrom transversely to the side edge of the stationary platform. A guideway structure **19** is supported on an upper edge of the brackets and a movable member in the form of a platform **20** is mounted in the guideway structure for movement therealong while supporting the first head portion, as later described.

The guideway structure has a planar base **21** extending between the brackets and being supported thereon. A pair of elongate guideways **22** and **23** for directing the movement of the movable platform are formed at opposite side edges of the planar base by flange portions which have edges walls **24** and **25** extending upwardly from opposite side edges of the planar base and ribs **26** and **27** extending inwardly from upper edges of the walls with flanges **28** and **29** depending from inner edges of the ribs to form an inverted J-shaped flange portion at each edge of the base, thereby defining the guideways. Stop or bumper members **30** are positioned at opposite ends of the guideways to maintain the movable platform within the guideway structure.

The movable platform is illustrated as a generally planar member having a pair of spaced ribs **31** and **32** depending from a lower surface thereof and positioned inwardly of the depending flanges and substantially parallel therewith. A plurality of axles **33** are mounted in each of the ribs adjacent a lower edge thereof, whereby spaced wheels **34** are rotatably mounted on the axles for travel within the guideways between the stops or bumper members. The wheels engage the base and are confined within the respective inverted J-shaped guideway thereby restricting the movement of the platform **20** to linear movement toward and away from the saw blade and substantially parallel to the stationary platform.

The head portion of the fish must be accurately positioned relative to the saw blade during severing of the head portion from the body portion, therefore, fish engaging members **37** are mounted on the movable platform adjacent a side edge **38** thereof for positioning the head portion of the fish relative to the saw blade. The side edge of the movable platform is adjacent and spaced from the side edge of the stationary platform, and the saw blade is positioned between the side edges **16** and **38**.

The fish engaging members are illustrated as a pair of spaced rods **39** and **40** extending upwardly from an upper surface of the movable platform adjacent the side edge thereof with the rods being positioned to engage pectoral fins **41** of the fish. The pectoral fins are adjacent the head portion thereby holding the head portion in a fixed position on the movable platform and the body portion in a fixed position on the stationary platform, both in a fixed position relative to the saw blade. The rods **39** and **40** are positioned to clear the saw blade when the fish is moved toward and beyond the saw blade.

An end wall **42** is mounted on and extends upwardly from the movable platform at a front edge **43** thereof. The end wall is positioned to be grasped by an operator (not shown) while moving the fish toward and beyond the saw blade. A chute **44** is positioned adjacent a rear edge **45** of the stationary platform for receiving the body portion after it is severed from the head portion. The chute is positioned below the stationary platform and leads to other processing equipment, such as cleaning operations.

In use, the movable platform is moved to a forward position having the rods **39** and **40** positioned forward of the saw blade. A fish which has been previously killed, as by being electrocuted, is placed between the rods with the head portion on the movable platform and the body portion on the stationary platform. The tail **46** of the fish is grasped and the fish is pulled toward the stationary platform to engage the pectoral fins firmly against the rods, thereby holding the fish in a fixed position relative to the rods and the saw blade. An operator grasps the tail and the end wall **42** and moves the movable platform forward while pulling on the tail until the fish engages the saw blade at a location adjacent the pectoral fins and spaced slightly toward the tail.

After the head portion and body portion are severed, the body portion is moved into the chute for further processing. The fixed partition **17** maintains the severed portions of the fish at separate locations, whereby the head portions may be suitably disposed of and the body portions placed in the chute.

The fish head removing apparatus has a particular advantage at a wide range of body weights, so that a fish from eight ounces to three pounds may be production dressed without any adjustment. The location of the rods **39** and **40** is such that the engagement of the pectoral fins therewith properly positions the head portion to be severed relative to the saw blade thereby indexing each fish, of any in the wide range of sizes, by the engagement of its pectoral fins.

The spacing between the rods permits large fish bodies to pass between them until engaged by their pectoral fins and also engages the pectoral fins of the small fish all without any adjustment of the spacing of the rods. This permits a plurality of sizes of fish to be intermixed and production dressed in a substantially continuous operation.

PROCESSING CARP FOR HUMAN CONSUMPTION

Carp is considered undesirable for eating by most people for the following reasons:

(1) The unappetizing mud vein in the tissue which runs down both sides of a carp just under the skin.

(2) The darkness of meat which is due to an excess of blood cells.

(3) The presence of fine bones, commonly called hair bones, running longitudinally of the carp and which are in such great numbers as to be too numerous to fillet out by ordinary methods.

These characteristics not only make the carp undesirable for eating when filleted by the usual methods, but also make it undesirable for use in forming fish sticks or patties. The only state in which the carp has been marketed to any extent is in the ground state.

A process for treating carp to eliminate these three undesirable characteristics, and thus to make it commercially acceptable as food has been developed by *D.J. Biechele, Jr.; U.S. Patent 3,690,899; September 19, 1972*, and consists of these steps: (a) the mud vein is removed after cutting to expose it; (b) a cold salt solution leaching operation is used to remove excess blood cells from the tissues which ordinarily give the meat its dark color; and (c) a special scoring operation is used to cut the fine longitudinally extending bones, which cannot be removed by ordinary filleting, so that they will be in such short pieces that they will not be troublesome.

The best method of carrying out this process will be described with reference to the accompanying schematic drawings in which: Figure 3.3a is a longitudinal side elevational view indicating most of the body portion of a carp and illustrating diagrammatically its bone structure. Figure 3.3b is an enlarged transverse sectional view of the carp taken along line **2–2** of Figure 3.3a. Figure 3.3c is a perspective view illustrating the transverse scoring of a fillet of the carp to cut the longitudinally extending hair bones and slice the fillet into thin slices. Figure 3.3d is a perspective view illustrating the slices in a pressing or forming pan. Figure 3.3e is a perspective view showing the formed loaf of fish and a stick or cake cut therefrom.

A live carp of two pounds or above is secured and is killed by a sharp blow with a wooden or metal billy club. The head and tail are removed and it is skinned in the easiest manner. The carp has a bone structure as indicated in Figures 3.3a and 3.3b which consists mainly of the backbone **10** and the attached transverse ribs **11**, and the longitudinal hair bones **13** disposed outside the ribs, with the mud vein **12** running the full length of the fish from head to tail. The carp is then filleted by means of cutting and lifting out the backbone with the ribs and cutting into two halves.

However, the long hair bones, which extend substantially the full length of the fish, will not come out with the skeleton which includes the backbone and ribs. After filleting of the fish and cutting into two halves, the mud vein is removed by by cutting along the vein to expose it and pulling it out of the tissues by hand. In doing this, the carp is cut longitudinally along each of the two filleted halves, one of which is indicated by the numeral **15** in Figure 3.3c. However, each half will still have the secondary bone structure in the form of the longitudinally extending hair bones embedded therein and adjacent the outer side thereof.

The next step is to treat the fillet **15** to improve its color. For this purpose, it is leached in a saline solution which is prepared by using about one pound of salt to fifteen gallons of water. The fillet is soaked in the salt water for a period of about three to five minutes and the temperature of the solution is maintained low to prevent development of any harmful bacteria and the enzymes that tend to give the fish undesirable characteristics. The solution temperature should be below 45°F and preferably in the range of 33° to 45°F. After soaking, the fillet is removed and rinsed with fresh water.

FIGURE 3.3: PROCESSING CARP FOR HUMAN CONSUMPTION

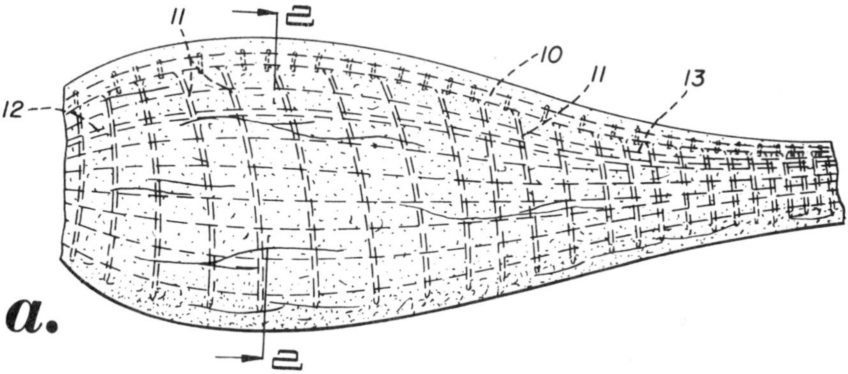

a.

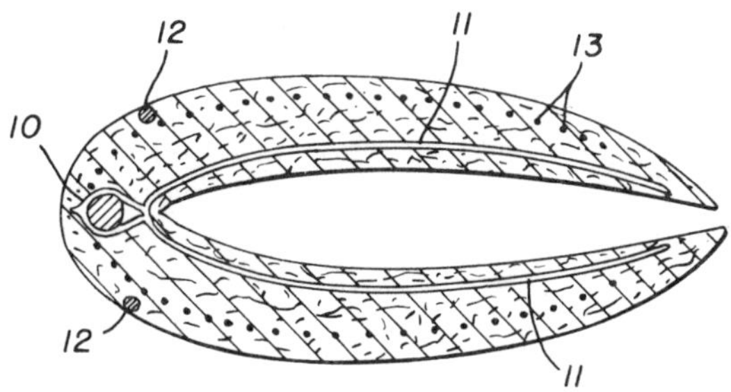

b.

(continued)

FIGURE 3.3: (continued)

c.

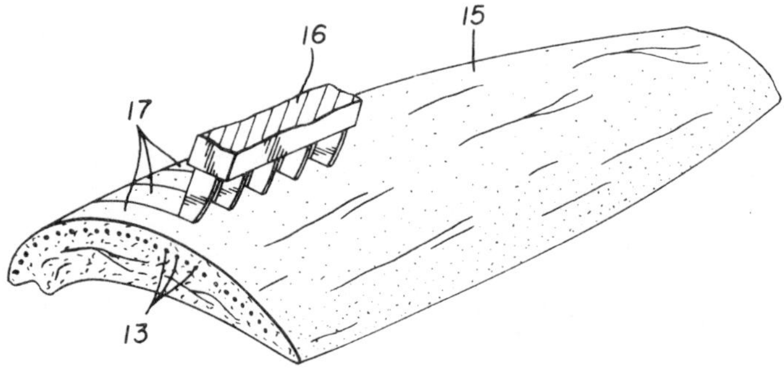

d.

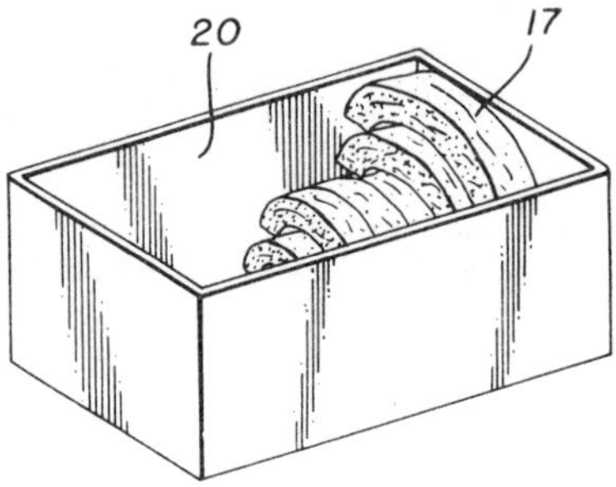

e.

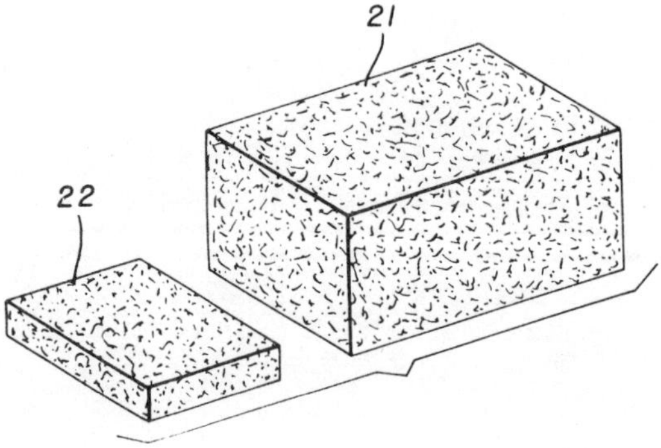

Source: D.J. Biechele, Jr.; U.S. Patent 3,690,899; September 12, 1972

Next, the fillet is cut transversely into steaks or strips of 1/16 to ¼ inch thickness. Simultaneously, the hair bones are cut into small pieces of corresponding lengths so that such minute pieces of bone will not be harmful or noticeable when eating the finished product. This operation is preferably accomplished as indicated in Figure 3.3c, in which a multiple-blade scoring knife is indicated at **16**, although a single-bladed knife could be used. The fillet is placed onto a cutting board, belly-down, to expose its outer surface with the bones **13** close to the surface. If the scoring knife is used as indicated, it is preferred that the blades be set ⅛ inch apart. The fillet is scored transversely from one end to the other to produce the strips or steaks of ⅛ inch thickness and to cut the hair bones.

The steaks or slices **17** are then placed in a pressing or forming pan **20**. If possible, the steaks or slices should be placed flat in the pan as indicated. Then pressure is applied to press the steaks or strips into a loaf **21**, as indicated in Figure 3.3e, from which patty size cakes or sticks **22** can be cut. The size of the forming pan will determine the area of the finished product and its thickness may be selected in slicing the loaf. A suitable size is 4" x 5" x ⅛" thick. The loaf is preferably subjected to sharp freezing while in the forming pan and after the freezing is removed for slicing or cutting into patties or sticks.

The patties or sticks can be battered and dipped into breading to make them ready for frying or they can be stored in this form, if frozen, for future use. The above described process is unique and simple, but is very effective for removing from the carp these undesirable characteristics which have made it unattractive for human consumption. It results in a commercially acceptable edible product which is much more appetizing than carp in the ground state which has been the only form on the market in the past.

HOMOGENEOUS FROZEN FILLET BLOCKS

L.B. Knutrud; U.S. Patent 3,615,720; October 26, 1971; assigned to Frionor Norsk Frossenfisk A/L, Norway has developed a process for the production of deep-frozen fish fillet blocks that are later to be divided into smaller units. In the production of such blocks the fillets are placed in a mold having an open top or, if desired, in a packing inserted into the mold, after which the mold with the fillets therein is placed in a freezing apparatus, for instance a plate freezer. The mold may be in the form of a pan, or it may consist of a frame resting on a bed, for example an aluminum tray. The packing may be in the form of two box halves, a bottom and a top half that are placed on top of each other and at the sides are surrounded by the frame. The fish fillet blocks may have different sizes, but they are ordinarily relatively flat, and a frequently used block size has a length of about 60 cm, a width of about 50 cm, and a height of about 7.5 cm.

As indicated such fish blocks are divided into smaller units of a size between 20 grams and about ½ kg. These units may be subjected to a further treatment, and especially for the smaller units it is usual to subject them to a breading and, if desired, to cook them in oil before they are again packed and frozen.

It is of great importance for the further treatment of the large fish blocks that the blocks are homogeneous, that is that voids in the blocks are avoided as far as possible. Such voids would entail substantial variations in the weight of the divided units, waste of raw material due to trimming of the blocks, and sometimes an aftertreatment of the units to fill recesses in the surfaces thereof. Voids in the block occur especially when

(1) the individual fillets are not appropriately or carefully placed,

(2) the fillets are especially large, and

(3) the fish is in rigor mortis so that the fillets will not yieldingly adapt themselves to each other.

With respect to the laying of the trimmed fillets in an appropriate way in the mold, for example in cartons erected in steel frames, much effort is devoted to training and supervising the workers, but due to changing personnel, unskilled labor and the demand for fast laying, it is impossible to eliminate the voids entirely. With respect to large fillets, attempts are made to neutralize the tendency for voids to be formed along the sides by splitting the fillets. It is customary to solve the problem with respect to the formation of voids due to the fish being in rigor mortis by letting the fish stand until they are more pliable. However, it will be appreciated that this is an unsatisfactory solution, as any delay in freezing results in a depreciation of the quality of the fish.

The object of this process is to provide a method for the production of deep-frozen fish fillet blocks as indicated above, by which voids in the frozen blocks can be avoided in a simple way and without any substantial increase in production costs, even though the processed fish may be in rigor mortis, and without the necessity of imposing rigid requirements with respect to careful laying and to the size of the fillet pieces placed in the mold.

This is achieved due to the fact that before the fish block in the mold is passed to the freezing apparatus it is subjected to a vacuum which is subsequently eliminated. In this way all the substantially enclosed voids within the block and between the block and the mold walls or the packing walls respectively, that is, the voids that are not in open communication with the space around the mold, are eliminated, since the air pressure in these voids will be reduced to near zero, the result being that the voids will collapse when the vacuum is eliminated. Therefore, the blocks will be substantially free of voids on the bottom side, along the lateral faces, and within the block.

However, the blocks will usually still have irregularities or recesses in their upper surface. In order also to eliminate these recesses, it is further suggested that a substantially plane surface should be kept in pressure contact with the upper surface of the fish block during the elimination of the vacuum.

Figure 3.4 illustrates a plate **1** of aluminum and a rectangular frame **2** resting on the plate. In this frame a cardboard box **3** may first be placed, whereupon the fillets **4** are placed. Then a cover **5**, which may be a box similar to the box **3**, may be placed over the fish fillet block.

According to the process the mold of the plate and the frame, as well as the boxes and the fillets therein, are then placed in a vacuum apparatus, which is illustrated in the drawing in a highly diagrammatical way and is indicated as consisting of a table **6** and a hood **7**, which can be moved pivotally or otherwise with respect to the table and placed over the mold and the fish fillet block so as to seal against the table by means of gaskets **8**, thereby forming a closed chamber **9** that may be placed under vacuum by means of an evacuation conduit **10**.

FIGURE 3.4: APPLYING VACUUM TO FILLET BLOCK BEFORE FREEZING

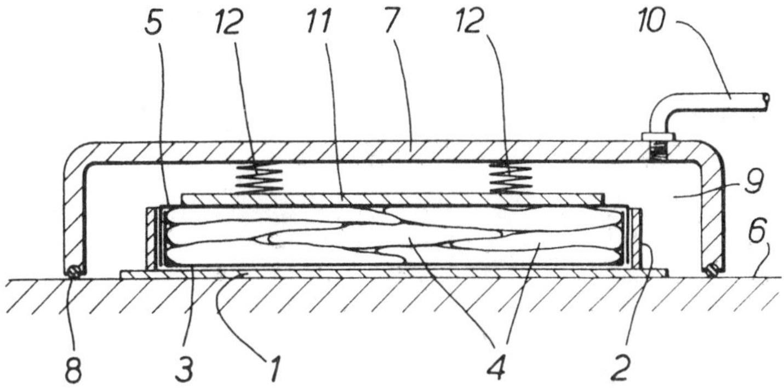

Source: L.B. Knutrud; U.S. Patent 3,615,720; October 26, 1971

When the hood is placed over the fish block, a plane rigid plate **11** connected with the hood is forced against the upper surface of the block. The force by which the plate is kept in contact with the fish block should preferably be constant, irrespective of any movement of the fish block, and therefore preferably a pneumatic pressure is used to force the plate against the fish block. However, for the sake of simplicity springs **12** for forcing the plate downwards are illustrated in the drawing. It may be noted that in tests with the method excellent results have been achieved by using an ordinary vacuum apparatus of the kind used for vacuum packing of food articles.

FISH ROE

Preparation of Ikura

T. Sugihara, C. Yashima, H. Tamura, M Kawasaki and S Shimizu; U.S. Patent 3,759,718; September 18, 1973; assigned to Riken Vitamin Oil Co., Ltd., Japan use protease to decompose the supporting tissue in order to separate the granular eggs of salmon (ikura) and trout.

The eggs of salmon and trout consist of the ovary, egg granules and the membraneous or supporting tissue which envelops the individual granular eggs. It has been found that the supporting tissue is quickly decomposed by the action of protease, while the membrane of granular eggs remains unspoiled, and thus individual granular eggs are separated. This process for preparation of ikura is characterized in that the granular eggs are taken out of the ovary of salmon and trout, then the mass of blood and other filth is quickly washed off in fresh or salt water. Subsequently washed eggs are taken from the water and steeped and agitated in a protease solution in order to decompose the supporting tissue and ovary membrane. The thus-obtained granular separated eggs are salted by a known process. Any protease may be used, for example, alkaline protease, neutral protease, or acid protease.

The enzyme solution is prepared by dissolving the protease in fresh water or a solution of salt water having a concentration up to saturation, and preferably in the range from 2 to 10%. Moreover the solution may be adjusted, depending on the property of the enzyme, to a pH of 3 to 10 by adding a buffering agent.

The temperature of the enzyme solution is desirably in the range less than 70°C which corresponds to the coagulation temperature of the protein contained in eggs of salmon and trout and more than 15°C for effective enzyme action, as well as maintaining the flavor and taste of the ikura. The treating time in the protease solution varies from 2 to 60 minutes depending on the activity of enzyme added, and the temperature of the enzyme solution.

By this process, separated eggs can be obtained with high yield in a short time without applying any physical pressure; much less care is required insofar as the freshness and maturity of the eggs are concerned, and long experience and skill is not required as compared with the conventional processes. Furthermore according to this process, the production yield is more than 85%, while it is 50 to 60% in the conventional processes.

Example: Eggs of silver salmon of 1 kg weight are quickly washed to remove blood and filth in fresh water. Subsequently, the washed eggs are soaked in an enzyme solution (pH 6.6) prepared by dissolving 300 grams of common salt and 5 grams of neutral protease in 10 kg water. The solution is kept at a temperature of 38° to 40°C and after agitation for 5 minutes, the separated granular eggs are scooped up in a net, and then cooled as well as washed in a large amount of solution containing 3% sodium chloride which is kept at a temperature of 15°C.

After being steeped in a saturated solution of salt for 15 minutes, the treated eggs are left on the wire net frame for two hours in order to remove water, the ovary and other filthy residue which remain. Then the surface of the granular eggs are coated with pure vegetable oils (quantity of oil to be coated is 2% of eggs by weight) whereby 852 grams of the product (so called ikura), are obtained (the yield of the product is 85.2%). There is no difference in the color, flavor and taste between the product obtained by this process and those made according to conventional processes.

Improving the Color of Fish Roe

Fresh salmon roe is reddish in color. However, red-colored salmon roe tends to turn brown. This discoloration leaves the salmon roe with an unappetizing appearance. The rate at which the brown discoloration of the reddish colored salmon roe takes place increases the higher the temperature of storage of the roe. For example, brown discoloration of the roe is noticeable within a day or a few days when the roe is stored at room temperature. When stored at sub-freezing temperatures, the brown discoloration is noticeable after several months. Salmon roe is often stored at subfreezing temperatures for many months from the time it is packed until it is marketed in order to maintain the edibility of the roe. Although the edibility of the roe can be preserved by such subfreezing storage, roe may nevertheless turn brown during such long storage. One method of preserving the red color of salmon roe has involved treating the roe with nitrite, for example, potassium nitrite. This is effective in keeping the reddish color of salmon roe. However, there have been reports which link nitrites to producing harmful effects on the body.

S.W. Yip; U.S. Patent 3,852,489; December 3, 1974; assigned to New England Fish Company has developed an improved process for preserving the color of fish roe, using a safer additive comprising an edible citrate and/or sulfite, bisulfite, hydrosulfite or meta bisulfite as the color-setting material in place of a nitrite.

Although the process may be utilized with red colored frozen salmon roe, in the preferred practice skeins of red salmon roe, while still fresh, are initially cleansed by being rinsed with fresh water, and thereafter rinsed in a salt (NaCl) solution, preferably at a concentration in the neighborhood of about 3 weight percent. The skeins are then drained to free them of excess brine. They are thereafter placed in a brine solution containing the color-set additive or additives and are agitated in the solution until the salt contained in the roe reaches a minimum of about 5 weight percent, and preferably a salt content within the range of about 8 to about 10 weight percent. The brining time for reaching this concentration depends upon the size of the skeins and upon the maturity of the roe, the time increasing with size and with maturity.

Although not essential, the packed roe can be matured, prior to low-temperature storage, by being allowed to stand at room temperature for approximately a day or two, or even three to six days, depending upon the temperature. The temperature may desirably be between about 55° and about 75°F, being preferably held close to about 60°F. The time of maturation will bear an inverse ratio to the temperature. Near the higher portion of the temperature range, the maturation may be about 1 to 3 days, and in the lower portion of the temperature range, the maturation may be for about 4 to about 6 days.

The packed salmon roe is desirably coated with vegetable oil, as by applying such coating on the top surface layer of the product, after which the lids are applied to the boxes and tightly fastened down; and the boxed product is then subjected to a low-temperature environment, and held at such temperatures during storage and/or shipment. The low temperatures employed may be between about 10° and about 20°F, but should be considerably lower, preferably in the subzero range, if the products are being preserved for any extended period of time, for example, a temperature within a range of about −4° to about −10°F is especially effective.

In the normal course of marketing, and delivery to the consumer, the resulting salmon roe has its normal original appearance, or in many cases, an enhanced appearance, especially as to color, and is tasty, palatable and wholesome.

Unless otherwise stated in the description of the process, fresh salmon roe is subjected to the treatment steps outlined on the following page.

(1) rinsed with fresh water;

(2) rinsed with a 3 wt % NaCl solution, withdrawn therefrom and allowed to drain;

(3) placed in a saturated brine solution;

(4) agitated therein until the salt content thereof is 8 to 10 wt %;

(5) removed from the brine solution and allowed to drain; and

(6) stored in a suitable container at a temperature of about -10°F for a 4 month period.

Example 1: Roe is treated according to Steps (1) to (6) above. No color-set material is used. After 4 months of storage, visual examination of the roe shows that it is brown in color.

Example 2: A saturated brine solution containing about 0.35% sodium citrate and about 0.25 weight percent sodium sulfite by weight of water in the brine solution is used in Step (3) above. After a 4 month period, visual examination of the roe showed that it was red in color and appreciably comparable to the red color of the roe before it was subjected to cold storage.

It should be pointed out that the impregnation of the salmon roe with one or more of the color-setting or enhancing materials described is rendered very simple, effective, and uniform as to final concentration, when it is done as a part of the brining operation. It is believed that the citrate and sulfite function to prevent the formation of brown Heme pigment metmyoglobin in the roe. Further, the results are especially good when both a citrate and sulfite are employed in combination.

INDIVIDUAL FISH PORTIONS

Fish Steaks

Fish cakes have been made of ground or shredded fish, but such comminuted product loses much of the moisture, the flavor, and the nutrients which accompany the moisture, and so the fish cakes made of such material are of somewhat inferior quality. Moreover, they have no appreciable tendency toward cohesion, and tend to revert to shreds during handling and cooking.

D.S. Johnston; U.S. Patent 3,433,647; March 18, 1969 describes a process for making fish steaks of chunks or strips of fish from fillets and portions of fillets, the chunks being pressed together and overlapped so that they tend to adhere one to another, and while thus compressed in a column are cut or sliced off cleanly, one by one, from the column in measured quantities.

In the process of their severance from the column they are lapped one over another, and pressed together, this procedure being termed herein polishing, and when discharged they tend to remain in the form of a standard steak composed of a number of such cohesive chunks. Such steaks are suitable for freezing and storage or for direct marketing and use. They simplify packaging and merchandising by weight and size, are highly nutritive and possess most of the flavor and much of the smooth textural qualities of the original fillets. Because of the nature of this product, it is necessary to sever each steak cleanly from the column in a manner that will maintain a given plane of severance, to the end that the steak, when severed from the column, is maintained in its predetermined shape and size, with all chunks tending to cohere one with the others in an individual steak of a standard size and weight, such as a steak suitable for an individual serving.

In producing these fish steaks, whole pieces of fish are pressed into a mold cavity so as to fill the cavity to a predetermined depth under pressure sufficient to cause substantially uniform density of material throughout the cavity and cohesion between the pieces, but which pressure is insufficient to destroy the tissue structure within the pieces or to extrude natural juices from the flesh, which would impair its nutritional value and destroy its flavor characteristics.

In the preferred practice of the method, these pieces are fed en masse as a column, are advanced into the mold cavity under pressure, and finally are sheared off at the mold face while under such pressure, and while being subjected to a lateral polishing action which tends to lap one chunk over another and to insure their cohesion. Subsequent ejection of the molded product prepares the mold for subsequent reuse, whereby under rapid production conditions, the process may be conducted on a continuous basis.

The machine for producing the fish steaks incorporates a succession of mold cavities moved progressively across the discharge end of the feed tube or duct through which the column of fish pieces being fed under pressure against the cavity are guided. As a mold cavity moves progressively into and from registry with the duct, the chunks flow somewhat plastically as the cavity is filled, and its contents are bonded together under pressure sufficiently to remain intact as a molded steak after severance and upon ejection from the cavity.

A dry member, characterizing the process, is interposed between the mold cavity and the duct. This dry member or shear plate, as it is herein termed, rotates continuously about an axis defining the direction of feed, and serves both to sever the column of fish at the entrance to the mold cavity, and to polish or smooth the steak at the cavity face as the cavity is being filled and advanced beyond the discharge end of the duct. Such polishing or smoothing action, effected by the shear plate, works or spreads the surface materials of the different pieces or chunks across the surfaces of the neighboring pieces, and thereby further integrates the mass into a cohesive whole.

The clean severance effected by the rotative shear plate maintains the fish within the individual steaks in the form of chunks, and these chunks tend to adhere one to another. In consequence, the steaks, all of the same weight and shape, may be individually wrapped, or may be frozen in the form in which they were severed from the mass, or may be immediately cooked and consumed.

The chunks, being pieces of fish of appreciable size, contain the greater part of the moisture and the flavor, for the pressure to which they have been subjected is insufficient to break down their tissues, or to squeeze from them the juices, and so they still contain the nutriment which is characteristic of the particular fish, and are not lacking in these characteristics as a fish cake of shredded fish. They require no binder to hold them together in handling or cooking, for all constituent chunks have been pressed sufficiently tightly together. The patent contains detailed drawings and descriptions of the machine used in the process.

Fish Sticks and Other Preshaped Fish Portions

The production of such deep frozen products as fish sticks, breaded, shaped fish portions, etc., is customarily undertaken by first forming the products to a deep frozen block and subsequently, cutting or sawing the desired portions from the block. When producing such fish products from fillets, it is necessary to employ methods other than the deep frozen block since these fish products are produced from relatively coarse cut fish, i.e., fillet pieces and the like, and the direct forming of the products into the shape of the finished commercial goods is extremely difficult. The production of a deep frozen block involves extra expense in operational costs, packaging and storage, and cutting the block by means of saws or knives entails a substantial waste.

The finished products may be punched out instead of cut or sawed. When punching however, it is necessary to soften the goods to a temperature suitable for punching, but without risk of damage to the goods. Foodstuffs which must keep for long periods of time are usually frozen to a temperature of - 25° or -30°C in order to attain the necessary keeping quality of the products. When punching, which is preferably carried out on prefrozen plates, the goods are easily crushed at these temperatures, so the goods are usually softened by raising their temperature to approximately -5°C.

The softening must be sufficient to allow the goods to be readily worked but only to the point where they retain their shape when subjected to treatment. This also entails substantial operational and storage expense, and the cost of the freezing is also greater since the product must be twice frozen to deep frozen state.

E. Schjolberg; U.S. Patent 3,579,359; May 18, 1971 has developed a process in which the raw material is ground to a mass consisting of grain size particles, which is formed into a bond-like product and frozen to a consistency suitable for portioning into shaped products.

It has been proved that division or grinding of the raw material into a mass consisting of grain sized particles involves no special disadvantages with respect to the consistency of the commercial product or the keeping quality thereof. Since the raw material is formed, in ground state, to a band-like product, it is possible to carry out production in a continuous production line. The portioning to shaped products may be undertaken expediently by means of punching and the waste caused by portioning is minimal. Freezing prior to portioning is regulated according to the portioning method employed.

By punching the shaped products from the band-like product it is possible to vary the shape and size as desired, without being restricted to straight-sized products. After portioning, the temperature of the shaped products is further lowered by means of a final freezing operation to the desired keeping temperature, e.g., -30°C, and the deep frozen products may then be conveyed to optional further treatment, e.g., to a dip and breading machine from which the products may be conveyed to a packing line. This process offers significant advantages. In the first place it is possible to carry out the operation in a continuous process and the waste caused by portioning is reduced to a minimum while, at the same time, greater variation in the shape of the products is possible than was the case with previous methods. In the second place, the expense of freezing, operational costs, storage etc, is substantially reduced inasmuch as the necessity of intermediate storage of the deep frozen blocks or plates is eliminated, as is also the possible softening station for sheets which are to be punched.

A layout is illustrated diagrammatically in Figure 3.5, and the process is further described with reference to the drawing. The raw material, fish, is conveyed to a grinding machine **1** which grinds the raw material to a suitable mass of grain sized particles. From the grinding machine, the raw material thus treated is delivered onto a belt **2** which conveys the fish mass toward the right on the drawing. The fish mass is formed into a band of desired thickness by means of forming means **3**, which comprises the necessary forming shields and rolls.

Subsequent to the shaping means, various means for control are indicated by the number **4**, for example, transillumination for discovery of undesirable particles and removal thereof, and for patching of the mass to provide uniformity in the band form of the mass. Subsequent to the control, the belt conveys the band-like fish mass into a freezing channel **5**, preferably in the form of a radian freeze channel, and the fish mass is frozen therein to a temperature of -1° to -5°C. From the freezing channel the frozen mass is conveyed to a punching machine **6** which punches out the portions of desired shapes and dimensions.

FIGURE 3.5: METHOD FOR PRODUCING FISH STICKS

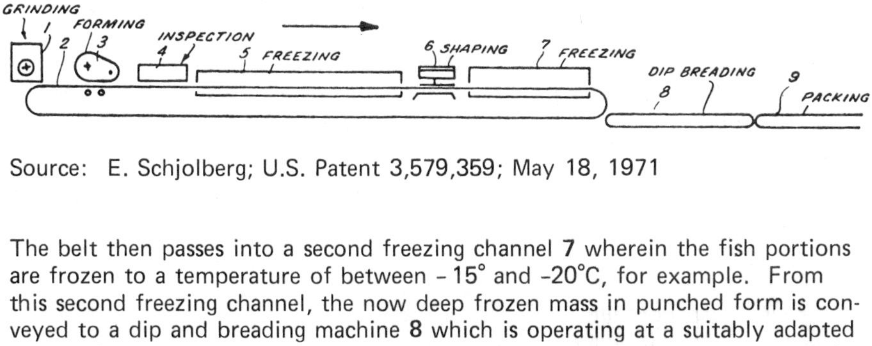

Source: E. Schjolberg; U.S. Patent 3,579,359; May 18, 1971

The belt then passes into a second freezing channel **7** wherein the fish portions are frozen to a temperature of between -15° and -20°C, for example. From this second freezing channel, the now deep frozen mass in punched form is conveyed to a dip and breading machine **8** which is operating at a suitably adapted speed. From the dip and breading machine the shaped products are conveyed to a packing line **9**. The belt is disinfected during its return.

By means of this process it is possible to produce the desired products in an inexpensive and efficient manner, the products having good shape-retaining qualities in finished state. In the case of fish the protein thereof will provide the necessary binding quality, and this binding quality is retained in that the raw material is ground to a mass consisting of grain sized particles instead of being finely ground to a cream like mass.

Small Seafood Pieces Bound with Comminuted Fish Scraps

Maintaining the physical integrity of a seafood product through heat processing, refrigerated storage and subsequent cooking is very difficult. It is almost mandatory that such products be stabilized against autolysis prior to extended storage. Stabilization requires the inactivation of naturally occurring enzymes and this is most conveniently done by heat. Heat, however, literally disassembles flesh portions such as fish fillets into muscle segments by destroying the connective tissue between segments.

In the preparation of fisheries products, it is often desirable to bind together small segments or particles of flesh to form a larger or differently shaped product. Generally the desired size and shape of the product is governed by both practical and esthetic considerations. It is of great convenience, especially in the marketing or serving of cooked fisheries products, that the product portions be of relatively uniform size and shape.

A host of different binding materials and techniques have been used in the past. Breading techniques have been used but are undesirable since the breading becomes soggy upon thawing and the product tends to fall apart. Natural gums such as guar and the like have been used but they tend to give the product a slimy appearance and feel. Proteinaceous materials such as egg albumen have been traditionally used as binders in all types of food preparations. Egg albumen tends to give a decided crusty appearance and texture upon cooking and often affects the flavor. Other naturally occurring binder materials used in the past include alginates, enzymes and proteins derived from milk and vegetable sources.

None of these materials is completely satisfactory for use in binding fishery products. Some lose their binding qualities upon heating. Some drastically change the the texture of the final product. Others add distinctive tastes or aromas to the fishery product; especially those products which have delicate or subtle flavors.

L.R. Ronsivalli and R.J. Learson; U.S. Patent 3,712,821; January 23, 1973; assigned to the U.S. Secretary of the Interior have found that flesh from fish, mollusks and crustaceans may be bound together with a binder paste prepared by comminuting the flesh of these same animals. Relatively small flesh pieces are mixed with binder paste and shaped by mold pressing into a product having the desired size and configuration. This reconstituted product maintains its physical integrity during heat processing, refrigerated storage and subsequent cooking.

Broadly, this is a process for preparing products which will maintain physical integrity during processing, storage and cooking while substantially duplicating taste and texture of the natural meat. Product portions of uniform and predetermined size and shape are produced from smaller flesh protions of fish, mollusks, crustaceans and mixtures of these. Small flesh pieces are bound together with an adhesive binder paste prepared by comminuting portions of the meat of those marine animals. The binder paste conveniently may be prepared by comminuting flesh portions in a high shear type of mixer together with sufficient water to form a relatively thick, homogeneous paste.

Binder paste is then added to the flesh pieces and gently mixed until each flesh piece is coated with the binder paste. Product portions are then formed by pressing the coated flesh pieces in a mold or other suitable forming device and heated to about 150°F to seal or set the binder. The product temperature is not critical and is determined by the texture desired in the product and whether or not additional processing is to be done. Generally 120° to 180°F would be used for most applications.

The product portions may then be stored under normal refrigerated storage, frozen storage or room temperature storage after being heat sterilized or radiosterilized. After refrigerated storage the product portions may be thawed and cooked in a conventional manner without the crumbling and disintegration normally encountered. Adhesion of the binder paste may be enhanced by scratching or abrading the surfaces of the individual flesh pieces prior to mixing with the binder paste. Amount of binder paste used is not critical and is determined by the texture desired in the product and by the size of the flesh pieces used in the process. Generally, binder paste will comprise from about 5 to about 25% of the product.

Example 1: A number of binder pastes were prepared and evaluated. In each case, fish flesh was comminuted in a shearing-type mixer. It was found that binder pastes of good consistency and homogeneity were produced with mixtures of fish flesh and water in the proportions of about 1 part fish to about 1 to 3 parts water.

Adhesive properties of the binder pastes for marine flesh were determined to be primarily a function of the water content of the paste and the degree of comminution of the fish flesh. More dilute pastes had less adhesive power but were generally easier to mix with flesh segments. As the degree of comminution increased, it was possible to obtain progressively greater adhesive properties. Extended comminution of the fish paste produced a binder giving almost a rubbery texture to the product. By simple experimentation, it was possible to obtain any desired degree of adhesive properties and to substantially duplicate the texture of natural products.

It was found that scratching or abrading the surfaces of individual pieces or segments of flesh making up a composite product had the effect of increasing the binding qualities of the pastes. Products prepared by binding small flesh pieces into a larger fillet or loaf configuration could be heat sterilized, frozen, stored, thawed and cooked without flaking or disintegration. Evaluation of the product by test panels concluded that taste and texture substantially duplicated natural products.

Example 2: The ocean quahog is a clam found in plentiful supply off the northern coast of the United States. This clam is characterized by a very strong flavor making it unacceptable for general food use. Hence, at the present time it remains a relatively unexploited food source. However, these clams may be processed to form a very acceptable, or even premium food product by blending their flesh with that of a bland fish such as haddock or cod. The blended flesh particles are bound together or reconstituted to form a product having a texture closely approximating natural flesh. The processing is carried out as follows.

Quahog meat is finely diced or coarsely ground and is then mixed with similarly prepared flesh of a bland fish, preferably haddock or cod. A binder paste prepared by thoroughly comminuting fish flesh with sufficient water to form a thick paste is then added to the mixture of quahog and fish with gentle but thorough stirring. The resulting mixture is formed into appropriately shaped and sized portions. Flavor of the finished product is controlled by varying the proportion of quahog to fish. It has been found that a ratio of about 15 parts quahog to about 85 parts haddock produces a product virtually indistinguishable in taste and aroma from the traditionally utilized clam species. Texture of the product can be tailored to substantially correspond to that of natural meat by adjusting the degree of comminution of the binder paste and of the amount of binder used. Generally, the amount of binder paste necessary will range from about 5 to 20% of the finished product and most desirably will range from about 10 to 15%.

Example 3: Crab meat in the form of large segments of lumps, often known as backfin meat, is highly prized and commands top market prices. Extraction of this lump meat by hand is no longer economical so a variety of sophisticated machines have been developed for this purpose. Even so, a significant portion of the meat is recovered in shredded form marketable only at a much lower price.

In another inherently simpler approach, meat is recovered mechanically with no attempt being made to recover it in lump form. Basically this approach is to mechanically crush the cleaned, meat-containing crab shells to form a mixture of shell splinters and meat shreds. Meat is then separated from the shell by pneumatic techniques such as flotation or centrifugation.

Shredded meat recovered from Maryland blue crabs by mechanical means, such as described above, was reconstituted to form large segments or lumps in the following manner. A portion of fish flesh was mixed with water and was comminuted by intense mechanical mixing to form a thick paste. Ratio of water to fish flesh is not critical but about 1 to 2 parts of water per part of fish flesh forms a paste having about the right consistency. The variety or species of fish desirably is one having a bland taste so as not to noticeably affect the distinctive crab flavor. Cod and haddock are particularly desirable for this purpose.

Shredded crab meat was dispersed in the fish paste in an amount equal to about 7 parts of crab meat per part of paste. After gentle but thorough mixing, the preparation was mechanically formed into lumps or segments similar in size and shape to those naturally recovered from crabs. After cooking, evaluation panels could not distinguish any significant differences in odor, taste or texture between natural crab meat and the reconstituted product.

Texture of the reconstituted product is controlled both by the quantity of binder paste used relative to the crab meat and by the degree of comminution of the fish paste. In order to substantially duplicate natural crab meat texture, it has been found that the ratio of crab meat to fish paste should be from about 5 to about 12 parts of crab meat per part of fish paste. Particularly preferred is a ratio of about 7 parts of crab meat per part of fish paste.

Preshaped Frozen Fish Portions

A method and apparatus for forming a preshaped fish product is described by *I.H. Langlands; U.S. Patent 3,728,136; April 17, 1973; assigned to National Sea Products Limited, Canada*. The method includes the step of forming individual portions of the fish, each being of predetermined size and being at least in a partially frozen condition, into a predetermined shape by forcing the fish portions individually into a forming die cavity adapted to impart the predetermined shape to each portion. The volume of each portion prior to the forming step equals the volume of each corresponding formed food portion. The frozen preshaped edible fish food product has a smoothly contoured marginal outline, with the part adjacent the marginal edge being in a deformed condition relative to the part in the central region of the product.

The apparatus includes a forming station having a forming die all arranged to engage and form portions of fish fed thereto. A device for feeding individual portions to the forming station is provided together with mechanism for effecting actuation of the feeding device and the forming die in timed relation to one another to effect forming of the portions. In the preferred form of the apparatus, the forming station includes a fixed platen with the feeding device being arranged to feed each portion to a position intermediate the forming die and the platen when the die is in a retracted position.

When the forming die is advanced, the fish portions are engaged in pressurized relation between the die and the platen. The feeding device preferably includes an endless conveyor belt defining a path of travel which extends over the fixed platen. The means for effecting actuation of the feeding means includes a mechanism for indexing the conveyor belt in step-by-step fashion to bring successive fish portions intermediate the forming die and the fixed platen to permit them to be formed in sequential fashion.

The process provides ejecting means associated with the die for positively ejecting the formed portion when the die is retracted. Preferably, a web of synthetic plastics material is arranged such that it is interposed between the interior or cavity of the die and the fish portion during the forming step and the web moved relative to the die cavity to effect ejection. In one form of the process the web may be shifted forwardly relative to the die to effect ejection of the formed food portion.

Although it is contemplated that the process may be used with various kinds of meats, by far the greatest success to date has been achieved with the use of uncooked fish. The process is described as follows: The fresh fish is first frozen under pressure by any suitable prior art method and is then usually stored for any desired period of time. While still frozen, the fish is cut into regular shapes, the size of the portions being determined according to the weight desired. Generally speaking, each portion comprises a generally rectangular shape having length and width dimensions several times greater than the thickness dimension. The fish portions are then tempered, that is, their temperatures are allowed to increase somewhat thereby to reduce the amount of pressure required to effect shaping of the frozen portions.

Broadly speaking, the fish portions will be tempered until they have temperatures somewhere between about 10°F and the upper limit of the freezing range for fish which is about 28.5°F. If the tempering temperature is too low, excessive damage to the flesh of the fish occurs during the forming step and the flavor of the final cooked product will be impaired. If the tempering temperature is too high, the portions will thaw out during the forming stage and may lose the shape imparted thereto and require refreezing.

At least reasonably good results can be achieved when the fish portions are tempered anywhere in the range from about 10°F to roughly 25°F; better results can be obtained if the tempering range is narrowed down to temperatures from about 15° to about 25°F, with optimum results being achieved about the middle of this range, that is, when the fish portions are at a temperature of about 20°F just prior to the forming step.

The subdivided portions of tempered fish are then fed individually to a forming station and are there pressed into a predetermined shape. The shape of the forming die will, of course, depend upon the final shape desired for the fish portion, and the pressure required to effect proper formation will depend upon the size of the portion and the temperature at which it is formed. For example, in order to effect formation of a portion of cod flesh weighing approximately 3½ ounces and tempered to about 20°F roughly 600 lb pressure must be applied for each portion of fish.

After the portions have been formed to the required shape they are carried away from the forming station and are either refrigerated or are prepared for subsequent processing. Since the portions of the fish are still in a substantially frozen condition after the forming step they retain the shape imparted to them by the forming die. The patent gives figures and detailed descriptions of the apparatus for die-forming the fish portions.

BATTER-COATED SEAFOODS

Aqueous Starch Slurry for Batter

J.J. Ducharme, H.S. Black, Jr. and S.J. Leith; U.S. Patent 3,052,545; September 4, 1962; assigned to National Starch and Chemical Corporation have found that the use of amylaceous materials having a sediment volume in the range of 40 to 50 provides an improved deep-fried food batter which is entirely free from the undesirable properties associated with conventional mixes, such as lack of adhesion to the foodstuff.

As is known to those skilled in the art, amylaceous materials, such as starches, will vary in the rate in which they will settle out from their aqueous suspensions. Thus, certain starches, depending upon their manner of preparation and subsequent treatment, will tend to settle out at a much more rapid rate than other starches. A convenient method for measuring this phenomenon is by means of the sediment volume determination. As used herein the term sediment volume is defined as the volume of sediment, on a percentage basis which settles out after 18 hours from a 20.45% by weight aqueous suspension of the starch being tested.

The starches used in this process may be derived from such sources as corn, high amylose corn, wheat, potatoes, waxy maize, tapioca, sorghum, sago, or rice. It should also be noted that cereal flours containing starches, are fully equivalent to starches. These flours may be derived from such sources as wheat, corn or potatoes.

The batter mix compositions of this process display a degree of adhesion to deep-fried foods which is so tenacious as to actually render the breaded batter coating into an integral component of the foodstuff. This adhesion is, in fact, so intense that it is virtually impossible to remove the deep-fried, breaded batter from the foodstuff without actually pulling off a portion of the foodstuff along with the breaded batter. Inhibited starches comprise starches which have been modified so that upon heating in water to a temperature beyond the usual gelatinization (i.e., pasting) point of the particular starch, there results a material in which the ability of the individual starch granules to swell, burst, disintegrate and otherwise lose their identity as individual granules has been retarded or avoided in contrast to the manner in which ordinary granules lose identity upon cooking.

In order to prepare mildly inhibited starches which may be employed in the deep-fried batter mixes of this process, it is ordinarily necessary to react starch with a cross-linking agent. These include: aliphatic dihalides such as propylene dichloride, dichloropentanes (ordinarily available as commercial mixture), ethylene dibromide, glycerol dichlorohydrin and dichlorobutane; ether forming epoxy halogen compounds such as epichlorohydrin and epibromohydrin; certain polyfunctional reagents such as cyanuric chloride, phosphorus oxychloride, metaphosphates and polymetaphosphates; aldehydes such as formaldehyde and formaldehyde containing resins and prepolymers; and, mixtures of adipic or citric acid with acetic anhydride.

In general, these cross-linking agents may be defined as compounds containing at least two functional groups which can react with at least two hydroxyl groups of the starch molecule or molecules (in the granule form) thus altering the cooking characteristics of starch in the manner described earlier.

The inhibition reaction may either precede or follow the derivatization or other modification of the starch. With respect to the amount of cross-linking reagent needed for the reaction, this is determined by the granule swelling power (GSP) of the resulting cross-linked starch. Granule swelling power is a measure of the extent of granule inhibition, and may be defined as the amount of swollen, hydrated paste which is formed by the cooking in water under specific conditions, of one gram of dry starch as divided by the weight of anhydrous starch in the swollen paste.

The GSP is determined, in practice, by dispersing one gram of starch (anhydrous weight) in enough distilled water to give a total weight of 100 grams. Normally the starch is suspended in this water, stirred over a boiling water bath for 5 min, and then covered for the remainder of the cooking cycle. After cooking is complete, the sample is readjusted to a weight of 100 grams and transferred, quantitatively, into graduated 100 ml centrifuge cups.

The sample is then centrifuged at 2,000 rpm for exactly 20 minutes and the starch dispersion is removed as a clear supernate and a compacted swollen paste. The percent solids in the supernate is determined by evaporation of an aliquot. The wet weight of the swollen paste is determined directly after the decantation of the supernate and the amount of dry solids in the paste is determined by evaporation. The granule swelling power is then calculated by the formula:

$$\text{GSP} = \frac{\text{Wet weight of swollen paste}}{\text{Weight of dry starch in swollen paste}}$$

In order to function effectively as batter mixes, mildly inhibited starches should have a GSP value in the range of from 10 to 30, since within this range they appear to give the highest possible film strengths. Therefore, the quality of cross-linking reagent to be used in the inhibition process may be defined as that amount required to obtain a product having a GSP of between 10 and 30. As these reagents all differ in their reactivity with starch, the optimum proportions will be different for each reagent.

As regards the actual preparation of the batter mixes, the practitioner need merely prepare an aqueous slurry of the starch, which, of course, having a sediment volume in the range of from 40 to 50 and preferably being mildly inhibited to a GSP of between 10 and 30. This slurry may contain from 100 to 150 parts, by weight, of water per 100 parts of starch. The precise starch:water ratio will, of course, depend upon the type of foodstuff being prepared, the manner in which the batter mix is to be applied, as well as on the particular starch being utilized.

The starch slurry is then kept under agitation prior to its application to the foodstuff so as to prevent the starch from settling out. The batter mix may thereupon be applied to the foodstuff, which may be either fresh or frozen, by any convenient procedure such as dipping, spraying, or cascading.

After draining off the excess batter, the breading is applied and the foodstuff may then be deep-fried. It should be noted that it is unnecessary to heat the starch slurry prior to its application to the foodstuff. The practitioner may add such seasoning, seasoning extenders or food colors to the batter mix slurry as are desired; although in most cases, such additives are not needed. The resulting breaded, deep-fried foodstuffs made with the batter mix are, in all cases, characterized by the golden-brown color and smooth, even texture of their breaded coatings which are always notably free from any crumbling or disintegration during their subsequent handling and eating.

In an alternative method, the batter mix starches may be applied to foodstuffs by means of a dusting procedure wherein the dry starch is directly applied to the surface of the foodstuff. The breading is then similarly applied and the foodstuff may then be deep-fried.

The starch batter mix composition may thus be used in the preparation of all types of deep-fried fresh and frozen foodstuffs including fish, shell fish, fish sticks, fish fillets, fish steaks, whole fish, scallops, oysters, clams, shrimps, lobster parts, etc. The following examples will further illustrate the process. In this example all parts given are by weight unless otherwise noted.

Example: This example illustrates the use of the starch batter mix in the preparation of prefried frozen fish which require further cooking prior to their ultimate consumption. Frozen, raw, skinless codfish fillets were cut into 3" x 2" x ½" portions and dipped in an aqueous batter mix slurry which was being maintained, under agitation, at 70°F and which comprised a mixture of 110 parts of water and 100 parts of cornstarch which had been mildly inhibited with 0.02% by weight of epichlorohydrin so as to obtain a product having a GSP of 24.3 and a sediment volume of 45.

In determining the sediment volume of this starch sample, as well as for all of other starches described, the procedure employed comprised placing from 10 to 15 ml of distilled water into a 100 ml graduated cylinder. To this water there was then added 25 grams of the starch sample whose weight had been corrected for a moisture content of 10%, by weight. The resulting mixture was agitated whereupon additional water was added so as to bring the total volume to 100 ml. After additional agitation the cylinder was allowed to stand at a temperature of 72°F. After 18 hours the sediment volume was determined by reading off the level to which the starch particles had settled.

After the excess batter mix had been drained off, the fish portions were thoroughly coated with a bread crumb breading, lightly patted, and placed in a wire mesh frying basket. The breaded fish portions were next immersed in corn oil which was at a temperature of 385°F and were then prefried for 45 seconds whereupon they were removed from the hot oil and allowed to drain.

The prefried fish portions were then cooled in a refrigerator at approximately 40°F for a period of one hour and were thereupon refrozen at 0°F. Samples of the frozen, prefried codfish which had been stored at 0°F for 7 days were then prepared for consumption by being baked for 18 minutes in an oven set at a temperature of 350°F.

Additional samples were also prepared by baking for 11 minutes at a temperature of 450°F. After removal from the oven, both sets of the ready-to-eat samples were evaluated for general appearance and, more particularly, for the adhesion and cohesion characteristics of the breaded coatings. In both cases it was observed that the breaded coatings displayed a firm, pleasing texture along with a uniform, golden-brown color. Moreover, they possessed an outstanding degree of adhesion to the fish so that removal of the breading could be accomplished only with the accompaniment of a part of the fish.

As for their cohesive properties, it was noted that these fish could be handled and eaten without any crumbling or disintegration on the part of the breaded coatings. In contrast, when the above described procedure was repeated using, in this instance, an ordinary batter mix comprising 3.5 parts of water and 1.5 parts of a mixture containing 72.2% wheat flour, 19% nonfat milk solids, 2.5% dry egg white, 7% salt and 0.3% of pepper, the following results were noted.

Although the breaded coatings possessed a firm texture they displayed very poor adhesion to the fish and could be stripped away with very little effort. Moreover, their cohesion was entirely negligible as was evidenced by their ready crumbling and disintegration upon handling and eating.

Dry-Oxidized Starch in Batter

J.A. Antinori and M.W. Rutenberg; U.S. Patent 3,208,851; September 28, 1965; assigned to National Starch and Chemical Corporation have modified the procedure of U.S. Patent 3,052,545 previously described by using a starch which has been oxidized. Among the oxidizing gases that have proved most effective are ozone, chlorine, chlorine dioxide, nitrosyl chloride, nitrogen dioxide, and similar compounds. The amount of oxidizing gas employed in this procedure will vary, of course, with the particular gas used and with the type of amylaceous material treated. Thus, for example, it has been found that, as based on the weight of the material being oxidized, ozone may be used in amounts ranging from about 0.025 to 10%, whereas gases such as chlorine, chlorine dioxide, nitrosyl chloride, and nitrogen dioxide may be used in amounts varying from about 0.04 to 5%. The dry amylaceous material is agitated in any suitable container, and a stream of the oxidizing gas is passed through the agitated, dry mass for a period of from about 15 minutes to about 60 minutes. Again, the length of the reaction time will vary with the type of starch and the oxidizing agent used. After the reaction has been completed, the excess oxidizing gas is drawn out of the container, usually by flushing the latter with air. The oxidized amylaceous material is then usually screened or otherwise sifted and may be employed to prepare the batter mixes.

Example: This example illustrates the use of the starch batter mix in the preparation of prefried frozen fish which require further cooking prior to their ultimate consumption. Frozen, raw skinless codfish fillets were cut into 3" x 2" x ½" portions and dipped in an aqueous batter mix slurry which was being maintained under agitation at 70°F and which comprised a mixture of 100 parts water to 75 parts cornstarch (all parts by weight) treated for 30 minutes with 0.025% of ozone, as based on the weight of the cornstarch.

After the excess batter mix had been drained off, the fish portions were thoroughly coated with a bread crumb breading, lightly patted, and placed in a wire mesh frying basket. The breaded fish portions were next immersed in corn oil which was at a temperature of 385°F and were then prefried for 45 seconds whereupon they were removed from the hot oil and allowed to drain.

The prefried fish portions were then cooled in a refrigerator at approximately 40°F for a period of one hour and were thereupon refrozen at 0°F. Samples of the frozen, prefried codfish which had been stored for 7 days at 0°F were then prepared for consumption by being baked for 18 minutes in an oven set at a temperature of 350°F.

Additional samples were also prepared by baking for 11 minutes at a temperature of 450°F. After removal from the oven, both sets of the ready-to-eat samples were evaluated for general appearance and, more particularly, for the adhesion and cohesion characteristics of the breaded coatings. In both cases it was observed that the breaded coatings displayed a firm, pleasing texture along with a uniform, golden-brown color.

Moreover, they possessed an outstanding degree of adhesion to the fish so that removal of the breading could be accomplished only by removing a part of the fish. As for their cohesive properties, it was noted that these fish could be handled and eaten without any crumbling or disintegration on the part of the breaded coating.

Coating Followed by Rapid Hot Cooking Oil Dip

L.L. Libby; U.S. Patent 3,078,172; February 19, 1963; assigned to Tra-Way Corporation has provided a process for producing an uncooked frozen food product having exceptional preserving quality and excellent flavor.

The frozen food product is prepared by first coating the uncooked food, frozen or not frozen, with an edible sealing material such as flour and batter, coating the sealed uncooked food with an edible cooking material such as an oil and a fat at a temperature at which the cooking material is maintained in a fluid condition without degradation thereof and such that the food remains uncooked, and, if required, freezing the cooking material-coated food at a temperature sufficient to solidify the cooking material thereon.

The uncooked food is coated with a sealing material such as flour and/or batter for the purpose of coating it with a material which will prevent the natural juices from flowing out of the food. The sealing material is edible, and it also serves to impart flavor to the food upon being cooked.

The flour can be, for example, white flour of wheat, rye, or corn origin, or whole wheat flour. The batter consists of egg, milk and seasoning, with or without flour. The seasoning can be salt, pepper or any other material which will impart the desired flavoring effect to the food.

The sealing of food with flour and/or batter should be effected at a temperature at which the food is not cooked at all, since the purpose of this process is to prepare a frozen food product which is in an uncooked condition. After the food has been sealed with flour and/or batter, it may be coated with bread crumbs, cereal crumbs, flour or other coating materials. The coated food is then immersed in a cooking material such as oil or fat of the edible type.

Generally, the cooking material is liquid at temperatures ranging from about 110° to about 375°F and its setting point is about 32°F or higher. The coating operation is conducted at temperatures from about 110° to about 375°F for a period of about 5 to about 40 seconds, preferably about 15 to 30 seconds.

The actual time and temperature conditions employed vary with the food product to be treated and with the physical state of the food product. For example, a frozen uncooked fish will be subjected to different time and temperature conditions than fresh uncooked fish at room temperature. The temperature of the coating material is preferably such that the product to be treated has deposited thereon only a thin coating of cooking material.

If the cooking material is maintained at too low a temperature and a frozen product such as frozen fish sticks is passed through the cooking material, the temperature of the cooking material will be lowered and instead of the fish sticks having a thin coating deposited thereon, they will be coated with a thick coating of congealed cooking material.

While the conditions of treatment may vary over a wide range of time and temperature, the important consideration is that the period of treatment is shorter than the period of time required to effect appreciable cooking of the food at the prevailing temperature. In regard to the cooking material, a large variety of oils and fats can be used, such as, for example, hydrogenated shortening, corn oil, butter, peanut oil, cottonseed oil, soybean oil, beef fat, lard, etc.

In preparing the frozen food product for consumption it is recommended that it be cooked in an oven to a temperature of, for example, from 400° to 450°F and maintained at that temperature for about 15 to 50 minutes. The prepared food can also be cooked by broiling or in a skillet without the addition of any oil or fat. Thereafter, the food is ready for serving.

It is found that although the product of this process is in an uncooked condition, for example shrimp, it requires essentially the same period of time to cook as it takes in warming or reheating a frozen food product which has been previously cooked. Accordingly, from the standpoint of cooking, there is no disadvantage in using the frozen uncooked product.

Example: A frozen block of cod fillets was cut into strips of sticks measuring 3¾" x ¾" x ½". The sticks were coated lightly with finely milled flour and then submerged directly into a batter of milk, eggs and seasoning. The batter and flour coated sticks were then dropped into a drum containing browned bread crumbs which adhered to the batter and flour and thus produced a uniform coating of bread crumbs on the fish sticks.

The fish sticks were then immersed for five seconds in hydrogenated cottonseed oil (MP 105°F) at a temperature of 120°F. The fish sticks were immediately packaged and stored at a temperature of about −5°F. (Note that freezing was not necessary.) After the sticks were stored, a portion thereof was removed from their packages and placed in a shallow pan.

The pan was put into an oven which had been previously heated to a temperature of 425°F. The sticks were cooked for a 20-minute period without adding any cooking material other than what had been present as a result of the coating operation. The cooked fish had an excellent flavor which was comparable to fresh fish which had been cooked shortly after being caught.

Silica Gel Added to Cooking Fat

A.H. Nagel and T.P. Finucane; U.S. Patent 3,574,638; April 13, 1971; assigned to General Foods Corp. have developed a process of preparing French fried products comprising frying the food product at temperatures of 350° to 400°F in deep fat containing less than 2% silica gel, having an average particle size of about 10 microns. This process requires a smaller quantity of silica gel than the process of adding the gel directly to the batter mix, and since the surface is of chief importance, incorporation of the silica gel in the frying oil provides a vehicle for depositing the silica gel on the surface of the fried product in a uniform manner, and at a location where the greatest benefit is desired. More specifically, the uniformly coated fried product of the process prevents the exudation of fat from the outer surface of the fried product, and also retains the moisture in the interior of the fried product.

As a result little intermingling of fat and moisture occur upon reheating, thereby lessening the sogginess and enhancing even-browning, without burning. The prevention of fat exudation upon reheating also eliminates fat drippings. Thus the prior art problems of off taste, sogginess, fat dripping and lack of even-browning, are eliminated in the French frying and reheating steps of the process.

Although the silica gel may be added directly to the batter mix prior to frying, a much larger quantity would have to be added. Moreover, the direct addition of silica gel to the batter mix gives a fried product which has an apparent tendency to get soggy upon reheating, and such sogginess impairs the crispness and taste of the reheated fried food product. Another disadvantage of directly adding silica gel to the batter mix is that it produces a fried product which drips off fat upon reheating.

It is a critical feature of the process that the silica gel used in the cooking oil be present in an amount less than 2% based on the oil, and have an average particle size of not more than about 10 microns. The edible shortening fats or oils which may be employed may be liquid materials, semisolid materials or those which are solid at room temperature, either natural or chemically produced such as by hydrogenation. The oils may be cottonseed oil, corn oil, peanut oil, sunflower or other edible oils; the semisolid fats may be lard, butter, margarine or the like; the solid fats may be animal fats or they may be the liquid oils previously mentioned which have been hydrogenated, or mixtures thereof. The term batter as used herein includes conventional mixtures of flour or bread, salt, baking powder, eggs, and milk and/or water etc. The following is an example of the process.

Example: Micron size silica gel (Syloid 63) was added to liquid cottonseed oil at 1% levels. The oil was heated to 375°F and battered and breaded fish was fried therein for a period of 40 to 120 seconds. The deep fried lightly brown product was removed from the frying oil and chilled at 0° to 70°F to freeze the fish and solidify the fat. In general, the fat content of the piece produced will range from 6 to 10%. After chilling the product is generally packaged and distributed in the frozen conditions. However, here, it was reheated in an oven at 425°F for 10 minutes to give a nonsoggy, crisp, evenly browned, good tasting fried product.

Controlling Temperature of Batter-Coated Fish Portions Until Frying

The prevalent practice in production of "fish and chips" in the U.S. is to use frozen fish portions, thaw them, dip and coat them with batter, and then fry them. However, once the fish thaws to the point where juices can flow from it, the fish rapidly suffers a great loss of flavor and continues to do so. There is therefore a need for a process for preparing fish which will permit volume preparation of deep-fried fish without significant loss of flavor.

D.J. Malin; U.S. Patent 3,622,348; November 23, 1971; assigned to Arthur Treacher's Fish & Chips, Inc. has found that this may be accomplished by a process wherein initially hard frozen food is first raised to a temperature range of 20° to 32°F at which the food acquires an intermediate consistency. At an intermediate consistency, it is neither frozen solid nor soft and fleshy. It has a quality of stiffness but can be manipulated and cut with a knife. The food is then coated with a batter and inserted in hot cooking oil while it is substantially at the intermediate consistency. The food is permitted to sink and will remain submerged until it has been properly cooked. After the proper cooking, it will rise and float in the oil and should be substantially immediately removed therefrom.

The steps in preparing fish are as follows: Normally, initially hard frozen bulk-size fish will be shipped in a frozen solid condition to a restaurant installation. When the fish is received it will be stored in a deepfreeze which will be kept below 0°F. Each evening, after a day of operation and prior to the next day's operation, a sufficient quantity of bulksize fish for one day's operation will be removed from the deepfreeze and stored in a refrigerator, kept at approximately 25°F.

The following morning, after the large bulk pieces of fish have had their temperature raised to approximately uniformly 25°F, they are removed from the refrigerator and are cut into serving portions of approximately 2 to 4 ounces. The cut-up portions are then immediately returned to the refrigerator for withdrawal as they are needed for cooking.

As fish is ordered throughout the day by consumers, the serving portions are withdrawn from the refrigerator, dipped into a saline solution for a few seconds, and then dipped in and coated with a dusting compound. The saline solution cleanses the surface of the fish, removes any ice particles which would otherwise burst the batter, adds flavor, and provides a suitably damp surface to aid the adherence of the dusting compound. The dusting compound preferred is rice cones or granulated rice. The dusting compound serves to absorb the surface dampness of the fish and gives the fish a uniform, dry, outer surface which helps seal the moisture in the fish. The dusting compound also aids the adherence of the batter to the fish, helps the fish retain its crispness and inhibits heat loss from the fish.

After being coated with the dusting compound, the fish is dipped in a batter and coated with a relatively thick layer of batter to aid the moisture retention of the fish. After being dipped and coated with the batter, the fish is ready to be inserted in a hot cooking oil. It is important that during all these steps, the fish remains substantially at the intermediate consistency and therefore remains substantially in the temperature range of 20° to 32°F. By keeping the fish in this temperature range flavor loss is substantially prevented and still cutting and manipulation through the coating can be accomplished.

The fish are inserted into an oil having a temperature of 362°F. Although a different temperature might slightly alter the cooking time and result, the temperature is not critical. Regardless of the specific size of the serving portion, which may be any size in the approximate range of from 1 to 6 ounces, it will remain submerged in the hot cooking oil until it has been cooked the proper length of time. Only after it has been cooked so that it has the most desirable and commercially saleable taste, will it rise to the top and float in the oil. Smaller serving portions will rise earlier, and the larger portions will remain submerged for a longer period of time. However, all portions, regardless of when they rise, will be cooked to provide a uniform product which is cooked to perfection.

Starch Mixed with Oxidizing Agent

Adherence of the coating is a problem in the preparation of coated frozen food products, particularly fish fillets. Batter starches have been used for coating the fish to cause bread crumbs to stick to the frozen fish. It is well known that these products do not do a completely satisfactory job because the coatings tend to fall off or pull away from the coated food product.

J.C. Fruin; U.S. Patent 3,767,826; October 23, 1973; assigned to Anheuser-Busch, Incorporated has found that it is possible to increase the adhereability of the bread crumbs by coating the fish with a batter formed from a dry mixed, slightly oxidized starch product as the base for the batter. The starch is dry mixed with water for forming a batter, coated on the fish, and thereafter bread crumbs are applied to the batter coated fish. The crumb coated fish are fried, refrozen and packaged as consumer items. The crumb coated fish can be frozen and then fried just prior to serving by the consumer. The process, being dry mixing, uses conventional starch processing equipment and is safe and easy to practice.

A conventional starch, such as cornstarch which has been wet milled, is dry mixed with an oxidizing agent, such as calcium hypochlorite. Other oxidizing agents may be used, and other types of starches or starch derivatives can be used. The batter also can be used for food products other than fish. Both the oxidizing agent and the starch are dry when mixed. The dry starch may contain up to 20% moisture and still be suitable for this process. The process involves a slight oxidation treatment of the starch. This preferably is done in the starch plant and involves the dry mixing of the starch with a hypochlorite in the range of about 0.1 to about 5.0 parts hypochlorite to 1,000 parts starch by weight. If substantially less hypochlorite is used, the oxidation effect is achieved much more slowly; and if substantially more hypochlorite is used, the oxidation is so strong that a good batter starch is not produced. While hypochlorites are preferred as the oxidizing agent, other oxidizing agents can be used.

After treatment with the hypochlorite in the dry form, the treated starch may be allowed to age. Accelerated aging can be used and longer time and higher temperatures also are effective. As previously mentioned, the treatment is a mild oxidation treatment and involves the use preferably of calcium hypochlorite. Other alkaline earth hypochlorites can be used, but calcium hypochlorite is the most practical and economical. Other oxidizing agents include chlorites, bromates, peroxides, etc. The dry blending is important because it gives a faster and more certain reaction. It also is desirable that the starch be stored at room temperature or higher after treatment.

A reducing atmosphere causes undesirable action in a starch to be used as a batter starch. This is shown by tests in which the starch is placed in a desiccator, the air is removed by applying a vacuum, and SO_2 gas is added to replace the air. This produces a starch which does not have good batter starch characteristics.

The mild treatment caused by the calcium hypochlorite, the dry mixing of the oxidizing agent, and the level of calcium hypochlorite added are significant characteristics and are important in producing a starch batter which causes fast sticking power of bread crumbs to frozen fish. The mildly oxidized starch is mixed with corn flour and/or other substances, such as wheat flour, milk powder, alginates, cellulose gums, salt, etc., to form a commercial product. This product then is mixed with water by the user to form a batter for coating frozen fish. The exact compositon of the mix varies by proprietary product.

Example: 5 lb of granular dry cornstarch having 10% moisture was mixed with 0.005 lb of calcium hypochlorite. The mixing was done in a Twinshell mixer for 20 minutes. Thereafter, the hypochlorite treated starch is stored one week at room temperature. To make the batter, 150 grams of the treated aged starch is mixed with 150 grams white corn flour obtained by conventional milling techniques and 400 grams water is combined with this mixture of starch and flour to make the batter. The fish are a lot of frozen cod portions weighing 3 ⅓ ounces apiece. The fish are coated with the batter starch and then coated with bread crumbs. The crumb coated fish are deep-fat fried at 375°F for 3½ minutes.

The fried fillets are then placed on edge and slit with a knife, exposing the interior surfaces in a manner similar to opening a book, so that the inner surface is exposed. Flesh is scraped from the batter-breading coating and the amount of fish flesh adhering to the coating is estimated. The table shows the percent of sides of fish which have the range of adhesion of bread crumbs set forth in the table. In preparing the table, each side of the fillet is examined and the percent of coverage is estimated by a worker skilled in working in this art. The table on the following page shows examples of results obtained in deep fat frying of fish, dipping pieces of frozen fish in batter and then coating with breading prior to deep fat frying. The table shows the degree of adhesion of batter to fish in ranges of 95 to 100%, 90 to 95%, 80 to 90% and less than 80%.

Type of starch	Treatment	Age of starch when tested, days[1]	Adhesion, percent			
			95-100	90-95	80-90	0-80
Dent corn	None (control)	3	25	37	13	25
Do	0.10 $KBrO_3$	3	50	50	0	0
Do	None (control)	12	0	0	13	87
Do	0.10% $KBrO_3$	12	0	10	20	70
Do	0.05% $Ca(OCl)_2$	12	8	50	25	17
Do	None (control)	4	0	0	0	100
Do	0.10% $Ca(OCl)_2$	4	42	25	25	8
Do	0.05% $Ca(OCl)_2$	4	25	50	17	8
Do	None (control)	9	0	17	33	50
Do	0.015% $KMnO_4$	9	25	33	25	17
Do	0.10% Na ClO_3	9	33	42	17	8
Do	0.20% H_2O_2	9	0	42	8	50
Do	None (control)	1	0	0	12	88
Do	0.05% $NaClO_2$	1	8	42	33	17
Do	0.25% $NaClO_2$	1	8	33	42	17
Do	0.01% $NaClO_2$	1	8	25	42	25
Waxy maize	None (control)	1	0	0	8	92
Do	0.10% $(CaCl)_2$:	1	0	25	42	33
Dent corn	None (control)	2	0	0	0	100
Do	0.10% $Ca(OCl)_2$	2	42	42	16	0

[1] At room temperature.

OTHER CONSUMER PRODUCTS

Instant Fish-Potato Mixture

Little attention has so far been given to the preparation of instant fish products. Dried fish is a very old article of food but the techniques previously available for drying fish have been incapable of providing a product which could be rehydrated by addition of liquid to give a palatable form of reconstituted fish.

E.A.M. Asselbergs; U.S. Patent 3,201,260; August 17, 1965; assigned to Canadian Patents and Development Ltd., Canada has developed a process for the production of a dehydrated, cooked, fish-containing product comprising mixing together comminuted, cooked fish and a cooked carbohydrate material selected from the class comprising mashed cooked potato, precooked starches, precooked cereals and mixture thereby obtained and drying this thin layer to a low moisture content. It is essential in the process that the fish be cooked prior to comminution and the carrying out of the drying operation.

The carbohydrate material which is incorporated in the mixture may be any kind of precooked starch, for example potato starch or corn starch. Precooked cereals such as rice, wheat or barley can also be used. However, the most satisfactory carbohydrate material to be used in the preparation of the mixtures of the process is mashed, cooked potato.

When producing a mixture of fish and potato by the process it is essential that the potatoes are cooked and mashed prior to the formation of the fish-potato mixture. The manner of cooking and mashing the potatoes is not critical. The precooking of the fish is carried out in a generally conventional manner. The fish, which may be fresh or frozen, is cleaned and then cooked, preferably by immersion in boiling water for whatever length of time is necessary depending on the nature of the fish.

The ratio of the amount of fish to the amount of potato contained in the fish-potato mixture can be varied within wide limits. There is no practical upper limit to the proportion of potato that can be introduced into the mixture. However, the reduction of the fish content to a very low value will seldom be desirable because the protein content will then fall off and the product will eventually become indistinguishable from mashed potato. Theoretically, there is no upper limit to the proportion of fish in the mixture. However, with most types of fish the content of fish in the mixture should be not more than 90% by weight to obtain the most palatable products. If more than this amount is used there is a tendency for the product to become hard and rubbery in the subsequent drying operation and this makes it difficult to obtain a satisfactory product.

Thus it is preferable to form a fish-potato mixture which contains at least 10% of mashed potato. In general it is found that the best products from the standpoint of taste and nutrition and also for ready carrying out of the drying operation are those which contain from about 40 to about 60% by weight of fish, the balance being mashed potato.

The fish-potato mixture is formed into a thin layer which is then dried. For best results the thickness of this layer should not be greater than 0.015 inch. By operating with a layer of this thinness the dried product which is obtained is of small size and can therefore be more easily reconstituted. Thicker layers of the mixture can however be utilized, the practical upper limit of thickness being dictated by heat transfer considerations and by reconstitutability of the product. With thick layers the product has a large particle size and is less readily reconstituted.

The formation of the thin layer of fish-potato mixture can be effected by forcing the mixture through a suitable slot. For example, the mixture may be passed through the nip between two roller members consisting of rotatable heated drums. A layer of the fish-potato mixture is then formed on each of the drums and then drying of the layer is effected on the surface of the drums. The following example is given by way of illustration.

Example: Freshly dug potatoes of the Sebago variety grown in New Brunswick, were peeled in an abrasion peeler, and trimmed to remove defects such as eyes and spots. They were then sliced in "French-fry" portions consisting of strips of potato having a cross section of 0.375" x 0.375". Cooking of the potatoes was then carried out in boiling water at 212°F in a steam jacketed kettle of stainless steel with a weight ratio of water to potatoes of about 1 to 1. The cooking time was 6.5 minutes.

After draining off the water the potato strips were spread out on a perforated screen and air at room temperature blown over them. This had the effect of forcibly removing the moisture-saturated air surrounding the potato strips thereby effecting surface drying and fluffing of the potato strips to give a mealy appearance.

The potatoes were then fed to a conventional mashing machine, commonly called a ricer, comprising a hopper into which the potatoes were dropped and a conical screw conveyor which served to withdraw the potatoes from the base of the hopper and force them through the perforations in a perforated conical screen closely surrounding the screw conveyor. In the formation of a commercial product suitable additives, well known in the art, could be introduced at this point for ensuring a desired texture, flavor, color, stability and odor of the final product. The mashed potato falling through the perforated conical screen of the mashing machine was then mixed in batches with a quantity of comminuted, cooked fish, 10 lb of the mashed potatoes being mixed with 25 lb of fish in a slow speed mixer. The fish has been prepared by boiling frozen or fresh fillets of hake in water for 4 to 6 minutes. Vinegar, salt, pepper and nutmeg were added to the cooking water. After the fish was cooked the water was drained off and the cooked fish was comminuted using a standard type of comminuter with a fine screen (perforations of 1/32"). The use of a fine screen is important as it ensures that any bones passing through the screen are finely comminuted.

The mixture of cooked, comminuted fish and mashed potatoes, containing about 71% of fish, was fed to a double drum drier, the rotating drums of which had a length of 7.75 inches and a diameter of 6 inches. Steam under 30 lb pressure was supplied to the interior of each of the drums. The clearance between the drum was varied between about 0.001 and about 0.015 inch, four runs being carried out with clearances of 0.002, 0.006, 0.010, and 0.014 inch respectively. The drums rotated at a speed of about 2 revolutions per minute.

In each run a continuous but perforated layer of fish-potato mix was formed on the surface of each drum. In appearance the layer resembled a lace work or felted fibrous structure having a random orientation of the fibers. Knife scraper devices were used for removing the layers from the drums, each of the layers having been in contact with the associated drum for a period of about 8, 12, 16, and 20 seconds respectively, depending on the drum clearance, prior to its removal from the drum at which time the product contained about 8% moisture.

The layers of dried product thereby obtained were readily disintegrated by slight mechanical force into a mass of crystal-like particles resembling snow crystals. The moisture content could be reduced below the above-mentioned value of 8% by operating in an environment the humidity of which was more carefully controlled than that used for the above experiments. For commercial purposes moisture contents of less than 5%, preferably from 0 to 3%, are desirable. In any event the products of the process are to be dried to a moisture content of less than 15%.

The product of each run could be reconstituted by admixture with a suitable amount of milk, or a milk-water mixture, and eggs. The amount of fluid used in the reconstitution can be varied in accordance with taste and with the purpose for which the product is to be used. The texture and flavor of the reconstituted product were found to be excellent.

Fish "Chip"

E.N. Bacon; U.S. Patent 3,301,686; January 31, 1967; assigned to Nova Scotia Research Foundation has developed a fish product in the nature of a thin, crisp, wafer composed of fish flesh and resembling a conventional potato chip but having a distinctive taste.

Any suitable type of fish is employed as the starting substance. The fish are first cleaned in conventional manner by removal of the heads and entrails. The skin, bones, and other residue may also be removed during the cleaning stage, but alternatively, this step may be omitted. The cleaned fish are then partially dried by exposing them to atmosphere for a period of time or by subjecting them to a short, mild heating step to dry the gel-like secretion therein. The partially dried fish, preferably cut into lumps or chunks, are now subjected to a fiberizing treatment in water to produce an aqueous slurry of separated fish flesh fibers. Any suitable conventional hydropulper, beater or the like is employed.

Figure 3.6a shows a fiberizing device and Figure 3.6b shows a screening device. Figure 3.6a illustrates at **1** a suitable beating device comprising a vessel **2** containing a series of rotatable blades **3** mounted on a shaft **4** driven by a motor **5**. The vessel is charged with a quantity of water **6** and the fish pieces or chunks **7**. The charge comprises at least 90% water and 10% fish by weight. Preferably, the charge comprises approximately 98% water and 2% fish.

FIGURE 3.6: APPARATUS FOR PRODUCING FISH "CHIP"

a.

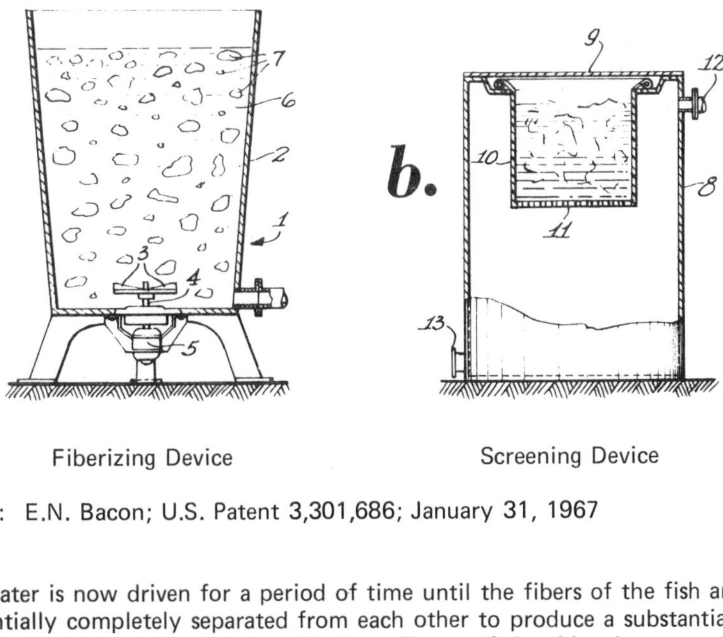

Fiberizing Device | Screening Device

Source: E.N. Bacon; U.S. Patent 3,301,686; January 31, 1967

The beater is now driven for a period of time until the fibers of the fish are substantially completely separated from each other to produce a substantially homogeneous slurry of water and fish flesh fibers. If the skins, bones, and other residue has not been preliminarily removed, the slurry is screened to remove this foreign matter from the fiber slurry.

The fiber slurry is now further screened to partially remove the moisture content thereof. This partial dewatering step may be effected in any suitable filter or screen device. Figure 3.6b illustrates a suitable screening device comprising a closed tank **8** having suspended from the top **9** thereof a cylindrical receptacle **10** having a perforated or wire screen bottom **11** The screen bottom may be 40 mesh. The tank is subjected to suction or vacuum through a line **12**. A drain plug **13** is provided to remove the separated water. The receptacle having a moist mat **14** of interlocking fish flesh fibers therein is then removed from the tank. If necessary to obtain further dewatering of the mat, a pressure plate carried by a shank may be employed to press the mat against the screen. Such a pressure plate may be employed during the vacuum screening step described.

It is important that the moisture content of the final mat be closely controlled to limit subsequent shrinkage and loss of fat therein. The moisture content of the mat should not be less than 25% and preferably in the approximate range of 40 to 50%. The mat may have a thickness ranging from a few thousandths of an inch to at least one inch depending upon the desired character of the final product. For instance, if a chip-like product is desired, the thickness may be a few thousandths, say, 0.005 inch.

The mat is now cooked in any desirable manner. For instance, it may be passed as a sheet between heated rolls and the cooked sheet then broken up randomly. Alternatively, it may be cut into any desired shape and the cut shapes fried to the proper degree of crispness in a deep fat of appropriate temperature.

It will be apparent that seasonings or flavorings may be added to the mat before cooking to modify the taste and add to the palatability of the final product. Other addition to the fish fiber may include flour, corn starch or ground cracker crumbs. The fish employed may be partially dried fresh fish or conventional forms of salt or dried fish.

In respect of the fish chips, patties or cakes produced in accordance with the process, it will be apparent that no binder is required since a coherent mass is formed by random mechanical interlocking of the fibers plus added adhesion caused by surface tension of the initially moist fibers.

Gelled, Extruded Fish Product

J.C. Pottie; U.S. Patent 3,493,391; February 3, 1970; assigned to Nova Scotia Research Foundation describes a procedure for comminuting cleaned, raw fish to produce a rigid gel, which can be extruded into a desired shape for cooking. The raw material employed is the cleaned, dressed, edible portion of fish. Preferably, but not necessarily, the fish employed are of the white-fleshed species. Fish fillets, or the cuttings from the filleting table, may be used. These cuttings usually contain a great number of bones. It is not necessary to remove these bones, which are a source of valuable minerals.

No other special preparation of the raw fish material is required. For example, the fish may be in the fresh condition as it comes from the cutting table; it may be completely thaw-dripped, or thaw-dripped with slight surface desiccation from standing. It may have been frozen for short or long periods, thawed to just before the thaw-drip begins, and then used. It may have been frozen for short or long periods, completely thaw-dripped, and then used. The important consideration is that the raw fish employed be free from any additive such as water or gelling agents.

The temperature of the raw fish at this stage and before subsequent treatment is relatively unimportant. For instance, it may be partially frozen, as indicated above, or completely at room temperature, say about 27°C or higher. The raw material is now subjected to a comminution treatment, which may or may not be preceded by a coarse grinding step. An important and critical feature of the process is that during the comminution and subsequent steps to be later described, no addition other than possibly a relatively minute quantity of flavoring or pH adjusting agents is made to the fish material. Thus, the raw material consists essentially of fish with or without a relatively minute quantity of flavoring or pH adjusting agents.

For instance, minute quantities of sodium chloride, monosodium glutamate, or the like, may be added to modify texture and flavor, without detracting from or reducing the substantially pure fish character or condition of the product. More specifically, it is important that the use of gelling, thickening, stabilizing, emulsifying, and the like agents be avoided. The use of added water during the comminution step, which would result in subsequent shrinkage of the product, is, of course, to be avoided.

The comminution may be carried out in any suitable type of conventional apparatus. Thus, a conventional machine such as a Waring Blendor, having an upright vessel with a rotatably driven multibladed cutting device mounted in the lower portion thereof, is quite satisfactory. It is important that the comminution machine is located in an atmosphere not exceeding 5°C whereby its contents reach and are maintained within a temperature range of 5° to 30°C.

Each fish muscle fiber consists of a bundle of fibrils, each of which in turn comprises a bundle of proto-fibrils. The fibrils comprise two proteins, myosin and actomyosin. These proteins are hydrophilic colloids which, when they are released or extracted from the fibrils by shearing action of the cutting knives, result in the formation of a gel.

The fibrils and proto-fibrils are oriented in such a way in the mass that the natural water content of the fish is imprisoned in the spaces between the fibrils. The action of the cutting tools also incorporates air into the gel. The addition of sodium chloride further modifies the texture by furthering the extraction of the myosin and the actomyosin from the fish muscle.

The comminution step may be continued until a substantially rigid plastic mass with partial gel formation has occurred or until substantially complete rigid gel formation has been effected. Ordinarily, the comminution step is effected within a period of two to seven minutes. The resulting plastic mass, in either semigel or gel form, is maintained at a temperature not exceeding its comminution temperature until cooked.

The mass may be placed in an extrusion device of conventional structure directly after formation and extruded in shaped bodies directly into hot deep cooking fat maintained at appropriate cooking temperature, and fried to the desired degree of crispness. Immediate extrusion into deep fat of appropriate temperature following semigel or gel formation seals in the flavor of the fish. The cooked shapes may, of course, be of any desired form with appropriate choice of extrusion nozzle.

The final product may comprise simply the uncooked plastic mass in semigel or gel condition, but it is important that, if cooking of this product is to be delayed, the product must be frozen and kept in frozen state for later cooking. The cooked product may be, for instance, in the form of a thin, rectangular crisp wafer about one inch wide and one or two inches long. Alternatively, it may be in the form of a crimped cylindrical crisp, about ⅛ inch in diameter, and from one to two inches long.

Lox and Eggs

Lox is marinated and smoked salmon. It is prepared by soaking the salmon sides in a strong marinating solution, which may be a strong brine solution or a water solution containing sugar and sour cream, for several, viz 10 to 14, days and then smoking the so-marinated fish sides in the presence of cool or hot smoke for an extended period of time lasting anywhere from 14 hours to 4 or 5 days.

The dish known as lox and eggs is well known, and its preparation also is well known. *M.P. Katz; U.S. Patent 3,012,896; December 12, 1961* describes a process for preparing a precooked frozen food composition consisting of onion, butter and lox which can be prepared in quantities sufficient for many portions of several sittings and which can be stored for long periods of time without any loss of flavor, natural juices, and taste characteristics associated with the freshly prepared combination of the ingredients. The following example will illustrate the process.

Example: The following ingredients are prepared:

Ingredient	Parts by Weight
Onions (red or Bermuda)	6
Butter	1
Lox	1

The onions are diced and washed. The butter and the onions are placed in a pan, and the mixture is slowly heated, while being stirred, until the butter is melted. The heating and stirring are continued simultaneously, until the onions become soft and sufficiently browned, as desired. The lox, which has been subdivided into pieces approximately 4 to 5 times the size of the onion pieces, is then added slowly to the onion and melted butter mixture under constant stirring. The heating is continued for approximately ¾ hour per approximately 2 lb of the mixture of lox, onions, and butter. The composition is then transferred into a container, preferably of a metallic foil material such as aluminum, and frozen.

The composition has been used in the following manner to prepare a lox and egg dish from a portion thereof. A portion of the frozen mass is separated and allowed to thaw out. The thawed portion is heated until the butter becomes molten, and a predetermined amount of shelled eggs, in accordance with the portion being prepared, is added to the mixture while it is kept at such a temperature that the butter remains molten and simultaneously kept under constant agitation.

The cooking is continued until the desired degree of preparation is attained. The finally prepared dish is such that it retains the flavor, natural juices, and taste characteristics and appearance of a freshly prepared dish, even though the hereinbefore described mass has been subjected to freezing and storage for a long period of time.

Fish Flavoring Material

I. Katz, C. Giacino, L.J. Strasburger and M.H. Vock; U.S. Patent 3,627,540; December 14, 1971; assigned to International Flavors & Fragrances Inc. utilize a small amount of 2,4-pentadienal for adding a seafood flavor to foods, or for improving the already existing taste of seafood.

The quantity of pentadienal utilized can be varied to provide a wide range of seafood-type flavors. As used herein, seafood-like flavors will be understood to mean flavors associated with marine animals including crustaceans such as shrimp, lobster, and the like; mollusks such as oysters, clams, and the like; and particularly fish, including smoked fish, dried fish, salted fish, and the like.

Thus, at the level of 50 parts per billion, the pentadienal begins to impart a very light, almost indefinable, sensation in aqueous solution. As the level is increased to one part per million (ppm) a definite pleasant salt water fish-like character is imparted to an aqueous solution. As the level reaches 2 ppm, a definite, good characteristic anchovy flavor is obtained and this flavor persists up through levels of 10 ppm and beyond, where the substance leaves a light, lingering aftertaste.

Foodstuffs flavored in this way include both solids and liquids prepared for oral consumption by human beings and by animals, particularly warm-blooded domestic animals such as are used for pets or raised for consumption. The seafood flavor characteristic of the pentadienal can further be varied by including one or more cyclic oxazo compounds, particularly Δ^3-oxazolines having the formula

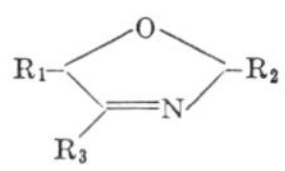

wherein R_1, R_2 and R_3 are hydrogen, alkyl, or alkylene and can be the same or different. It is preferred to use lower alkyl derivatives of oxazoline, preferably those having up to about three carbon atoms in the alkyl group or groups. A preferred oxazoline is 2,4,5-trimethyl-Δ^3-oxazoline.

The amount of oxazoline can be varied over a range to provide a variety of shadings of seafood flavor notes. The quantity of oxazoline utilized varies from about an amount equal to the quantity of pentadienal up to ten or more times the quantity of pentadienal. As components for use in this process, oxazolines provide a flavor redolent of cooked fresh seafood, particularly fresh fish.

The compounds can be combined with adjuvant ingredients as necessary to impart special characteristics. Such other ingredients include vehicles such as water, ethanol, propylene glycol, glycerol, glyceride oils, and the like, and carriers such as gum arabic, tragacanth, and the like to provide a base for a spray-dried flavor composition. They also include thickeners such as alginates, carrageenan, and the like to impart a heavier body to gravies and similar products and condiments and spices.

The pentadienal and optical oxazoline can be incorporated into foods such as soups and soup mixes, casserole dishes, canned and frozen vegetables, animal and pet foods and other veterinary products, sauces, gravies, broths, chowders, stews, simulated seafood products, dietitic products, spreads and dips, and the like. It is generally preferred that from about 0.5 to about 25 ppm of the pentadienal be comprised in the total foodstuff to be flavored. Amounts somewhat less or somewhat greater than this can be utilized, dependent upon the particular reaction product, the foodstuff to be improved, personal preferences, and local tastes, and the like. Likewise, the amount of oxazoline ranges from an amount equal to the pentadienal to 10 times as much.

Example 1: A fish flavor is prepared by admixing 5 mg of 2,4-pentadienal and 3 grams of cade oil with sufficient cod liver oil to make up a kilogram of material. 10 grams of this oil is admixed with 10 grams of vinegar and applied to a greens salad containing sliced radishes and pieces of fresh tomato. The salad so obtained has an excellent anchovy flavor nuance.

Example 2: A fish flavor is prepared by admixing 3 grams of cade oil, 20 mg of 2,4,5-trimethyl-Δ^3-oxazoline, 3 mg of 2,4-pentadienal, and sufficient cod liver oil to make 1 kg of finished oil. The fish flavor composition, in the amount of 0.9 cc is added to 7.3 grams of a soup base consisting of the following ingredients in the indicated quantity (parts).

Finely ground sodium chloride	35.62
Hydrolyzed vegetable protein (Nestle 4 BE)	27.40
Monosodium glutamate	17.81
Sucrose	10.96
Beef fat	5.48
Sethness caramel color (powder B & C)	2.73

The resulting mixture is added to 12 ounces of boiling water to create a soup having an excellent fish chowder flavor.

Another process for giving food a seafood flavor, developed by *L.C Chhuy and E.A. Day; U.S. Patent 3,773,523; November 20, 1973; assigned to International Flavors & Fragrances Inc.*, consists in adding a product made of an algae extract or enzymatic digest of seafood or seafood extract or seashell extract, a sulfur compound such as cysteine and, if desired thiamine and /or thiazole alkanols and other free amino acids and/or polypeptides. In addition, if desired, an oxazoline and/or 2,4-pentadienal and/or thiazole alkanol may be added.

In accordance with the process edible compositions are prepared comprising the product of the processing of a natural seafood extract or an enzymatic digest of seafood or a seashell extract or an algae extract with at least one sulfur-containing compound, in absence of a sugar and, optionally, in the presence of thiamine, or thiamine precursors which are thiazole derivatives, and other free amino acids and polypeptides.

Cysteine, cystine and methionine are especially preferred sulfur-containing amino acids. Especially preferred products resembling natural seafood in their flavor and aroma are produced by means of intimately admixing and heating mixtures of cysteine and/or methionine and/or their salts with seashell extracts and/or enzymatic digest of seafood and/or seafood extracts and/or algae extracts.

The amount of seafood extract or seashell extract or algae extract or enzymatic digest of seafood used in these compositions prior to processing may vary within wide limits. For example, liquid compositions may be prepared using as little as 1% by weight or as much as 40% by weight of seafood meat extract or algae extract or seashell extract or enzymatic digest of seafood based on the total weight of the composition. Dehydrated seafood or seashell or algae or enzymatic seafood digest product are prepared for preparing these compositions to obtain the best flavors. It is also preferred to extract the seafood or seashell or algae materials with solvents which can easily be separated from the desired extracts. Examples of such extracting solvents are water, alcohols, methylene dichloride, ethylene dichloride, methyl isobutyl ketone, glycols such as propylene glycol and aqueous saline solutions such as 5% NaCl.

When it is used, thiamine may be used as a free base, but is preferably employed as an acid salt. The preferred salts are halogen acid salts and, of these, the hydrochloride is especially suitable. The amount of thiamine used in the preparation of seafood flavors whether as the free base or as the acid salt generally varies from about 0.2% to about 20%, by weight, based on the total weight of the liquid composition, although useful products may be prepared with greater or lesser amounts.

Cysteine and cystine are sulfur-containing amino acids readily available as free amino acids or as salts, both acid and basic. Preferably, the acid salts, especially the halogen acid salts, such as the hydrochloride are utilized. For most purposes the desired flavor quality is achieved by the use of from about 0.2 to about 20% by weight of each of these amino acids as free acids or as salts, or a combination of them totaling the same proportion by weight based on the total weight of the liquid composition. Useful products may be obtained using somewhat greater or lesser amounts.

The inorganic sulfur-containing compounds include sodium sulfhydrate and sodium sulfide. Instead of sodium compounds, corresponding compounds of potassium, calcium and ammonium or substituted ammonium may be substituted. Any such compound employed should be capable of reacting to give the desired seafood note. Organic sulfur-containing compounds useful include methyl, ethyl, and propyl mercaptan, dimethyl sulfide and disulfide, and methyl ethyl sulfide. In the following examples, ingredients are given in parts by weight.

Example A: To 75 grams of dehydrated seashells [25:25:50 (weight ratio) shrimp shell, lobster shell, mussel shell finely ground mixture] is added a mixture of 25 grams of sodium chloride and 100 grams of water. The resulting slurry is refluxed at atmospheric pressure for a period of 1 hour. The resulting slurry is filtered and the presscake is refluxed for 1½ hours again with a mixture of 75 grams of water and 18.75 grams of sodium chloride. The supernatant liquid batches resulting from the above mentioned first and second filtrations are then combined to form a natural seashell extract.

Example B: To 75 grams of dehydrated pulverized shrimp is added a mixture of 25 grams of sodium chloride and 100 grams of water. The resulting slurry is refluxed at atmospheric pressure for a period of 1 hour. The resulting slurry is filtered and the press cake is refluxed for 1½ hours again with a mixture of 75 grams of water and 18.75 grams of sodium chloride. The supernatant liquid batches resulting from the above mentioned first and second filtrations are then combined to form a natural seafood extract.

Example C: To 75 grams of dehydrated pulverized dried seaweed [a 50:50 mixture of red algae and brown algae (Rhodophyta and Phaeophyta)] is added a mixture of 25 grams of sodium chloride, 100 grams of water and 150 grams of methylene dichloride. The resulting slurry is refluxed at atmospheric pressure for a period of 1 hour. The resulting slurry is filtered and the presscake is refluxed for 1½ hours again with a mixture of 75 grams of water, 18.75 grams of sodium chloride and 100 grams of ethylene dichloride. The supernatant liquid batches resulting from the above mentioned first and second filtrations are then vacuum evaporated to ¼ of their original volumes, removing the ethylene dichloride and methylene dichloride, and then combined to form a natural algae extract.

Example D: To 75 grams of dehydrated chopped and pulverized anchovies is added a mixture of 25 grams of sodium chloride and 100 grams of water. The resulting slurry is refluxed at atmospheric pressure for 1 hour. The slurry is filtered and the press cake is refluxed for 1½ hours again with a mixture of 75 grams of water and 18.75 grams of sodium chloride. The supernatant liquid batches from the first and second filtrations are then combined for a natural seafood extract which can be replaced by Anchovy Extract (Unilever, N.V.).

Example 1: The following ingredients are refluxed for 4 hours. The resulting mixture is then aged for 3 days and an aliquot withdrawn and dried. Based on the weight of the dry solid product obtained, sufficient gum arabic is added to the batch to provide a composition containing one part by weight of flavor solids and one part by weight of gum arabic. The composition is then spray dried to produce a seafood flavored product.

L-cysteine hydrochloride	0.88
Seashell extract produced in Example A	98.24
Thiamine hydrochloride	0.88

Example 2: The following ingredients are refluxed for 4 hours, aged for 3 days admixed in 1:1 ratio with gum arabic and spray dried to produce a dry solid having a seafood flavor. Before spray drying one part of sodium chloride is added.

L-cysteine hydrochloride	1.71
Seafood extract produced by the process of Example B	28.63
Thiamine hydrochloride	1.71
Water	67.95

Example 3: The following ingredients are refluxed for 4 hours, aged 3 days and spray dried to produce a dried product having a seafood flavor.

Water	53.74
Dibasic ammonium phosphate	1.06
Thiamine hydrochloride	0.63
L-cysteine hydrochloride	0.27
Potassium chloride	0.27
Phosphoric acid (85%)	0.21
Dibasic potassium phosphate	0.21
Calcium lactate	0.10

(continued)

Salt	5.30
Beef fat	21.20
Seashell extract produced by the process of Example A	17.54

Example 4: The following ingredients are refluxed for 2 hours and sufficient gum arabic added to provide a composition containing one part gum arabic and one part flavor solids by weight. The composition is drum dried to produce a dry product with a flavor resembling that of fresh flounder.

L-cysteine hydrochloride	2.00
Thiamine hydrochloride	2.00
Water	67.00
Algae extract produced by the process of Example C	29.00

Example 5: The following ingredients are refluxed for 4 hours.

L-cysteine hydrochloride	0.88
Anchovy extract produced by the process of Example D	18.00
Thiamine hydrochloride	0.88
Water	67.30

The resulting mixture is then aged for 3 days and an aliquot portion is withdrawn and dried. Based on the weight of the dry solid obtained, sufficient gum arabic is added to the batch to provide a composition containing one part by weight of gum arabic. The composition is spray dried to produce a fish-flavored product.

Stuffed Fish Fillet Product

In the fish industry it has long been a problem to find ways to utilize various types of fish which are plentiful and nutritious but heretofore too small or not sufficiently taste-worthy in their natural form to create a significant market demand.

J. Joaquin; U.S. Patent 3,718,481; February 27, 1973 has solved this problem by supplying a ready-to-cook fish product that can be made from a wide variety of flat type fish that can be stored in a frozen form after being assembled, and that can be easily prepared as a tasteful, satisfying and nutritious form for the consumer merely by heating the product to a predetermined temperature range sufficient to cook it properly.

The first step for making the product **10** involves the filleting of a flat type fish in the conventional manner. Such fish which are harvested in large quantities by commercial fisheries may include sole, sea perch, flounder and other such species which typically have a much greater width than thickness. In filleting such a fish **18**, as shown in Figure 3.7a, a knife **20** may be used to first cut the side of fillets of the unskinned fish starting near the tail and running close to the backbone toward the head of the fish. The skins of the fillets **12** and **14** are then removed by the knife so that each fillet is then skinless and boneless, as shown in Figure 3.4b. In any fish fillet that is cut from a raw fish and skinned in the aforementioned manner, the characteristics of the flesh varies through the fillet cross section. As shown in Figure 3.7c, the side **22** of each fillet from which the skin was removed has a layer **24** of flesh with a slightly more fibrous or muscular composition which enables the fish to curve or bend in both directions as it swims. In the next step of the method, both of the fillets are laid flat on a plane surface **26** with their skinned sides **24** facing up. At this point, a thin layer of flour **28** is sprinkled on the upturned surface of each fillet.

FIGURE 3.7: STUFFED FISH FILLET PRODUCT

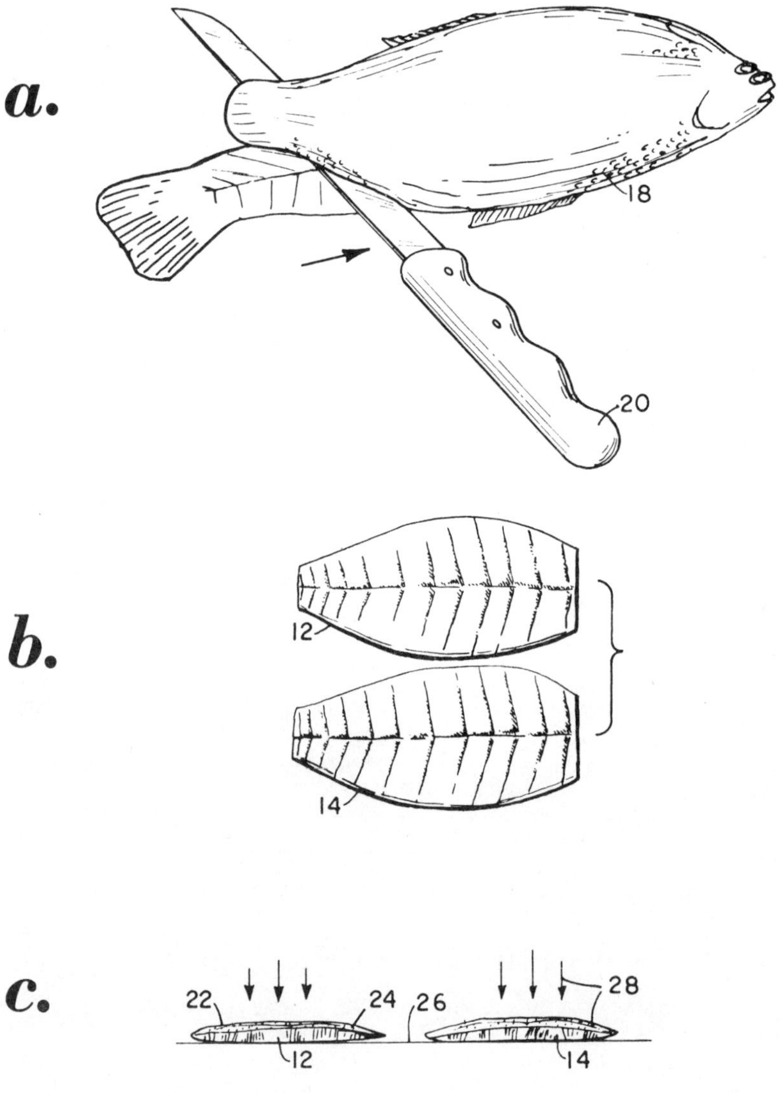

(continued)

FIGURE 3.7: (continued)

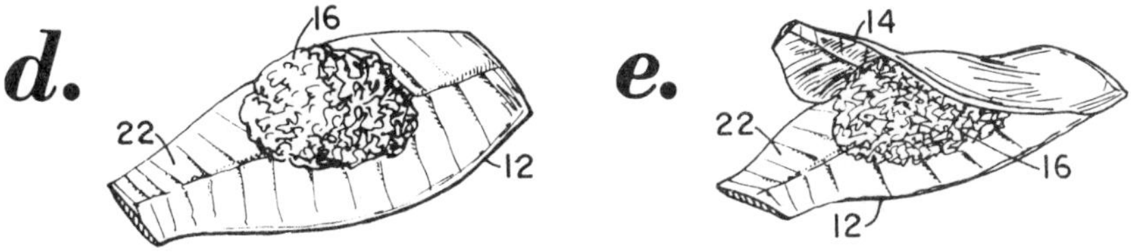

Source: J. Joaquin; U.S. Patent 3,718,481; February 27, 1973

In the next step, a quantity of stuffing material **16** is placed on the upper skinned side **24** of the first fillet **12**, as shown in Figure 3.7d. Different ingredients for stuffing material may be used including various types of edible vegetables, meat or seafood materials or combinations thereof. For example, curried rice, spinach, artichokes, celery and asparagus are highly satisfactory. Also more complicated mixtures of crabmeat, shrimp or lobster tails may be mixed with a suitable vegetable in any desired combination.

The stuffing may be prepared raw or it may be precooked, and it is preferably spiced or seasoned to provide the desired flavor. A prime requirement for the stuffing is that it must have the proper consistency so that it does not liquefy to any great extent or fall apart during subsequent steps of the method or during the storage or cooking of the product. This consistency problem is overcome in part by the careful selection of type and amount of ingredients in the stuffing and also by reducing the temperature of the stuffing to near or below freezing (e.g., 34° to 40°F), just before it is applied to the first fillet. Thus, a portion of stuffing when placed on the first fillet, will generally hold its shape without slumping and falling off of it.

In the next step of the method, shown in Figure 3.7e, the second fillet is placed with its skinned side down on the portion of stuffing. This fillet thus drapes over the stuffing, and its outer edge portions are arranged to extend down and engage the similar edge portions of the first fillet. The flour that was previously applied to the fillet surfaces and which has become moistened by the fillet now serves as a glutinous material to help provide a seal between the two fillets, thereby forming an enclosure between them containing the stuffing material.

The two fillets have thus formed an elongated, sealed package **10** for the stuffing whose exterior surface, if desired, can be coated with such well-known surface agents as flour, bread, or cracker crumbs. The fillet package at this point can be prepared immediately for cooking by placing it in a broiling oven pan for a preselected time and temperature. However, the package is particularly adaptable for cold storage where it may be frozen and kept in this condition for long periods until ready for use. The preferred cooking procedure is broiling, and this may be accomplished by placing the fillet package in an oven (e.g., 350° to 400°F) for 20 to 30 minutes.

During storage and particularly when the package is cooked, several important reactions occur. The fillets when subjected to the cooking heat tend to warp considerably, that is, the skinned sides of each fillet tend to become concave. However, since the fillets are assembled with their skinned sides against the stuffing, the natural warping only tends to force their sealed edges closer together. Since, during the entire cooking process the product stays together as a unit, and the juices and flavor of the stuffing are kept inside the fillets.

MOLLUSKS

CLAMS AND OYSTERS

Opening Bivalves Chemically

Commercially, oysters are opened by hand, using a knife, mallet, chisel, block, or combination of all these tools or knife alone depending upon the technique of the oyster shucker. In the commercial oyster house, the shucker in achieving speed sacrifices quality of workmanship. These oysters may be cut in other places besides in the eye. The oysters, which are completely processed by hand, are washed and then packed in containers which are refrigerated or frozen. The oysters obtained in this manner are fresh, whole and have lost no weight. The quality of the oyster can approach that of the oysters in restaurants but usually is lower.

Another commercial method is to steam the oysters. The oysters are steamed at temperatures of 240°F plus, at pressures of 12 psig plus. The oysters under this form of process are essentially cooked. There is a 60 to 70% shrinkage in the oysters with an accompanying weight loss.

In order to eliminate the manual work that has to be done on a tightly closed oyster shell and avoid shrinkage in the oyster meats which is attendant on the steaming process, *D.J. Welcker and R.L. Welcker; U.S. Patent 3,013,883; December 19, 1961* devised chemical means to open oysters. Basically, this process provides solutions into which the closed oysters may be dumped and within 1 to 3 minutes, the oysters will open their shells. The meat may then be removed without steaming or manually prying open the oyster shells. The chemicals in the bath must be acceptable to the oysters so that they will crack their shells open and not immediately close and stay closed.

The oysters may be made to want to open and to pump water. The chemical changes must be fast acting, nonpoisonous and must not alter the physical appearance of the oysters or their taste. The method provides solutions of varying concentrations of water-soluble carbohydrates, nonpoisonous and edible for human consumption. The oyster can be fattened and enlarged to such an extent that it almost bulges out of the shell. This fattening step takes place with the addition to the solution of polysaccharides which when the salinity of the water is controlled and the water temperature controlled, will result in a quickly fattened oyster.

To insure a supply of oxygen, the solutions were aerated using equipment normally found in aquarium systems. To serve as food, soluble carbohydrates were used. Polysaccharides were most easily obtainable and the following were used. All tested were found acceptable: glucose, dextrin, maltose, dextrose, and sucrose (separately and in combination).

A proprietary product known as Cartose (a dextrin base commercially employed as a baby food) gave excellent results, was easy to use and gave a solution that looked and smelled like the natural meso-polyhaline waters.

The element or metal of the salt employed must be soluble and its ions loosely coordinated and capable of stabilizing the pyrophosphate structure of the adductor muscle. The salts used were the chlorides of aluminum, copper, zinc, magnesium, manganese and calcium, and the preferred compound was magnesium chloride.

This method of opening oysters has proven 100% effective on all sizes and shapes. The inherent property of the utilization of glycogen enables this process to be used to process the oyster most economically and justifiably at the time of the year that oysters are the largest and have the greatest food value.

With the shells thus opened, the oysters are easily removed by knife without danger of cutting the body and incurring weight loss due to bleeding. Such an over-all operation of chemically opening and hand shucking the opened oysters is very rapid. This is considering that a manual operation is to follow the chemical operation.

As previously stated, the oyster is in no manner physically altered by treatment with magnesium chloride singly or with polysaccharide. The same cannot be stated for the present mass production of steaming where there are terrific shrinkage and weight losses as high as 70%.

Shucking Oysters

Except when in undisturbed enjoyment of its natural habitat, a live oyster keeps its shells so tightly closed as to require the application of great force to pry the shells open. When an oyster is removed from its shell, it immediately begins to "bleed," losing much of its juices and liquid with consequent loss of weight and flavor. Ideally, therefore, an oyster should be eaten immediately upon its shell being opened and the meat removed therefrom. Hence the popularity of "oyster bars."

These factors have posed problems in connection with the economical marketing of oysters at a distance from the sea regions where they abound. The oysters must be shipped in their shells under refrigeration, they must be forcibly opened, and, in order to preserve yield and flavor, they must be immediately consumed.

F.S. Lapeyre, J.M. Lapeyre, L.E. Demarest and R.F. Couret; U.S. Patent 3,007,801; November 7, 1961 describe a method of shucking oysters whereby bleeding is eliminated or minimized and the original weight, flavor and value of the oyster is thus preserved. The process also provides a machine whereby mutilation of the oyster meat is eliminated or minimized and whereby the adductor muscle or "eye" is preserved as an integral part of the oyster meat.

The process includes the following steps: first, the freezing of live oysters within their shells; second, the subjection of the oysters to shear stress through impact or other means while the oyster begins to thaw, in order to break the hinge ligament or cartilage, to detach the adductor muscle from the shells, and to force the shells apart; third, the separation of the loose frozen meats from the empty shells.

The shells will then separate from each other delivering a meat which is still substantially frozen solid. The process offers the further advantages of individually frozen meats, rather than a frozen block of oyster meats as would be the case if the oysters were frozen only after the removal of the shells. The consumer may therefore thaw only the number of oysters required at one particular time. The patent contains a diagram and detailed description of the apparatus for shucking the frozen oysters.

Clam Evisceration

As is well known in the art, the processing of clams involves the steps of removal of the meat from the shell, referred to as "shucking" of the clams, followed by the removal of the belly contents from the shucked clams. Despite prior art efforts, processing of clams on a commercial basis is still done by procedures depending on manual labor. In the case of debellying, operators manually tear the belly away and wash the debellied clam. This is obviously an operation too tedious, time consuming and costly to be acceptable under present economic conditions.

R.K. Finley, D.J. Langlois and R.H. Nicholson; U.S. Patent 3,659,315; May 2, 1972; assigned to HCA Food Corporation have developed an efficient and simplified method for separating the visceral material, i.e., the belly and contents, from whole clams. The process is based on the fact that clean clam meat can be recovered in essentially viscera-free form by subdividing the shucked clams and subjecting the pieces to agitation in a hot aqueous medium for a controlled period of time, with the result that the visceral material, being exposed by subdivision of the clam, is disrupted hydraulically and distributed through the aqueous medium by solution and dispersion.

As a result of the agitation step, the visceral material is transferred to the liquid medium and, through the solution and dispersion of the visceral material, results in a thickening of the aqueous medium. The aqueous medium remains as a fully liquid body which can be readily separated from the pieces of clam meat, with the clam pieces then being recoverable as clean, essentially viscera-free pieces by simply washing away such residual liquid and fragments as may remain on the surfaces of the pieces.

The method is applicable to clams generally and as used herein, the term "clam" is intended to include the generic class of clams referred to in the art as Pelecypoda and the wide variety of species thereof. Examples of such species include surf clams *(Spisula solidissima),* soft shell clams *(Mya arenaria),* mahogany or black quahogs *(Artica islandica)* and regular quahog clams *(Venus mercenaria).*

In practicing the process, the clams are first shucked, i.e., manually or mechanically removing the meat from the shell, in accordance with techniques well known in the industry. If desired, the clams may first be heat treated to partially open the shell and facilitate separation of the meat from the shell.

After the clams have been shucked, the uneviscerated clam meat is chopped or cut into small pieces or slices. In general, the size of the pieces is not critical. However, the chopping operation should be such that at least a part of each clam, including the belly portion or visceral material, is subdivided into pieces smaller than the belly portion. Stated differently the chopping operation should be such that the "peritoneum" of each clam is at least partially intersected so that the viscera or belly content is exposed in part. In this regard it has been found that small pieces on the order of about one-quarter inch by one-quarter inch to about one inch by one inch are particularly advantageous. Slices on the order of about one-quarter inch by two inches in length may also be used.

The pieces of uneviscerated clam meat are then dispersed in an aqueous medium and subjected to agitation for at least 5 minutes. The temperature of the aqueous medium should be maintained in the range of about 100° to 150°F, with preferred ranges being on the order of 110° to 130°F. To a considerable extent the process is based on the fact that in subjecting the shucked pieces to agitation under the controlled conditions set forth herein, the visceral material is disrupted and transferred to the liquid medium as a solution and dispersion.

As indicated above, the pieces of clam meat should be subjected to agitation for at least about 5 minutes. Preferred time periods are on the order of from about 5 to 30 minutes. While agitation must be maintained throughout this processing step, the particular mode or apparatus employed to maintain the agitation is not critical. For example, agitation may be effected by a conventional bladed stirrer or by injecting high pressure steam and/or air through the aqueous medium, etc.

The concentration or percent by weight solids, i.e., the chopped clam pieces, of the aqueous dispersion or slurry may vary widely with particularly advantageous results being obtained at concentrations in the range of about 30 to 70% by weight of the total system.

After the clam pieces have been subjected to agitation in the manner described above, they are separated from the aqueous medium and are washed under conditions to remove any residue of the visceral material. Preferably this step is conducted by passing the recovered clam pieces onto a vibrating perforated screen provided with spray or jet washers directed on the meat as it passes along the screen. However, again the particular equipment employed is not critical and any conventional washing equipment may be used.

If desired, the washing step may be conducted with water at ambient temperature so as to reduce the temperature of the clam pieces. The washed clam pieces are recovered and further processed, as by canning, in the conventional manner.

Example: One hundred pounds of shucked uneviscerated surf clams was ground to pieces of approximately one-quarter inch by one-quarter inch size in a conventional grinding machine. Approximately 100 gallons of water was preheated in a 150-gallon stainless steel jacketed vessel to 130°F using 100 psig steam. The 100 pounds of ground clam meat, including the visceral material, was then immersed in the heated water in the stainless steel vessel and suspended by agitation. Steam was introduced into the outer jacket of the stainless steel vessel to maintain the temperature of the aqueous slurry at 130° ± 2°F. Agitation was maintained by a three-bladed propeller stirrer.

The suspension was then agitated for about 8 minutes, after which the entire contents of the vessel were drained by gravity from the vessel and delivered onto a vibrating screen, with the liquid phase passing through the screen and the clam meat remaining thereon. The screen was equipped with 85 psig spray washers which were directed onto the meat as it passed along the screen. The essentially viscera-free clam meat delivered at the end of the screen was charged to a cylindrical washer to remove any remaining sand or grit. The clean eviscerated clam meat was then canned in the conventional manner followed by thermal processing.

Comparison was made of the canned product of this example with that from clams that were debellied manually, by experts in the field. The comparison showed that the product of this example was superior to that obtained from manually debellied clams in that the latter product contained an observable proportion of visceral material while the product of this example was essentially viscera-free. Further, the clam meat of this example was fully equal in all respects to that obtained from the manually debellied clams.

The method of the previous patent has been very successful. Under some conditions of commercial practice, however, that method has proved to have one disadvantage, i.e., the cutting of the clams creates exposed tissue surfaces which are absorptive and tend, therefore, to take up small quantities of the aqueous medium into which the belly material has been dissolved and dispersed, the end result being that these surfaces are discolored by the small quantities of belly material.

While mere presence of a small absorbed quantity of belly material is not in itself deleterious, particularly when the clam meat is to be distributed in the form of product such as chowder, the discoloration caused thereby has proved to be a significant practical drawback.

To prevent such discoloration, *R.K. Finley, D.J. Langlois, R.H. Nicholson and H.F. Porter; U.S. Patent 3,794,746; February 26, 1974; assigned to Doxsee Food Corporation* have changed the order in which the process is carried out and perform the evisceration on whole clams before they are chopped for canning. The following example illustrates the process.

Example: Employing a 300-gallon kettle equipped with a 3-blade, 12-inch propeller agitator, 900 pounds of shucked, whole, uneviscerated surf clams containing 5% by weight residual fresh water from the shucking operation were pumped into the empty kettle promptly after shucking and without preliminary treatment other than shucking. When all of the clams had been delivered to the kettle, 2,000 pounds of fresh water which had been preheated to 160°F was delivered into the kettle over a period of 3 minutes, providing a suspension of clams in water with a weight ratio of water to clams of 2.4. Agitation was then started, with the propeller agitator rotating at 350 rpm, and continued for 6 minutes. At the end of that time, the entire contents of the kettle was pumped out and delivered to a shaker screen equipped with an overhead washing assembly arranged to deliver a washing shower of fresh water at ambient temperature. As delivered from the shaker screen, the clams remained in whole condition save for being essentially completely free of visceral material.

The temperature of the suspension pumped from the kettle was 120°F. The liquid of the suspension was observed to contain the belly material in solution and suspension. Though the belly material was clearly observable in the liquid of the suspension in the kettle, causing the liquid to be opaque, gray-green and somewhat mucilaginous, with observable small particles and fragments, no visible residue of this liquid or its dissolved and suspended content remained on the clams as delivered from the shaker screen.

The clean, whole, eviscerated clams discharged from the shaker screen were delivered to a conventional grinder, the ground clam pieces having a maximum dimension in the range of one-half to three-quarter inch, and the clam pieces were washed with fresh water at ambient temperature in a flume washer, then dewatered conventionally and passed over an inspection screen. The pieces were inspected for color and it was found that the exposed surfaces of the pieces, including the cut surfaces, were of normal color, exhibiting no off-color from absorbed visceral material.

Cutting Open Oysters Mechanically

T.S. Reinke; U.S. Patent 3,631,567; January 4, 1972 has developed a device for separating the valves of a partially opened oyster and removing the oyster, comprising diverging surfaces over which the valves are pushed to fully separate them, and cutting members which engage and follow the inner surfaces of the valves as they are separated and which cut the muscle which connects the oyster to the valves.

The oyster is first treated, as by electric shock or immersion in hot water, to partially separate the two valves, a very small separation being all that is required, and means are provided for forcing to fully open condition the two valves of such a partially opened oyster. With reference to Figures 4.1a, 4.1b and 4.1c, these means comprise a base **2** from which there extend upwardly two spaced members **4, 6** each of which has flat exterior walls which converge upwardly from the base to a sharp linear ridge, one member having walls **8, 10** which converge to ridge **12** and the other having walls **14, 16** which converge to ridge **18**.

These spaced members **4, 6** are aligned with each other and the corresponding side members **8, 14** and **10, 16** lie in the same plane, and the spaced members extend to the same height above the base whereby the linear ridges are also aligned, and are spaced apart such a distance that the adjacent edges of the aligned walls of the members are closer than the lateral width of the valves of an oyster of normal legal size, as clearly shown at A in Figure 4.1c. It will be apparent that if such an oyster, having been partially opened, is placed between and bridging the two members with the inner surfaces of its two valves engaging the sidewalls of the members, and is then forced downwardly, the downwardly diverging wall members will force the two valves apart to an extent determined by the extent of downward movement of the oyster.

Means are also provided for engaging the inner surfaces of the two valves as the oyster is forced downwardly in the described manner, to cut the muscle which attaches the oyster to the valves, thereby completely separating or shucking the oyster from the valves. Such means comprise two vertically extending cutters **20, 22** which are pivotally connected at their lower ends to the base and extend upwardly therefrom in the space between the two oyster opening members **4,6** with their upper ends just below the ridges **12, 18** and being slightly less in width than the space between the members.

FIGURE 4.1: OYSTER SHUCKER

a.

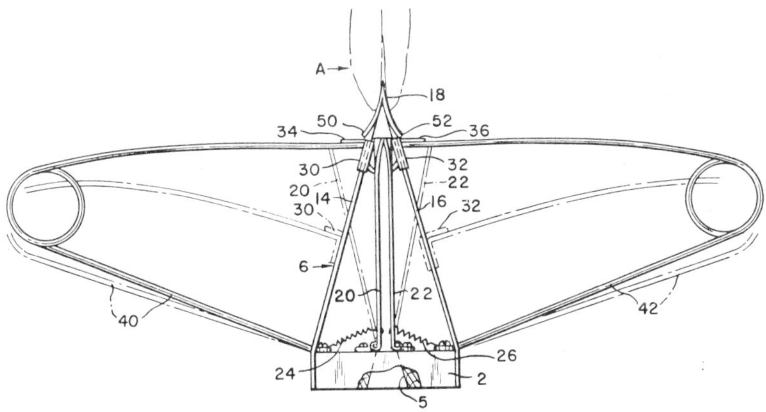

End View of Oyster Opening and Removing Device

b.

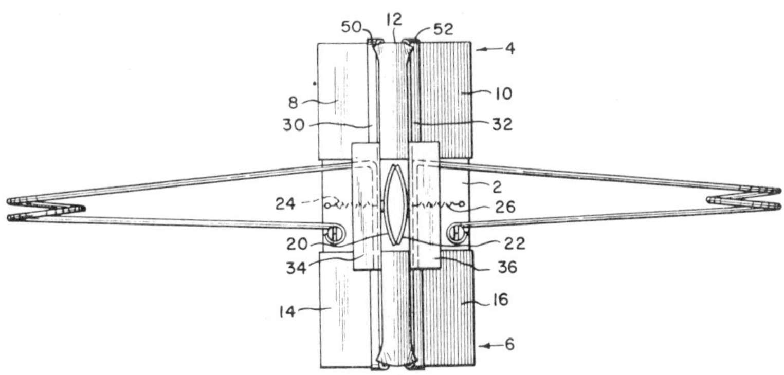

Top View of Oyster Opening and Removing Device

(continued)

FIGURE 4.1: (continued)

c.

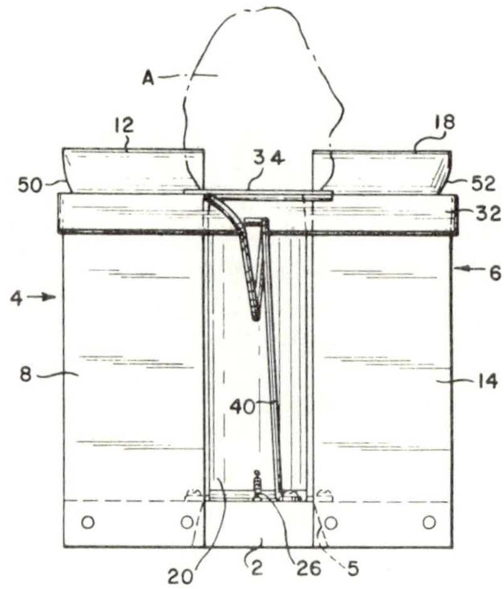

Side View of Oyster Opening and Removing Device

Source: T.S. Reinke; U.S. Patent 3,631,567; January 4, 1972

These cutters are preferably formed of sheet metal, having sharp upper edges, and at their upper parts, each has a shallow outwardly convex cross-sectional shape, with approximately the curvature of the inner surface of one of the valves of an oyster. These cutters are constantly and resiliently urged outwardly away from each other by spring means such as those shown at **24, 26** in the figures. It will be seen that if the sharp, upper edges of these cutters are moved along the inner surfaces of the two valves of an oyster as the oyster is forced downwardly over the members **4, 6** they will cut the muscle which holds the oyster to the valves.

In the use and operation of the device, an oyster is first partially opened in any suitable way, such as by an electric shock or immersion in hot water, and is then positioned at the upper part of the device bridging the gap between the two members **4, 6** with the inner surfaces of the lower or outer edges of its valves engaging the oppositely diverging sides of the valve opening members.

The oyster is now forced downwardly, causing the downwardly diverging surfaces to progressively separate the valves. At the same time the sliding members **30, 32** are also forced downwardly by pressure of the valves of the oyster on flanges **34, 36** and as these sliding members move along the downwardly diverging surfaces they move farther apart and as they do so the resilient means attached to each of the cutting members pulls the cutting members apart causing their upper cutting edges to follow the inner surfaces of the valves to thereby cut the oyster from the valve.

Opening Bivalves by Heat and Water Jets

Bivalve mollusks, such as clams, oysters, scallops and mussels, have two main shells or valves which are joined together by a hinge and held shut by adductor muscles which function to open and close the valves which, when open, pump in seawater containing food, etc. One class of bivalve, which includes oysters and scallops, contains one adductor muscle and is known as monomyrian. Another class, which includes clams and mussels, contains two adductor muscles and is known as dimyrian. When removed from the water the adductor muscle(s) of a bivalve mollusk becomes taut and pulls the two shells together against the force which is exerted by a ligament in the hinge.

The oldest method of removing the meat from the shells of bivalve mollusks is, of course, the manual one of inserting a knife between the lips of the two shells and severing the adductor muscle from the shell and the ligament adjacent the hinge then opens the shell so that the meat can be removed. It is well known that this method involves substantial skill and the application of considerable strength to the knife.

From the viewpoint of a processor of bivalve mollusks for human consumption, there are two divisions of the species of bivalve mollusks; one division has a siphon that is tough and long enough to be removed (soft shell clams, goeducks and horse clams) and the other division which either has no siphon or one which is too short to be removed (oysters, scallops, mussels, quahogs and skimmer clams). Removing the meat from a bivalve mollusk which possesses a siphon of appreciable size requires a different operation from a bivalve mollusk which has no siphon.

F. Hanks, Jr. and W.C. Grieb, Jr.; U.S. Patent 3,594,859; July 27, 1971 have developed a process for efficiently shucking bivalve mollusks which may with relatively slight adjustment be changed from shucking soft shell clams of varying sizes and having sizeable siphons, to shucking oysters of varying sizes which do not have such siphons. The apparatus for carrying out this process uses a production line comprising a chilling station, a loading station, a siphon cutting station, a heat shocking or cooking station and a shucking station plus means for moving the mollusks through these stations.

The chilling station comprises a tank containing water in which the clams or other shellfish are subjected to a chilling treatment while their siphons are extended in order to render them immobile during their movement through the loading station and the siphon cutting station. This treatment is satisfactorily provided by immersion in water at approximately 35°F for approximately 12 hours. However, it has been found beneficial to preliminarily place the clams in warm water (approximately 80°F) which contains as little oxygen as possible.

While in this bath, the clams should be placed as close together as possible in order that they take up all of the oxygen possible. During this preliminary treatment of the clams they keep their siphons extended in search of oxygen; and due to the relative absence thereof, they rapidly pass into a state of semi-suffocation. Then the chilling treatment of the clams with siphon extended is more effective to render them immobile during the time they are conveyed to, and move along, the rollers to the loading station.

Any suitable conveying means, such as a conventional endless belt conveyor, may be utilized to move the chilled clams (with siphon extended) from the chilling station to a hopper as defined by a pair of spaced and downwardly convergent panels which are disposed above the parallel, angularly declining, oppositely rotating rolls.

In the case of a clam, the siphon is at the end which is heaviest, and the widest point of the shell is past the center point on the opposite end. Because of this, the movement of clams in single file fashion along the rolls causes them to automatically assume a position with the siphon extending downwardly through the space between the rolls. The clam will be in an on-end position with its longitudinal axis vertical.

For clams and oysters, it has been found that the parallel, angularly declining, oppositely rotating rolls may be approximately 4 inches in diameter and with their axes spaced apart approximately 4¾ inches, thereby leaving a space between their most adjacent points which is approximately ¾ inch in width. However, roll diamaters of anywhere between 3 and 6 inches are contemplated with appropriate variation in spacing from say, ⅝ to 2 inches, according to the size of the bivalve mollusk.

With rolls of the described construction and arrangement, a rate of rotation of approximately 36 rpm has been found very satisfactory; however, this is subject to considerable variation. Accordingly, the means for driving the rolls may comprise a single variable speed electric motor which may, through suitable gearing, impart rotational speeds anywhere in the range of approximately 25 to 100 rpm. The rolls move the bivalves through a siphon cutting station where the siphons of the soft shelled clams may be cut off.

Figure 4.2a is an elevational view of a heat shocking station wherein the oysters or desiphoned clams are caused to further open their shells, this station being constructed and arranged to accomplish cooking if such is desired. Figure 4.2b is an elevational view of the shucking station which receives the oysters or clams (with their siphons cut off) and in either heat shocked or completely cooked condition, and performs the actual shucking operation, whereby the meat of the bivalve mollusk is removed intact. Figure 4.2c is a transverse elevational view, partly in section, which is taken on the line **8–8** of Figure 4.2b. Figure 4.2d is a detail view illustrating the manner in which the water jets are made to move in such manner as to follow the contour of the interior of the shell during the severing of the adductor muscle therefrom.

As the chilled clams (which now have had their extended siphons removed either wholly or in part) or the chilled oysters move from the siphon cutting station, they continue their single file movement on and along the parallel, angularly declining, oppositely rotating rolls **10** and into the heat shocking (or cooking) station **D**.

A clam initially moves along the parallel, angularly declining, oppositely rotating rolls with its heaviest portion with extended siphon lowermost; but upon severing the siphon, the change in center of gravity causes it to commence to change its previous on-end position and this orientation continues as the clam moves into and through the station **D** until its hinged portion is uppermost and its separated edges lowermost.

FIGURE 4.2: OPENING BIVALVES BY HEAT AND WATER JETS

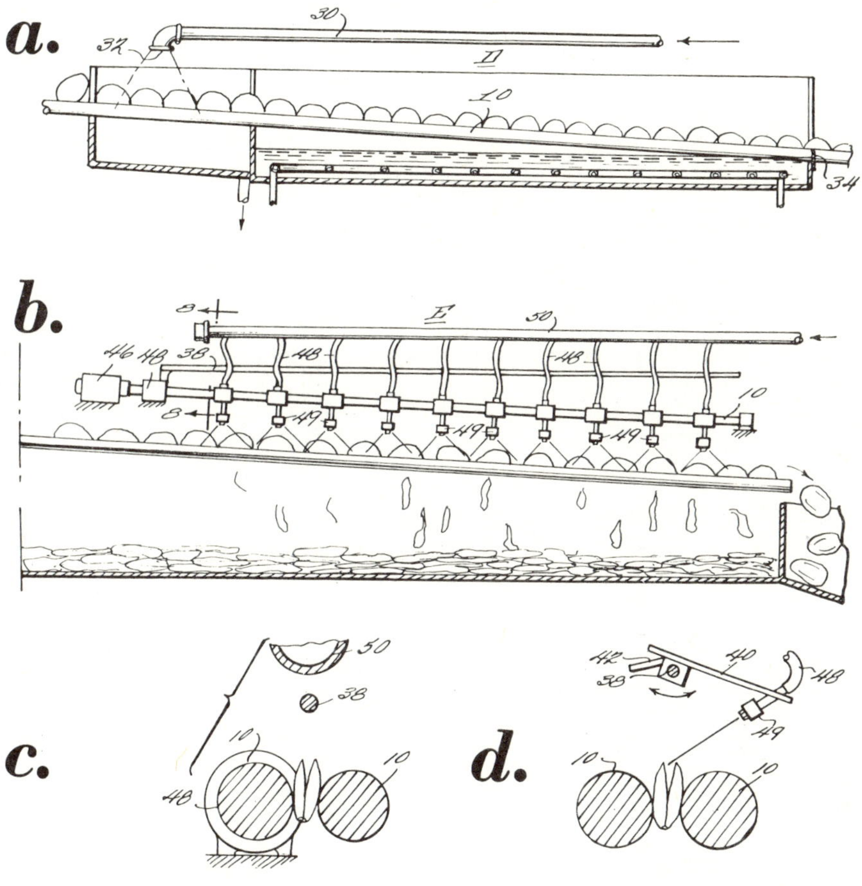

Source: F. Hanks, Jr. and W.C. Grieb, Jr.; U.S. Patent 3,594,859; July 27, 1971

For heat shocking purposes, distinguished from actual or complete cooking, the station **D** may be relatively short in length and simply comprise a hot water pipe or conduit **30** having a discharge outlet **32** which is disposed above the rolls **10** (Figure 4.2a) and adapted to discharge hot water, preferably at 212°F onto the clams as they move progressively therethrough, and the rolls are provided with wipers **34** for preventing the flow of the hot water longitudinally thereof.

With the desiphoned clams moving at a rate of say, 2 inches per second, it has been found that the time they are within the heat shocking station need be only approximately 10 to 15 seconds in order to obtain the desired result, i.e., further opening of the shells, and a proportional longer time in the case with oysters which have thicker shells and no siphons to be removed. However, if the bivalve mollusks are to be actually cooked before the meat removal step is undertaken, a substantially longer and higher heat conserving area may readily be provided.

The shucking station **E** is the most important single component in the production line and in it are the high pressure jets of water which completely sever the adductor muscle(s) from the shell and permit the removal of the meat of the bivalve mollusk wholly intact. In this portion of the apparatus (Figures 4.2b, 4.2c and 4.2d) a suitably journaled oscillating rocking shaft **38** is disposed above and longitudinally of the rolls **10** with its axis occupying a vertical plane extending through the pass between the rolls.

Attached to the rocking shaft and secured for movement therewith are two sets or series of laterally extending arms of equal length, one series **40** extending from one side of the oscillating shaft and the other series or set **42** extending oppositely with respect thereto, as shown in Figure 4.2d.

Disposed adjacent the oscillating shaft is an electric motor **46** which, through a suitable transmission unit **48**, drives the shaft in a back and forth arc of approximately 20°. Secured to and carried by the outer end of each of the laterally extending arms is a downwardly directed nozzle **49** and each of these nozzles communicates with a flexible tube **48** which is connected to a manifold **50**, the latter being connected to a source of supply which is adequate to furnish water to a manifold (gauge) pressure in the nature of between approximately 100 and 900 pounds per square inch and at a rate of between 1 and 2 gallons per minute per nozzle.

The nozzles may be of any well known high concentration type, and it has been found that the most satisfactory nozzles possess a slot shaped orifice with the sides thereof tapering at approximately 80° to provide a concentrated flat and fan shaped or arcuate spray.

Depending upon the speed with which the rolls move the desiphoned clams (or oysters) forwardly, and other variable factors, the total number of alternately extending lateral arms with their downwardly directed nozzles may vary appreciably, but it has been found satisfactory to utilize between 20 and 30 with the spacing between them approximately 3 to 4 inches.

With such an arrangement the trajectory forms an inverted 90° angle, and the flat and fan shaped or arcuate spray pattern is effectively directed into the shell of each individual clam or oyster from opposite sides thereof and is so adjusted as to strike the shell along the contour of the inner face thereof and contacting the meat of the bivalve as little as possible.

The jets are arranged in successive sets with each set emitting progressively increased pressure so the more difficult shellfish to shuck will eventually reach jets of sufficient pressure to cut their adductor muscle and remove their meats.

All of the flat and fan shaped or arcuate jets of water emanating from the nozzles are offset; and due to the arrangement of alternately extending lateral arms **40** and **42** and the motion imparted by the oscillating shaft **38**, the jets of water at point of impact follow the curvature of the contour of the inner face(s) of the shell and the meat of the clam (or oyster) falls by gravity between the rolls and into a suitable collection facility; and the shells move forwardly along the rolls for ultimate disposal.

Shellfish Opener for Home or Restaurant

Shellfish are a prized eating delicacy, but they present a very real problem to the gourmet; they are very difficult to open. In order to gain access to the edible meat of the animal, it is necessary to open the shell which surrounds the meat. The animal has very strong connective tissue which secures the two halves of its shell to one another, and this connective tissue resists attempts to open the shell. Ordinarily, the connective tissue holds the shell parts closely against one another with an appreciable force so that it is difficult to pry them apart even a little bit.

Even after the shell parts are pried apart slightly, the connective tissue still resists full opening of the shell. Even if that connective tissue is penetrated at a localized area, the length of the seam between the two shell halves will still have substantial amounts of connective tissue active therebetween, so that the shell will still strongly resist opening. It is necessary, if the shell is to be opened, that the connective tissue between the two shell halves be cut over all, or at least a very substantial proportion, of the length of the shell seam.

Shellfish of the kind under discussion here are of various types, the shells of which have different shapes. Moreover, even shellfish within a given type will vary somewhat in shape and may vary quite radically in size. Accordingly, it is very difficult to devise apparatus which will effectively perform the shell opening operation and which can be used with shellfish of different types or even with shellfish of the same type which are of different sizes.

Another problem involved is the time required to effect the opening of the individual shellfish. Each clam or the like has but a small morsel of edible meat. In order to make a dish for a person, it is necessary that a number of such shellfish be opened, and when a group of people are planning to eat together the number of shellfish which may be opened are correspondingly multiplied. Since each shellfish must be individually opened, the time required to carry out the opening operation becomes quite appreciable. Any apparatus which cannot expeditiously accomplish the shell opening operation is of little or no use.

There is another requirement for devices designed to open shellfish. They must be very readily cleanable, since shellfish residue, if permitted to stand, soon becomes quite odoriferous, and an opener which cannot readily be cleaned would soon become a very unwelcome appliance in a kitchen.

J.V. Valentino; U.S. Patent 3,748,692; July 31, 1973 has devised a shellfish opener whch can be used rapidly and effectively by persons with limited amounts of skill and which will be safe in its manner of operation, one which can be used either in a home, where shellfish are opened only relatively infrequently, or in a restaurant, where shellfish opening may be a very major operation, one which can readily be adapted to function with different types and sizes of shellfish but which is designed to be manipulated in the same fashion no matter what the type or size of the shellfish may be and one which is simple and sturdy in construction and which may readily be assembled and disassembled for purposes of cleaning.

The shell opener comprises a support adapted to engage and hold a shellfish so that the seam between the shell halves is exposed over at least a very substantial portion of its length. Mounted on the support is a carrier which carries a knife. The knife is adapted to overlie the seam of the shell and to be movable relative to the carrier toward and away from the shell in a precisely controlled manner, thereby initially to penetrate the shell seam and cut the connective tissue between the shell halves.

Thereafter the carrier is moved, and this causes the knife to move along therewith, thereby to sever the connective tissue over a sufficient length to permit the shell to be opened. Because the initial penetration of the seam is accomplished in a precisely controlled manner, the initial penetration is achieved without fragmenting the shell, this being desirable so that the shellfish meat can be eaten without danger. The subsequent cutting movement of the knife is carried out rapidly and with a relatively gross type of motion, but because the knife has already been properly aligned with the shell, having previously penetrated the seam thereof, this again can be carried out without danger that fragments of the shell will be formed.

A single handle is provided, manual manipulation of which accomplishes both the movement of the knife toward and away from the shell for penetration purposes and the further movement of the knife to complete the opening operation. The support is formed of a pair of adjustably positioned parts which grasp the shellfish between them and hold the shellfish in proper position. The grasping surfaces of the support parts are desirably formed of a spongy material, such as sponge rubber, so as to facilitate the shell engagement without damaging or breaking the shell and thus causing shell fragments to mix with the meat of the animal. The knife may be mounted on its carrier so as to have at least a limited degree of swinging or pivotal motion relative thereto, in order to permit it to adapt itself to the shape of the particular shellfish then being acted on. The patent includes drawings of the device and a detailed description of how it works.

Opening Oysters by Infrared Light

W.B.G. Ouw and A.L. Johnson; U.S. Patent 3,755,855; September 4, 1973; assigned to Long Island Oyster Farms, Inc. have developed a fast and inexpensive method for opening oysters and other shellfish which reduces significantly the effort and skill required by persons opening the shellfish and presents little or no potential for injury to the laborer. Furthermore, the product obtained from this process is an uncooked, still cool raw oyster, on the half shell if desired.

This process is thus usable on oysters intended for the most discriminating consumer. Oysters treated by this process have even been found to be frequently still alive after treatment and have been observed to remain so for periods up to half a day after treatment, as indicated by a noticeable heartbeat.

Broadly, the process comprises the steps of: (1) exposing the flat shell of the shellfish, in the region where the adductor muscle is attached to the shell, to a concentrated beam of infrared light for a time sufficient to weaken or sever the muscle attachment and render the flat shell easily removable, and (2) removing the flat shell by hand or other means.

Although it is preferred to expose the flat shell of the oyster, particularly when half shell oysters are desired, this process also works on either or both shells of the oyster or other shellfish. The main objective is to apply infrared light to the shell in the region where it overlies and is attached to the adductor muscle which holds the oyster closed.

The skill which has in the past been required of persons opening oysters has been required due to the fact that the oyster exerts such a strong force to keep its shell closed. In view of the strong force, it has in the past taken considerable manual dexterity and no small amount of strength to open an oyster, generally with a sharp knife. Due to the irregular shape of the oyster shell and the fact that they are generally wet, the job of sliding the knife quickly and strongly between the two shells of the oyster has required skill to avoid slippage and cutting of the hands and strength to pry open the shell. Further, particularly advanced skill has been required to achieve the speed required to open oysters at an economically feasible rate for a commercial operation.

By the above described method of opening oysters, the muscle attachment is severed or weakened to the point where the oyster can be opened by ordinary unskilled labor and such opening requires no great manual dexterity or physical strength. Rather, it is only required to slip an instrument such as a knife blade between the oyster shells and gently twist to remove the flat shell and break the ligament bond at the hinge between the two shells. In some instances, particularly with very fresh oysters, the muscle is severed and the oyster gaps, thus allowing the shell to be lifted off without need for a knife or other opening implement. In referring to the shells of an oyster, it should be remembered that such shellfish generally have an outwardly convex shell half and a rather flat shell half.

Figure 4.3a is a view in partial cross-section of an oyster. Figure 4.3b illustrates the application of a beam of infrared light to an oyster. Referring more particularly to Figure 4.3a, numeral **1** indicates the meat of an oyster. Numeral **2** designates the outwardly convex half shell of the oyster; **3** denotes the flat half shell of an oyster, and **4** indicates the region of this flat half shell which overlies and is attached to the adductor muscle **5** of the oyster. Numeral **4'** indicates the ligament joining the two half shells.

Referring more particularly to Figure 4.3b, **10** indicates an oyster positioned flat side up and held in place by any conventional means such as ring **11** on a surface **12**; **13** indicates the housing of an infrared lamp which is powered by electric power from a conventional electric cord **14** and socket (not shown); **15** indicates the infrared lamp bulb; **16** indicates an ellipsoidal focusing reflector and **17** indicates a focused beam of infrared light emitting from the infrared lamp bulb and striking the flat shell of the oyster in the region where the adductor muscle is attached.

As an example, applicants have found that a forced air cooled, 1¾ inch focal length, 1,000-watt infrared lamp having a beam width of 5/16 inch diameter at the focal length will function well in the method of this application. Such a lamp operates on 110 volts AC at 60 cps. The lamp is provided with an ellipsoidal reflector which focuses the infrared light into the above mentioned beam. Such a lamp is available under the trade name Infradex.

Applicants have found that using the above described apparatus oysters can be effectively treated by exposing them for about 2.5 to 20 seconds per oyster, preferably from about 5 to 15 seconds, obtaining a significant reduction in the force required to open the oyster without cooking the oyster. The longer exposure times in these ranges are, of course, required for thicker shelled oysters, while the shorter times may be used for smaller, thinner shelled oysters.

Although significant reduction in the amount of force required to open the oyster is apparent immediately after infrared treatment, best results are achieved with least effort if there is a short delay, up to about a minute, before opening the oyster.

FIGURE 4.3: OPENING AN OYSTER WITH INFRARED LIGHT

a.

b.

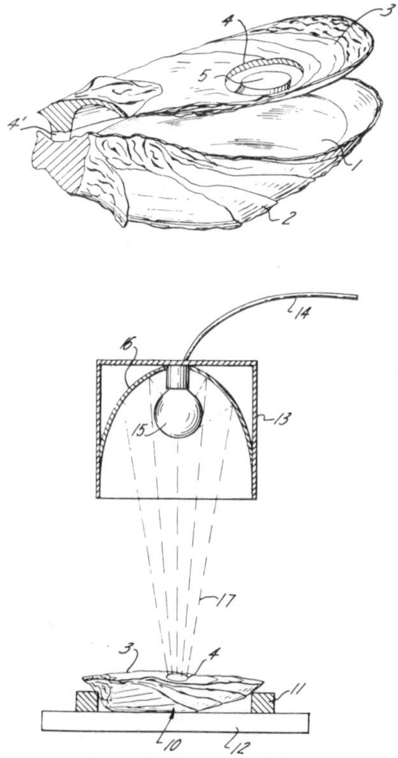

Source: W.B.G. Ouw and A.L. Johnson; U.S. Patent 3,755,855; September 4, 1973

Treating Tainted Bivalves

J.M. Dennis; U.S. Patent 3,418,138; December 24, 1968 describes a method for treating bivalves, especially oysters and clams, one object of the process being a treatment of oysters taken from polluted waters that restores the oysters to a normal or even better edible condition than when normal oysters have been tonged or dredged from their beds. Figure 4.4 gives a diagrammatical view of the complete apparatus for restoring normalcy to oysters taken from polluted waters.

FIGURE 4.4: APPARATUS FOR TREATING TAINTED BIVALVES

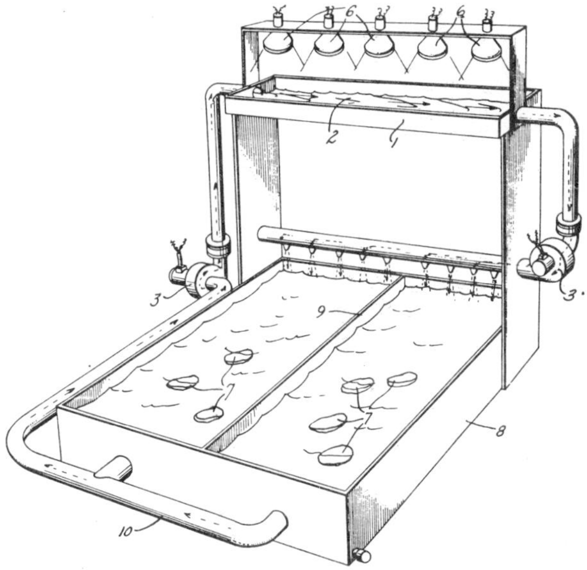

Source: J.M. Dennis; U.S. Patent 3,418,138; December 24, 1968

In the drawing, the numeral **1** designates a shallow rectangular water conducting open trough or tank, through which a very shallow sheet **2** of normal sea or simulated seawater is directed, by any means such as a deliverer or rotary pump **3**, there being ultraviolet lamps **6** placed thereabove to act on this flowing sheet of water, to allow this treated water to cover and act upon the oysters **7** that have been taken from polluted water.

This ultraviolet ray treated salt water after a determined period of time, from 1 to 336 hours, is sufficient to restore the natural health and even better flavor to such treated oysters. Tank **8** containing the seawater and oysters is divided into two compartments by the partition **9**. To provide the necessary circulation of the seawater the conduit **10** is in communication with both compartments of the tank and circulation is assured by the rotary pumps.

The temperature of the treated water is between 34° and 90°F, and the depth of the water being treated in the sluiceway or chute is from approximately one-sixteenth to twelve inches. The gallon per minute flow of the treated water may vary from 1 to 50 gallons per minute.

The bivalve when taken from the polluted seawater is first placed in a bath of ultraviolet treated simulated seawater or the natural seawater normally found where bivalves are dug, tonged or dredged for. Then this bath is directly under the outlet of the sluiceway so that the water that has been ultraviolet ray treated flows onto and in the water with the bivalves, preferably oysters, and this treatment goes on for 1 to 336 hours, or may vary in accordance with the amount of pollution that was in the water from which the oysters were tonged or dredged, until they have been restored to healthy condition, this being determined by shucking and testing oysters at stated intervals.

The oysters thus treated, when the same are restored to a healthy condition, in many cases have a better apppearance and even better flavor than the oysters tonged or dredged from nonpolluted waters. Clams taken from the sand and water that is polluted are treated similarly to the oysters.

A process developed by *D.D. Woodridge, J.C. Kitchel and W.R. Garrett; U.S. Patent 3,641,982; February 15, 1972; assigned to National Shellfish Processors, Inc.* utilizes an open end system of continuously flowing water directly from the source from which the bivalves have been removed and back into the initial water source.

This open end or continuous flow system initially has a pump to pump the water from the water source through a radiation sterilizer, such as a gamma or nuclear radiation irradiator, or special industrial or commercial ultraviolet water purifiers, so that sterilized water is provided which is substantially similar to the waters from which the oysters have been removed. This water is fed into large tanks at several levels in the tank where oysters may be located on removable shelves. The tanks also have means other than the action produced by the entering water for producing movement of the water in the tank.

The water leaves the tank from the bottom and passes through a second sterilizer essentially identical with the first sterilizer and back into the original water source. The second sterilizer has the advantage of returning substantially sterilized water back to the original source rather than allowing water recontaminated from the oysters to reenter the water source. The system may also be provided with means for movement of the trays of oysters such as by shaking, and for controlling the water temperature as desired, although this has not been found necessary in the warmer climates.

Figure 4.5 shows a flow diagram of the process. Referring now specifically to the figure, a flow diagram is shown in which the process may be seen to be an open end system in which water is pumped from the water source from which the oysters or the like have been taken and enters a pump **10** from pipe **11**. The pump in turn pumps the water through a pipe **12** into a sterilizer **13** which purifies or sterilizes the water passing therethrough. The sterilizer must be of a type that will sterilize the water without changing its content such as introducing any new substances therein which might result in the oysters refusing to depurate because of the noxious substances.

For instance, it has been found that chlorine sterilizers leaving a residue of chlorine in the water result in the oysters refusing to depurate. Accordingly, a radiation sterilizer is preferred. One such sterilizer might be a gamma radiation irradiator for very large operations which would make such a sterilizer feasible. Such irradiators have been found very effective in purification of contaminated fluids. A second type of sterilizer is an ultraviolet water purifier of a commercial or industrial type. These purifiers are commercially available for uses from amounts as small or smaller than 100 gallons per hour to in excess of 1 million gallons per day, as required, and may be obtained complete with power supplies, safety monitoring controls and also including such things as alarms and multiple wiper assemblies for keeping quartz jackets clean. One such commonly available ultraviolet water purifier is manufactured by Ultradynamics Corp. in a variety of sizes.

FIGURE 4.5: STERILE FLOWING WATER FOR OYSTER DEPURATION

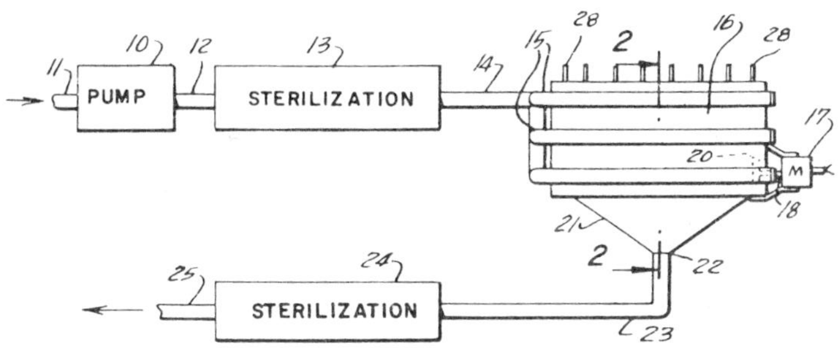

Source: D.D. Woodbridge, J.C. Kitchel and W.R. Garrett; U.S. Patent 3,641,982; February 15, 1972

Once the water is sterilized, it leaves the sterilizer in the same condition as entering except substantially sterilized. The water leaves the sterilizer in pipe **14** which enters into a three-layered pipe input system **15** wrapped in a U-shape around a tank **16**. These tank input pipes **15** spray the sterilized water across the top of the water in the tank and at two layers beneath the water, which keeps the water moving rapidly within the tank. Tank **16** has additional means for providing movement of water in the tank illustrated by a motor **17** attached to it with a shaft **18** passing through the tank and having a propeller **20** connected to the shaft inside the tank.

The shaft, of course, has shift seals thereon to prevent water from leaving around the shift. The tank has a funnel or pyramid shaped bottom portion **21** and a drain **22** leading out of the tank into pipe **23** and into a second sterilizer **24**.

This sterilizer may be essentially similar to sterilizer **13** and sterilizes the water that has been recontaminated by depuration of the oysters prior to this water being fed back into the original water source by pipe **25**. As can be seen, an open end depuration system provides for the continuous flow of water into and out of the system during the depuration of oysters and the like.

Converting Normally Inedible Clams to Edible Clams

In recent years there has been growing concern in the sea clam industry about the diminishing availablility of surf clams off the eastern coast of the United States. Because of the rapid depletion of the supply of surf clams which threatens the entire clam industry, numerous studies were undertaken to locate new beds of surf clams in areas of the Atlantic Ocean along the eastern seaboard which had not previously been investigated.

A research program was initiated in which the Fish and Wild Life Service of the United States Department of Interior outfitted a research boat for conducting exploratory work to survey parts of the Atlantic Ocean that have previously not been explored for clams. During the course of this research work there was found an apparently inexhaustible supply of mahogany clams, but surf clams were not found in quantities sufficient to replenish the diminishing supply.

The mahogany clams, primarily found in waters of 120 to about 200 feet deep, were rejected because it had long been known that they are normally inedible. The meat obtained from these clams has a foul and fishy taste, rendering it unusable as a food product. Furthermore, processing of these clams is complicated by the fact that the clam shell has a black fuzzy growth which discolors the meat and further imparts an undesirable color and odor to the meat during steam or water treatment. Thus, while a new abundant supply of clams has been found, no techniques were available for converting such clams into a usable food product.

A.S. Wendt; U.S. Patent 3,471,300; October 7, 1969; assigned to HCA Food Corporation describes a process for converting normally inedible mahogany clams into edible clams suitable for human consumption by removal of undesirable odor and taste from its meat by heat treatment to free the matter which causes the undesirable taste and odor. Figure 4.6 is a diagram of an apparatus suitable for use in carrying out the process.

Initially, it is necessary to remove the black fuzz growth on the shell of the mahogany clams. This can be accomplished mechanically as by employing a set of brushes which rotate, e.g., wire brushes. It is essential that the black fuzz growth be removed prior to the subsequent step of steam or hot water treatment because the use of steam or hot water in the presence of the black fuzz growth causes black dye to discolor the clam meat and impart an undesirable taste and odor to the meat.

After removal of the black fuzz growth, the clams, if desired, may be treated with steam or hot water to partially open the shell and facilitate separation of the meat from the shell. This treatment must be carefully controlled to keep the meat raw. As used herein, the term "raw" means before protein matter in the meat has partially or completely coagulated. The steam or hot water treatment should be for a period of not more than about 10 to 20 seconds. When the meat is cooked at this stage, the belly contents impart an undesirable taste and odor to the meat and render it unusable for food purposes.

The next step involves complete removal of the meat manually or mechanically from the shell, i.e., shucking the clams. This step may be carried out according to the techniques well known in the clam industry which have heretofore been in connection with processing of surf clams. After the clam meat has been shucked, it is necessary to remove the belly contents from the meat because mahogany clams have a large belly whose contents are not edible. The removal of the belly contents may be accomplished by slitting the "peritoneum" (skin that holds the belly to the meat) in several places, manually tearing the belly away and washing the belly contents, using jets of cold water. This treatment with cold water results in removal of belly contents and any undesirable belly odor and taste in the meat. After the belly contents have been removed and the meat washed, the meat, which is still raw, is ground with a grinder or cutter or the like.

According to this process the undesirable odor and taste associated with mahogany clams and other clams is removed by heat treatment of the clam meat to volatilize the undesirable odor and taste causing matter. Preferably, this phase of the process is carried out by steam treatment in an apparatus such as described in Figure 4.6. That apparatus is employed in the following manner.

A suitable quantity of raw clam meat which has been debellied and washed with cold water is placed in a wire basket **10** provided with a handle **11** or the like, the basket supported inside of a heating column **12** provided with outlet **14** for recovering clam juice obtained during the steam treatment. The bottom of the heating column is provided with a sparger **15** through which dry steam passes into the heating column.

FIGURE 4.6: APPARATUS FOR TREATING CLAMS

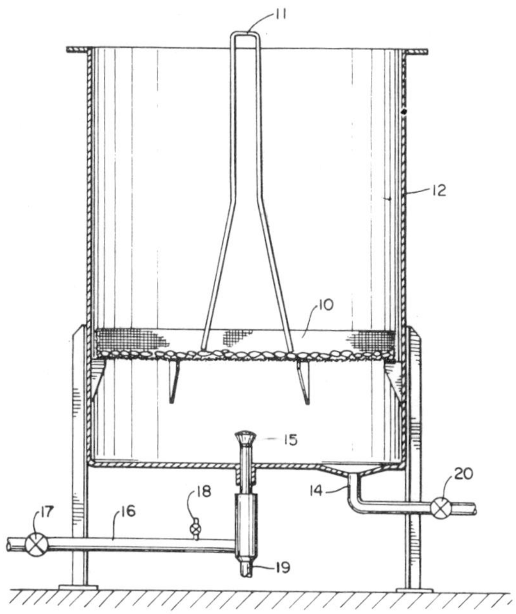

Source: A.S. Wendt; U.S. Patent 3,471,300; October 7, 1969

The steam is fed into conduit **16** having steam valve **17**, bleed valve **18** and condensate outlet **19**. The condensate outlet collects excess moisture in the steam and provides dry steam in the heating column. The use of dry steam reduces the quantity of condensate collected through the outlet by opening valve **20** and thereby reduces dilution of the collected clam juice with water.

In carrying out the process, dry steam is bled into the heating column where a suitable quantity of clam meat has been placed in the basket. Surprisingly, it was found that when the clam meat was subjected to steam treatment the matter present in the meat which causes the undesirable odor and taste is removed by volatilization. The foul odor characteristic of the normally inedible clams, very marked at the start of the steam treatment, was easily detected by smelling the vapors discharged from the top of the heating column.

After steam treatment for about 7 minutes the undesirable odor was no longer detected rising from the top of the heating column. This steam treatment is sufficient to volatilize all the unpleasant odor and taste causing matter from the meat. This phase of the procedure finally converts the meat into an edible food product.

Juice from the steam treatment of the clam meat condensed in the heating column and this juice was collected through outlet **14** in a beaker or other suitable receptacle. At the start of the steam treatment and throughout the entire steam treatment the juice was pearly white. The juice was sampled repeatedly from its initial collection to the completion of the steam treatment and it had no unpleasant odor or taste. This indicates that the matter in the clam meat which causes the undesirable taste and odor is volatile and was removed by steam treatment or other volatilization procedures.

After the steam treatment, which resulted in clam meat free from any undesirable odor and taste, the meat was processed in the usual manner by placing it in a can together with a brine solution. The can was capped and retorted at 240°F for about 1 hour to sterilize the meat. If desired, the brine solution may conveniently contain the clam juice obtained from the steam treatment step. It was found that the use of the clam juice in the brine solution enhances the flavor of the cooked clam meat.

By using the steam treatment described above, it is possible not only to use mahogany clams as a food product but also other species of clam, e.g., red clams. Further, the process can be employed to treat clams found in polluted sewage and oil beds which have acquired an undesirable odor and taste, e.g., to treat surf clams found in such polluted beds.

Pasteurizing Oysters

D.C. McMillan; U.S. Patent 3,615,726; October 26, 1971; assigned to Olympia Oyster Company describes a method of precooking oysters only to a pasteruization temperature by the use of microwave or processing frequencies of radiant energies, intensity and amplitudes such that oyster meat may attain a predetermined internal temperature sufficient to control bacterial growth and provide good keeping qualities, but a temperature such that off-flavors or smells foreign to those of freshly shucked oysters do not develop.

If oyster meat is exposed to microwave energy of the latter two frequencies and particularly the processing frequency for periods approximating a minute and a half, the interior temperatures will be elevated to a temperature about 140°F. Such elevation of temperature will permit the oyster meats to retain their firmness and fresh taste. At this temperature, the meats will be pasteurized against rapid bacterial deterioration and will fix flavor qualities and firm tissues. Substantially lesser attained elevated temperatures will not produce keeping qualities, and substantially greater temperatures will produce flabby structure and a "cooked" flavor.

It is a characteristic of microwave cooking of foods that the interior is heated more quickly than the exterior portions. If radiation is closely controlled, these internal temperatures will be attained and restricted.

When the oyster meats are discharged from the oven, they usually are breaded and packaged after being sharp frozen at temperatures approaching -40°F. They are held thereafter at keeping temperatures of around -10°F. They will have a shelf life of many months, as long as one year, rather than being measurable in days. Oyster meats thus subjected to a restricted internal pasteurizing temperature and external temperature of much less amplitude do not give up very much of their moisture content. The internal moisture is driven outwardly toward the peripheral surface of the oyster meat. This produces a viscous external covering that is sticky and adherent. This external covering is unbroken and defines a sealing element about each oyster meat against airborne bacteria and other sources of external contamination or deterioration.

It is desirable to utilize the viscosity and adherent quality to produce a dry firm seal or sheeting about each oyster meat. It is for this reason the oyster meat is breaded as soon as it is sharp frozen, after applying a coating of finely divided dry cracker crumbs or other particulate dry vegetable matter that will not introduce undesirable tastes to the product. This particulate matter absorbs and holds the excess moisture content of the covering and inhibits one coated oyster meat from sticking to the other. One advantage of this viscous covering is that when the external surface of the oyster meat, which attains a putty coloring upon discharge from the oven, is breaded with finely divided toasted crumbs the change in coloration, due to pasteurization, is masked by the brown or tan coloration of the toasted crumb coating material.

After the precooked oysters have had their temperature raised from -40°F, which is the approximate sharp freezing temperature, to a holding temperature approximating -10°F, it is desirable that the oysters be maintained spaced from each other, as in separate encapsulations. It is preferable that the encapsulating enclosures be made by two opposed sheets or films, having spaced pockets formed in one of the sheets.

Into each pocket a single precooked frozen and breaded oyster is lodged, similar to candy lodged in trays, and the pockets are sealed by the overlying sheet or film of sealing material. These sheets of sealing material are usually vacuum sealed or sealed in an inert gaseous medium to reduce oxidation. It is not essential that a subnormal atmospheric pressure be created within the package.

Enzymatic Separation of Undesirable Tissue from Clams

In present day practices, edible parts of marine creatures may be obtained from their shell enclosures by a variety of methods or means. Among the common commercial methods and apparatus are mechanical, electrical, and heat treatments. One of the most successful processes for shucking clams, for instance, includes subjecting the shell to extremely high temperature for a very short time to effect a release of the edible clam from the shell. While some of the processes and techniques may experience varying degrees of success, most of the prior known processes required at least some hand labor to treat the individual marine creature in order to remove the edible flesh and even may require further treatment to produce desirable edible flesh as the final product.

For instance, conventionally shucked clams at present must have the belly, located on the large end of the tongue of the clam, removed by manual squeezing. Workers tear the tongue from the remainder of the clam flesh and, with a squeezing-milking action, tear off that portion of the tongue in which the belly is embedded, removing a considerable portion of usable, desirable edible flesh along with the undesirable, though edible, belly in the process.

During the squeezing operation, sand and other earthy substances which may be entrapped in the belly may be forced deeply into the tissues of the tongue. This foreign material requires extreme washing treatment to remove even after clams are ground, minced, diced or sliced. The extreme rigor of the washing together with the large volume of water used tends to remove valuable soluble solids from the clam tissues, thus lowering the nutritive value and reducing recovery of usable flesh.

The purpose, therefore, of the process developed by *G.B. Fehmerling; U.S. Patent 3,729,324; April 24, 1973* is to separate the desirable edible flesh of a marine creature from the undesirable tissue. This is accomplished by the immersion of the fish or shellfish in an aqueous bath containing a combination of enzymes. The enzymes that must be used in combination are the proteolytic, glycoside hydrolase and lipolytic enzymes. Optionally, pentosanase, pectolytic and cellulolytic enzymes may be added. All of these enzymes may be obtained from any recognized source such as plant, animal or microorganism sources and may include the marine creatures as such a source.

Enzymes of the particular class found to be necessary for this process have been previously identified and are readily available to the art. All the following classes and subclasses of enzymes as well as further specific examples of each may be found on pages 672 to 785 of *Enzymes* by M. Dixon and E.C. Webb, 2nd Edition, (1964), Academic Press, New York City, hereinafter referred to as *Enzymes.* Examples of proteolytic enzymes are papain, ficin, bromelin, which find their source in plants. Other proteolytic enzymes, obtainable from animals, include trypsin, chymotrypsin and pancreatin.

The amount of the proteolytic enzymes to be used in the aqueous bath ranges from 1 to 75%, and preferably 5 to 15%, of the total weight of the enzymes in the bath. The glycoside hydrolase enzymes useful in accordance with the process are, for example, only hexoseoxidases and amylases, both alpha- and beta-amylase, as well as amyloglucoside. The glycoside hydrolase enzymes should be present in the bath in an amount between 1 to 75% total weight of enzymes;

preferably the amount of glycoside hydrolase enzymes should be varied between 5 to 25%. These enzymes cause the breakdown of 6 carbon, or multiples thereof, carbohydrates such as starch, hemicellulose, etc.

The lipolytic enzymes are those esterases capable of hydrolyzing fat and particularly include lipase/esterase. Any lipolytic enzyme may be used as is disclosed in *Enzymes* and may include, for example, glycerol ester hydrolase, lysolecithin acylhydrolyase, etc. The amount of the lipolytic enzymes used in the aqueous bath may vary from 1 to 75% and preferably should be between 5 and 50% by weight of the enzymes. The foregoing combination of enzymes provides a synergistic combination which effectively attacks and degrades the various connective tissue between the undesirable and desirable edible portions.

As a further facet of the process, it has been found that the optional addition of pentosanase enzyme further reduces and degrades the connective tissue that had been previously attacked by the glycoside hydrolase enzymes. Tests have demonstrated that the use of the pentosanase enzyme permits lower dwell times for the marine creatures in the enzyme bath. An added benefit of such a shorter time is that there is a minimal adverse effect upon the desirable edible tissue by the other enzymes in the bath when the pentosanase enzymes are incorporated into the bath.

Any pentosanase enzyme may be used as disclosed in *Enzymes* and may include kinases, ketolisomerases, epimerases, etc. Specifically, the pentosanase enzyme may be fructokinase. Pentosanase may be also referred to as gum hydrolysis or gumase. The amount of the pentosanase enzyme that is to be present should vary between 1 and 50% of the total weight of the enzymes present.

Optionally, either or both (a) any pectolytic enzymes such as pectinase/pectinesterase/pectin methyl esterase and (b) any cellulolytic enzymes may also be added to the above enzymes with or without the pentosanase. The amount of pectolytic enzymes that may be used may vary from 1 to 20% of the total weight of enzymes present in the aqueous bath. The amount of the cellulolytic enzymes that may be used may vary between 1 and 10% by weight of the total enzymes.

The aqueous bath into which the marine creatures are to be immersed should have a concentration of the enzymes in the combination as previously described of 0.003 to 3% by weight. The temperature and the time at which the marine creatures should be immersed in the aqueous bath of enzymes may vary considerably and need only be sufficient to dissolve at least partially the connective tissue that exists between the undesirable edible tissue and the desirable edible flesh so that the undesirable edible tissue will be easily removed by simply rinsing or water spraying the marine creatures so treated.

More precisely, however, the more effective time for immersion has been found generally to vary between about 30 seconds to 2 hours. The preferable time usually is 1 to 5 minutes for marine creatures that have been shelled; for fish the preferable times are 10 to 90 minutes. The temperature of the enzyme bath may vary greatly but preferably is from 75° to 175°F and most desirably 115 to 130°F. The pH of the enzyme solution is not critical, but may vary from 3.5 to 12.0 in use with various shellfish, such as abalones, conches, mussels, scallops, oysters, clams and the like, to remove the undesirable edible tissue from the desirable edible flesh after the shells have been removed.

With respect to clams only, it has been further found that it is preferable, though not essential, to divide the clams as by chopping, slicing or cutting into smaller portions. The final size is not important but with the larger clams, several divisions are desirable. Commercially, clams in strips of about one-half inch are often used and desired in industrial applications.

The dividing of the clam effectively permits a greater weight of clams to be packed into the same volume. For instance, clams passed through a ¾-inch commercial chopping plate fill a 3-gallon can with 28 pounds of clams, while the undivided 4 to 5-inch diameter clams fill the same can with 23 to 24 pounds of clams.

It should be understood, however, that while the slicing of the clam flesh will facilitate the removal of the undesirable edible portions, whole clam flesh may be similarly processed. If the clam were to be immersed in the enzyme solution whole, without being divided, additional time would be required though generally only several additional minutes.

Example 1: Clams were shucked in a high temperature oven and washed in a conventional circular washer under sprays of water. Twenty-five pounds of whole washed clams were sliced into ½-inch strips in a conventional clam slicer. Sliced clams were placed in a perforated stainless steel tub. The tub of clams was immersed immediately in an aqueous enzyme solution containing 0.026% enzyme mixture. The volume of solution was 6 gallons. The enzyme solution contained 40% proteolytic enzyme, 25% glycoside hydrolase enzyme and 35% lipolytic enzyme.

Temperature of solution was 125°F. The temperature of solution was maintained by circulating water at 125°F through the jacket of the stainless steel kettle in which the enzyme solution was contained. Agitation of the solution around and through the tub containing the clams was maintained by bubbling compressed air through the solution.

Reaction time was 2½ minutes. The perforated tub containing the clams was removed from the enzyme solution and sprayed by a jet of water at 40 psi from a hose. All the undesirable edible tissue in the form of belly material immediately disintegrated and washed through the perforations in the tub leaving behind the desirable edible clam flesh. The drained clam flesh was weighed and found to be 21 pounds, 1 ounce; this is equivalent to 84.30%. By conventional squeezing methods of belly removal, the recovery is approximately 70 to 72%.

For comparison, 18 pounds, 1 ounce of conventionally squeezed clam flesh was ground in a conventional clam chopper and canned under commercial practice simultaneously with clams prepared by the process. The enzyme treated clams were thermally processed in cans along with conventionally handled clams in the same pressure retort. After cooling and removal from retort, the processed clams were examined by "blind" comparison by experts in canned clam evaluation. No difference was noted in color, texture, flavor, aroma and drained weight between the enzyme treated and conventionally processed clams. The same evaluation was conducted on samples of both types of pack after accelerated storage for 10 days at 37°C (98°F). Again no difference was noted between the samples.

Example 2: The exact procedure of Example 1 was followed except that the enzyme solution contained 10% pentosanase enzyme and 30% proteolytic enzyme. The glycoside hydrolase and lipolytic enzyme concentrations remained the same. The immersion time was reduced to 1 minute and 45 seconds with even slightly improved product in color and flavor due to the clams having been subjected to the enzyme solution for an even shorter period of time.

Example 3: Clam tongues, that portion of the clam in which the belly is contained, were separated from the remainder of the clam flesh manually. The separated tongues with bellies attached were sliced in one-eighth inch strips in a conventional clam slicer. The sliced clams were subjected to an identical treatment as described in Example 1 except that reaction time in the enzyme solution was reduced to 1 minute and 5 seconds, and the enzyme solution contained 35% proteolytic, 20% glycoside hydrolase, 5% pectolytic, 30% lipolytic, 5% pentosanase, and 5% cellulolytic. The washed and drained clam strips were weighed and the recovery was 50.5%. Conventional hand squeezing for belly removal recovery is 32 to 37%.

The drained strips were chilled for 2 days, battered and breaded in conventional equipment with regular commercial battering and breading materials. The breaded strips were then frozen in a blast freezer in 2-pound cartons. After a 5-day storage, the enzyme treated clam strips and strips from conventional handling were deep fat fried in identical baskets in the same deep fat fryer at 375°F. Blind evaluations for flavor, color, aroma, and texture were conducted by experts and laymen. No difference could be determined between enzyme treated and conventionally handled clam strips.

With respect to the application of this process to clams, it has been found as a desirable modification to the process to wash the clams prior to preheating and immersion into the enzyme solution. The washing of the clams is preferably after they have been divided and the inner tissue exposed to the washing medium. Water is the preferable medium and may be sprayed, poured or otherwise applied to the divided clams or they may be immersed into a bath of water for a short time. The water temperature is not critical but generally may be below 150°F. Tap water temperature is entirely adequate. Time is not important in this preferred washing step and may only be sufficient, depending upon the means of applying the water, to wash the exposed tissue faces.

It is believed that this washing step effects the removal of a substantial portion of the naturally occurring enzymes is exposed cut cells and open interstices of the clam. Such enzymes may adversely affect the composition of the subsequent enzyme bath. Surprisingly, these naturally occurring enzymes present in the marine creature will tend to detract from the color, texture and taste of the clam, particularly if the divided clam is preheated to meet the temperature of the enzyme bath.

By washing as much of the variable naturally occurring enzyme from the clam, particularly when divided, a more uniform and effective subsequent treatment may be obtained from an enzyme bath of known and continuous desirable concentration. A uniform product and higher yield have resulted. It should be understood that this unique washing step is separate from the initial washing of the shucked clam which is designed to remove foreign matter and does not affect the internal enzyme concentration. The following example illustrates the application of this concept.

Example 4: One hundred pounds of heat shucked whole clams were chopped in a commercial food chopper with a ¾-inch plate. The chopped clams were washed in a perforated bottomed washer with a screw conveyor to move the flesh through the water into a tumble perforated reel with water sprays of tap water at a temperature of 60°F. Water from the perforated tumble reel was caught in a pan beneath the reel and pumped back to the first washer to conserve water. The clams passed from the reel to a perforated draining pan and then to 3-gallon lacquered cans.

Closed cans were refrigerated in a cold room under cracked ice to a center can temperature of 35°F which required 6 hours. Twenty-five pounds of the above refrigerated chopped clams with bellies in were placed in a perforated stainless steel tub. The tub was immersed immediately in a water bath containing 75 gallons of water at 128°F. The tub was raised and lowered in the warm water for 3 minutes. The tub of clams was removed from the warm water and immediately immersed in an aqueous solution containing 0.026% enzyme mixture, total volume being 6 gallons. The enzyme solution contained 20% proteolytic, 25% glycoside hydrolase, 50% lipolytic and 5% pentosanase.

The temperature of the solution was 125°F. The solution temperature was maintained by circulating water at 125°F through the jacket of the stainless steel kettle in which the enzyme solution was contained. The solution was agitated around and through the clams by bubbling compressed air through the solution. Reaction time was 2 minutes at which time the clams were removed from the solution and sprayed by a jet of water at 40 psi from a hose.

All of the undesirable edible tissue in the form of belly material immediately disintegrated and washed through the perforations leaving behind the edible clam flesh. The drained clam flesh weighed 21 pounds, 6 ounces, equivalent to 85.5% recovery. The color of the clam flesh was identical to fresh clams conventionally treated.

SCALLOPS

Gaping the Shell with an Oxyacetylene Burner

Before a scallop shell can be separated from the flesh within it, the shell must first be severed from the powerful adductor muscle. Experience in the prior art has shown that an application of heat in the local area of muscle attachment is effective to sever the shell and muscle and to open, or gape, the bivalve.

This well known process is performed simply and efficiently with a multitipped oxyacetylene burner in a process developed by *R.W. Nelson, R.F. Mackin and W.I. Tretsven; U.S. Patent 3,594,860; July 27, 1971; assigned to the U.S. Secretary of the Interior.* In the process of preparing a scallop for consumption as food, the shell is opened and the adductor and catch muscles, or meat, are separated from the shell and viscera. The shell and viscera are then discarded. When performed manually, the shucking-eviscerating process is tedious and time consuming. Mechanized shucking and eviscerating, however, speed the process to a pace consistent with the growing consumer demand for scallop flesh, reducing labor costs while producing a product of improved cleanliness and structural integrity.

Shucking and eviscerating in the apparatus of this process are performed in four distinguishable steps; gaping, breaking, eviscerating, and separating. In the first step a scallop in a horizontal position is heated from opposite the attachment area of the adductor and catch muscles on the upper half shell. A high thermal intensity oxyacetylene flame used for this purpose causes muscle separation and gaping in a very short time. Gravitational force draws the scallop flesh away from the horizontal upper half shell during this heating process until the shell gapes, so that the short flame exposure produces a minimal cooking effect.

After gaping, the second or breaking step is performed. By physical breakage at the hinge, the upper and lower half shells are separated from one another. The upper half shell is discarded. Although more complex structures might be used equivalently for performing this step, it is performed simply and efficiently by conveying the upper half shell against a rigid shucking bracket. The shucking bracket snags the upper half shell, breaking it free from the lower half shell as the scallop advances to the third step.

In the third step the scallop is eviscerated by a water suction process. Inverted in a tank of water, the scallop flesh is sprayed with water jets to loosen the viscera from the shell and muscle. The loosened viscera are drawn into a water suction intake and subsequently discarded. Turbulence generated by the incoming and outgoing water streams agitates the muscle with a cleansing action that loosens sand and residue, yet has little effect on the attachment of the muscle and shell.

After evisceration, the fourth step of separating the muscle and lower half shell is performed. Separation is achieved by heating the scallop shell in the same manner as in the original gaping step.

An apparatus for automatic, mechanical separation of edible flesh from the shells and viscera of scallops is shown in Figure 4.7. Unlike the flesh of other bivalve mollusks, such as oysters and clams, only the adductor and catch muscles of the scallop are consumed as food. Culinary preference requires an additional step of separating viscera from muscles when a scallop is shucked. This additional step is performed thoroughly and efficiently in combination with a complete shucking process by the structurally simple apparatus shown in Figure 4.7.

Four distinguishable steps in the shucking process are performed by the apparatus **10**. Beginning at Station A, a whole, live scallop **12** is loaded onto an individual tray **24** of a continuous conveyor **16**. The scallop passes through a very high temperature flame which loosens the adductor and catch muscles **42** from the upper half shell, causing the scallop shell to pop open, or gape. At Station B the upper half shell is physically separated from the lower half shell. With the muscle still firmly attached, the lower half shell is inverted as it proceeds to Station **C**. At Station **C** the scallop enters a water bath **52** where water jets strike the flesh, loosening the viscera **44** from the adductor and catch muscles. The viscera is then drawn away from the muscles and into a discharge pipe **60** by a hydraulic suction pump **62**, leaving only the cleaned muscles and half shell. Completely eviscerated, the scallop proceeds to Station **D** where the remaining half shells are separated. The patent contains more detailed descriptions and diagrams of the steps involved.

FIGURE 4.7: SCALLOP SHUCKING AND EVISCERATION USING OXY-ACETYLENE BURNER

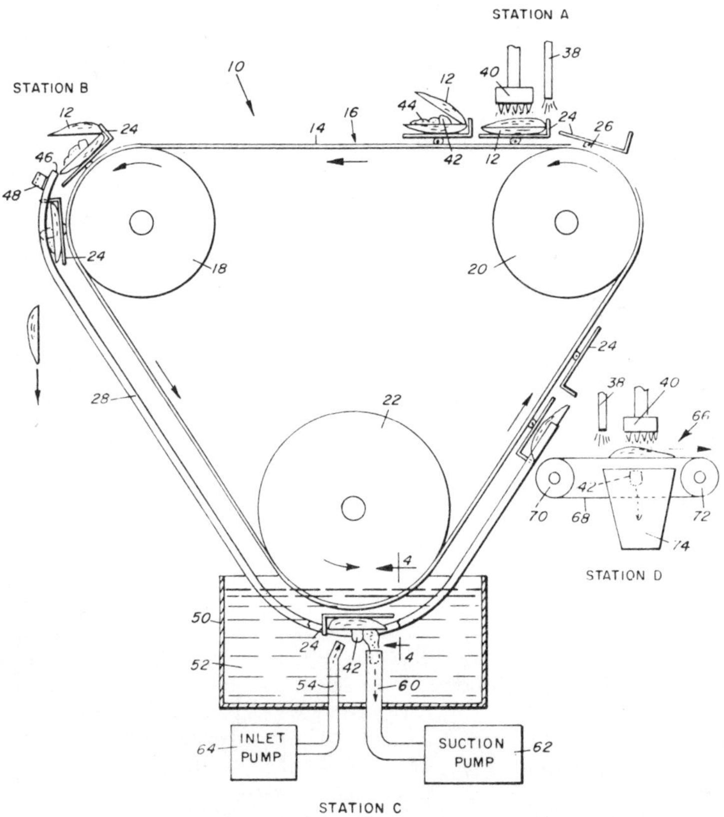

Source: R.W. Nelson, R.F. Mackin and W.I. Tretsven; U.S. Patent 3,594,860; July 27, 1971

Scallop Eviscerator with Rotating Rollers

Two types of scallop eviscerators now in use have rotating rollers to effect the separation of the viscera from scallop meat. One eviscerator, invented by C.M.

Bunnell, consists of a mechanical arrangement comprising a cylindrical drum within which turns a rotor which drives a number of rollers which turn by friction, under spring loaded pressure, against the interior drum surface and an eviscerating surface on a portion of the interior surface of the drum comprised of alternating textured and smooth areas.

Another roller type machine, now in experimental use by several different corporations, some of which are installed on ocean boats, uses the principle of a sloping bank of adjacent, rotating rollers any adjacent two of which in a regularly reversing cycle grab the viscera between the touching surfaces of any two rollers whose surfaces are moving downward and thus pull it off of the scallop meat.

The patent granted to *C.M. Bunnell; U.S. Patent 3,621,515; November 23, 1971* describes improvements and additional components which are adaptable to either type of roller eviscerator above described but specifically to the latter type.

These prior art machines depend upon the rollers to perform three separate functions: (1) to position the scallop by turning it so that the viscera becomes pointed downward toward the point of tangency between the two rollers; (2) to exert a downward pressure on the scallop meat and viscera by friction with the roller surface so as to bring the short tab ends of the viscera to that point of tangency of the rollers where they can be grabbed; (3) to grab the viscera between the rollers and pull it off the scallop meat.

One objection to these machines is that they are only partly effective in function (1) because they use rollers that are nontextured and when two smooth surfaced rollers turn against each other, it is quite apparent that they tend to turn the scallop meat in both directions at the same time which nullifies their turning ability and reduces the turning to happenstance rather than certainty.

Another objection to these machines is that they cannot perform function (2) efficiently unless the meats are of a uniform diameter and unvarying degree of firmness and even then not without an attrition in both the number and size of the meats.

Another objection is its limited ability to grab the slippery viscera. As the viscera breaks, each tab end becomes successively more difficult to grab. Another objection is the limited holding ability of two wet slippery roller surfaces pressing the slippery viscera between them and frequently severing the delicate tissue by the pressure exerted in the effort to hold the viscera firmly.

Also an objection is that each meat must travel over 40 to 100 rollers on the chance that the conditions will be right for the viscera to be completely removed, and many are not completely cleaned but have tab ends left which make necessary hand work in their removal or a less attractive product. Another objection is the inefficient removal of the slimy eviscerate from the rollers, as this slime complicates the efficient grabbing and holding of the viscera by the rollers. Other objections could also be made about the efficiency of the machines.

One improvement offered by the Bunnell apparatus is the provision of a means of rotating the viscera to be grabbed which does not depend alone on the friction of the rollers against the scallop meats to effect this action. Another improvement is in the provision of a positive means of turning the scallop meat to bring the viscera to a position at the bottom of the meat between the rollers where it can be forced to the point of tangency of the two rollers.

The fundamental principle involved here is one which includes an additional component and a change in the physical makeup and the function of the component rollers in the eviscerators currently in use. In these prior art machines, any two rollers, rotating against each other at that point in a reversing cycle in which the surfaces of the rollers are turning downward, and with a scallop meat and viscera between them, depend upon the two turning rollers alone to remove the viscera as described in the introduction to the first three objections. The process described here eliminates the hazard of drawing the scallop meat too far down between the rollers, where, depending on its softness and size, it may or may not be eaten by the rollers.

Elimination of the hazard is accomplished by using much smaller rollers of a diameter which permits no appreciable downward pull on the scallop meat itself because the angle formed by the tangents to the two rollers at the point on which a scallop meat rests is very obtuse rather than acute. A force from above is substituted to achieve function (2) of the rollers described in the introduction to the first objection, to hold, firmly and intermittently, the scallop meat and viscera against the eviscerating surfaces of the rollers and force the viscera to the roller tangency point. This component consists of a pressure control finger rack assembly.

Figure 4.8a presents a vertical view downward of the scallop control finger rack assembly. This is shown in its position superimposed over the rollers. **1** is a scallop control finger. These fingers are supported on a scallop control finger shaft **2**. At either end of the line of fingers a scallop control finger rack **3** is also supported on the shaft. The scallop control finger shaft is supported on a roller chain **4**. Together these parts make up the scallop control finger rack and chain assembly.

At the right end of the scallop control finger rack is a curved arm which is supported on a scallop control finger rack level control guide **5**. This establishes the level at which the lower end of the scallop control finger rests above the top of the roller surfaces. Not shown are clear vinyl spacers on the scallop control finger shaft to hold the fingers one-eighth inch apart nor are the chain guides. The scallop control finger shaft is of No. 7 gauge stainless steel. It is inserted into the chains at each end by enlarging the hole in the side rails of the chains to accommodate it and a hole not shown is drilled in each end of this shaft through which is inserted a retaining pin to prevent the shaft from slipping out of the chain.

The machine which followed the prototype contains six compartments, each with a scallop control finger rack and chain assembly containing thirteen racks with each rack containing sixteen scallop control fingers. **8** represents a compartment partition.

Figure 4.8a also shows a bank of rollers which slopes downward away from the viewer. This bank contains two types of rollers: **6** is a roller composed of one-half inch stainless steel shaft core onto which is cast a glass fiber reinforced polyester resin sleeve of a thickness of slightly over 1/64". Bonded to this with urethane resin is a coating of silica sand screened to 1/32" size and 1 grain thick. Over this sand is a thin coat of urethane resin to round the edges of the sand and seal the surface to a uniform textured but nonabrasive surface. The slick coating of tough resin makes it long wearing and easily cleaned.

FIGURE 4.8: SCALLOP EVISCERATOR

a.

b.

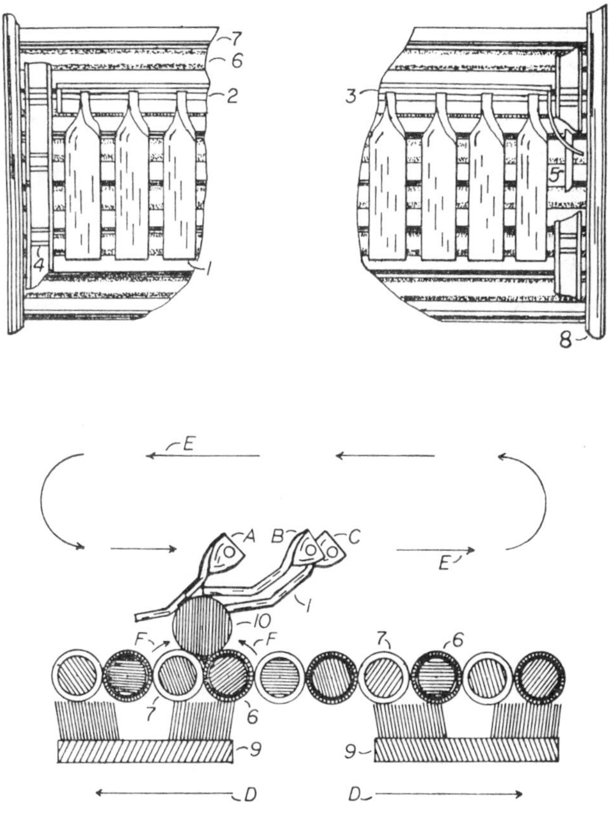

Source: C.M. Bunnell; U.S. Patent 3,621,515; November 23, 1971

7 is a roller which has a soft urethane surface consisting of a 1/16" thick sleeve of shore hardness 50 A slipped onto a ½" stainless steel shaft core. Because of the combination of a soft surface against a hard surface with recesses into which the viscera is pressed as it goes through the rollers, there needs to be no critical setting of the rollers against each other because the viscera is not pinched in two. It is also apparent that with one relatively rough and one smooth surface both turning against the lower surface of the scallop meat, the rough surface will exert more turning influence on the meat than does the smooth roller surface and thus turns the meat in a positive manner.

Figure 4.8b is a schematic diagram showing the relative positions of the major components of the apparatus and the direction of movement of each as it performs its function. A scallop control finger **1** shown at its at-rest height above the roller tops as established by the scallop control finger level control guide **5** in Figure 4.8a, at position **A** and traveling in direction **E**, contacts a scallop meat **10** which rests on the tops of and between rollers **6** and **7** which turn as indicated in directional arrows **F**. The viscera shown hanging from the bottom of the scallop meat as the scallop control finger moves to position **B** is forced by the weight of the scallop control finger to the point of tangency of the rollers, the surfaces of which are turning downward and is thus caught between the rollers and pulled through to the bottom.

Here the viscera is removed from the rollers by brushes **9** which rotate on vertical spindles, not shown, mounted on a carriage, not shown, but which moves the brushes back and forth as indicated by directional arrows **D**. As the scallop control finger moves to position **C** the sharp lower edge of its trailing end exerts a downward force on the right side of the scallop meat turning it clockwise as viewed unless the viscera is still holding the scallop meat steady as it is being removed. This prepares it for removal of other tabs of viscera if present.

In the machine there are standard floor cleaner brushes, four to a compartment, which in their travel clean the entire surface of the roller sleeves within each compartment working on the bottom of the rollers. Not shown is a low pressure water spray to the top of the roller area which is a component of the prior art machines. This helps to prevent sticking of the viscera to the roller sleeve surface. Figure 4.8b also shows the alternating arrangement of the two kinds of rollers **6** and **7**. The outside diameter of these rollers is ⅝". It can be seen that because of the cylindrical shape of the calico scallop meat it naturally fits down between the small diameter rollers so that even very short tab ends of viscera can be caught easily without danger of pulling the meat into the rollers.

Each scallop control finger rack moves slowly from the top of the bank of rollers where the meats with viscera attached are introduced to the machine down the slope of rollers at a regulated height above the rollers to the lowest end of the bank where on an endless chain it is lifted in accordance with directional arrows **E** and returned to the top of the bank of the rollers where it repeats the journey. In the course of this movement every scallop meat upon the bank of rollers is individually affected by one or another of the scallop control fingers in accordance with the action described. With thirteen racks of scallop control fingers in each chain assembly it can be seen that both the number of times the pressure is applied and the period of pressure against each scallop meat can be adjusted by an electronic speed control operating the gearmotor drive for this system.

Extracting the Scallop Adductor Muscle by Heat Shock and Mechanical Agitation

Historically, the shucking and eviscerating of scallops has been performed by hand on a piece-work basis. A 1961 amendment to the Fair Labor Standards Act of 1938 for the first time made minimum wage provisions applicable to the seafood industry and had a direct effect in raising the labor costs of hand shucking and eviscerating to very high levels. The increased labor costs particularly affected the feasibility of hand processing calico scallops which are found in abundance off the Atlantic Coast of Florida. The adductor muscle of the calico scallop is very small and shucking and eviscerating by hand are so time consuming in relation to yield, that hand processing these scallops became economically impractical.

E.D. Willis has, accordingly, developed a mechanical method for separating the scallop adductor muscle (the only part commonly used for human consumption) from the shell. The process and apparatus are described in *U.S. Patents 3,619,855; November 16, 1971; and 3,665,555; May 30, 1972; both assigned to Willis Bros., Inc. and U.S. Patent 3,662,431; May 16, 1972.*

This method of extracting the adductor muscle of an unshucked scallop is applicable to scallops generally, including sea scallops, *Placopecten magellanicus;* bay scallops, *Aequipecten irradians;* and calico scallops, *Aequipecten gibbus.* While the different varieties of scallops vary in size, their similar physiology allows them to be processed according to the methods of the process.

Scallops, after being harvested by dredging or the like, may have sand or other foreign matter on the exterior of their shells. Consequently, it is sometimes desirable to wash the exterior of the scallop shells with water to remove the foreign matter before processing according to the process. It is also sometimes desirable to mildly agitate the scallops prior to processing them, particularly when the scallop is processed within a short time after it is harvested from the sea.

In accordance with the process, unshucked scallops are immersed in a hot liquid bath for a time period sufficient to at least partially release the adductor muscle from the shell, but insufficient to substantially denature the muscle protein. Preferably, the hot liquid in which the scallops are immersed is water held at 180° to 200°F to provide a heat shock to the scallop. At these bath temperatures, the preferred immersion times are from 6 to 9 seconds. Optimum results are obtained by immersing the scallops for about 7 seconds in 195°F water. Under the preferred conditions, the adhesion of the adductor muscle to the shell is substantially lessened, but the scallop muscle does not absorb enough heat to be cooked.

The scallop muscle and shell are subsequently mechanically agitated to cause the release of the muscle and attached viscera from the shell. While the agitation must be vigorous to ensure release of the adductor muscle from the shell, it is desirable to avoid excessive agitation or crushing which would fragment the shells. Preferably, the mechanical agitation separates the two valves of the bivalves to ensure release of the muscle and viscera from the shell, but does not further reduce the size of the shell so that the bulk of the shell material can be easily screened to separate it from the muscle and viscera.

The valves and any large shell fragments resulting from agitation of the bivalve are separated from the muscle and viscera, preferably on a vibrating screen that has openings which will pass the muscle and viscera but will not pass the unfragmented valves, comprising the shell of the bivalve, which are discarded.

Muscle and attached viscera from the vibrating screen are preferably contacted with a stream of water to wash them and preferably to also carry them from the vibrating screen by fluming. During the agitation of the scallops and the separation of the muscle and viscera from the shell, the muscle and viscera may contact the outer side of the shells and pick up sand, etc. which is conveniently removed by the water stream. The muscle and attached viscera are strained from the stream of water and subsequently floated in a brine solution to permit small shell fragments to separate therefrom and to settle.

The scallop meats are then placed on the upper surface of an inclined path formed by a plurality of rollers having their axes transverse of the path. The initial orientation of scallop muscles on the path is immaterial, as the cylindrical adductor muscle is soon oriented by the rollers so that the axis of the muscle is generally parallel to the rollers. In this position, the muscle rotates about its axis, generally in the direction of the rotation of the lower of two adjacent rollers forming a nip, and exposes the viscera around substantially its entire periphery to the nip formed by each pair of adjacent rollers.

The viscera is pulled from the muscle and through the nip formed by adjacent rollers by rotating the upper portion of adjacent rollers towards each other. The rollers exert a frictional pulling force on the viscera in a downward direction through the nip of the rollers while at the same time restraining movement of the scallop muscle in that direction.

The rollers are oscillated (counter rotated) so that they alternate between pulling viscera from the muscle and advancing the muscles down the path. The degree of rotation of the upper portion of the rollers towards each other is preferably at least 360° for rollers at the top of the path, but the degree of rotation of rollers located lower in the path can be reduced below 360°.

The viscera are removed from the surface of the rollers at a point below the inclined path. Thus, the rollers physically separate and also effectively segregate the muscle and viscera. Preferably, the removal of the viscera from the rollers is accomplished by directing a stream of water under pressure at the underside of the rollers forming the inclined path.

The advance of scallop muscle down the inclined path is controlled by intermittently rotating the upper portion of adjacent rollers away from each other to permit the lower roller of each pair of adjacent rollers to advance the muscle down the path, and by spraying a flow of fluid on some of the muscles to affect their rate of movement down the path. Both the frequency and degree of rotation of the upper portion of the rolls away from each other affect the rate of movement of the scallops. Increasing the frequency and the degree of rotation of the upper portion of the rolls away from each other both tend to decrease the residence time of the muscle on the path.

The spraying of fluid on the muscles to affect their rate of movement down the path is preferably effected by spraying water over a substantial area of the path. By directing the water spray upwardly in relation to the path, movement of the muscles down the path can be impeded. Directing the water spray downwardly in relation to the path decreases the residence time of the muscles on the path. A further benefit of the use of a water spray on top of the rollers is the tendency of the water to form a continuous film that helps ensure that the scallop muscles rotate about their cylindrical axes and expose their entire peripheral surface to the pulling action of the rollers.

Viscera removed from the adductor muscles by the repeated pulling action of the nips formed by the rollers is cleaned from the rollers by water from spray pipes. The spray emitting from these pipes contacts the underportion of the rollers. The viscera are forced off the rollers by the water, drop into a pan, and are flumed therefrom by the water supplied by the rotating spray pipes. The viscera may be collected from this water stream and used in fertilizer, etc. Scallop muscles after traversing the length of the roller path fall into a collecting tray. These scallop muscles are substantially free of viscera and are a commercially marketable product. Detailed drawings and descriptions of the apparatus are available in all the patents.

Method for Shucking and Eviscerating Scallops Which Improves the Yield of Edible Muscle

A process for shucking and eviscerating scallops and an apparatus for carrying out the process are described in three patents to *R.T. Wenstrom and T.S. Gorton, Jr.; U.S. Patents 3,662,432; May 16, 1972; 3,665,554; May 30, 1972; and 3,683,458; August 15, 1972; all assigned to Slade Gorton & Co., Inc.*

Preliminary to the shucking operation, provision is made for cleaning and sorting the scallops dredged from the ocean bottom. The contents of the dredge are deposited in the sorting device which is adapted to remove all of the mud and other debris including undersize scallops and to leave only the marketable size scallops which are delivered to the shucking apparatus for processing. In practice, the separating operation lends itself to efficient conservation wherein provision is made for quickly separating the smaller scallops from the larger ones and to immediately return the small scallops to the sea.

Also, the scallops of merchantable size are separated from the empty shells, sand, rocks and both small and large miscellaneous sea life that is brought on deck together with the scallops, the separated scallops being delivered to the shucking apparatus for processing. Because of the speed of separating if done at sea, many of the fish, crabs, snails and other sea life may be returned overboard in condition to live.

In accordance with the described process for shucking scallops, the first step is to subject the scallop to a substantial shock which may be effected by dropping the scallop from a substantial height onto a receiving support. The initial shocking operation does not break or open the shell. However, it does prepare the scallop in some manner not understood for the next step in the operation which comprises immersing the scallop in a bath of hot water or subjecting it to another form of heat for a predetermined length of time and at a predetermined temperature sufficient to assist in relaxing the scallop muscle and to aid in separating the meat and viscera from the shell during a subsequent operation but not enough to cook the scallop.

The next step is to subject the scallop to a second shocking operation by again dropping it onto a support. This latter operation will effect separation of the two half shells of the bivalve and release of the meat and attached viscera.

The shells and the meat may then be guided onto a wide mesh screen conveyor which permits the meats to drop through the mesh onto a second conveyor to be carried to a storage point, and the shells carried on top of the screen conveyor are guided into a third conveyor for discharge at a suitable storage point. While disposed on the screen, the shells and the meats are subjected to a spray of water which assists in causing the meats to pass through the wire mesh.

In modification A of the process, the second shocking operation is performed by subjecting the scallops to a blow by rotating paddles mounted in a confined space such as a chamber comprising a box open at the bottom so that the scallops are opened and confined by the side walls of the box. The shells are opened, and the shells, released meats and viscera then fall onto an inclined wide mesh screen by which the shells and the scallop meats and viscera are separated.

The scallop meats and attached viscera are then delivered to a conveyor having resilient pusher elements which cooperate with an expanded metal plate having an undulating surface or closely spaced projections which effect separation of the viscera from the meat by a tearing and pulling or peeling operation.

In modification B of the process, both shocking operations for initially relaxing the scallop muscles are performed by projecting the scallops against the side wall of a box, the scallops being immersed in a bath of hot water after the first shocking operation. After the second shocking operation, the shells are separated, and the meats and attached visceras fall on an inclined wide mesh screen through which the meats with the viscera attached may pass. The shells which cannot pass through the screen are carried away on a conveyor.

The meats and visceras are then delivered to an apparatus by which the viscera is progressively removed from the meat in an efficient manner, the viscera being discarded and the cleaned scallop meat discharged onto a conveyor leading to a storage receptacle or other desired location for packing.

In still another embodiment of the process, the scallops are subjected to centrifugal force which also serves to shock and relax the muscles of the scallops immediately prior to subjecting the scallops to the hot water bath preparatory to a second shocking operation to effect opening of the scallops and separation of the meat from the shells.

Referring now to Figure 4.9, which more or less diagrammatically illustrates a preferred form of apparatus, **10** represents a hopper into which the scallops **12** to be shucked are deposited, and **14** represents a conveyor or carrier chain having a plurality of spaced pockets **16** and arranged at a relatively steep incline for carrying a regulated amount of the scallops from the hopper to a substantial height. The scallops thus elevated are carried over the upper end of the conveyor and are permitted to fall freely through a vertical chute **18** onto an inclined baffle plate **20** disposed at the lower end of the chute. The height of the free fall may vary from 20 to 60 feet depending on the size, shape and condition of the scallops, a preferred height being about 40 feet.

The baffle plate may be pivoted at **22** to permit adjustment of the incline by a slot and bolt connection **24** as illustrated. The baffle may comprise a metal plate **26** faced with a rubberized or other resilient material **28** of a relatively sharp blow to the scallop without breaking the shell. Upon stroking the semirigid baffle plate, the scallops will bounce off the plate and fall into a second hopper **30**, a stationary baffle **32** being provided to guide the scallops into this hopper.

FIGURE 4.9: OPENING SCALLOPS BY SHOCK

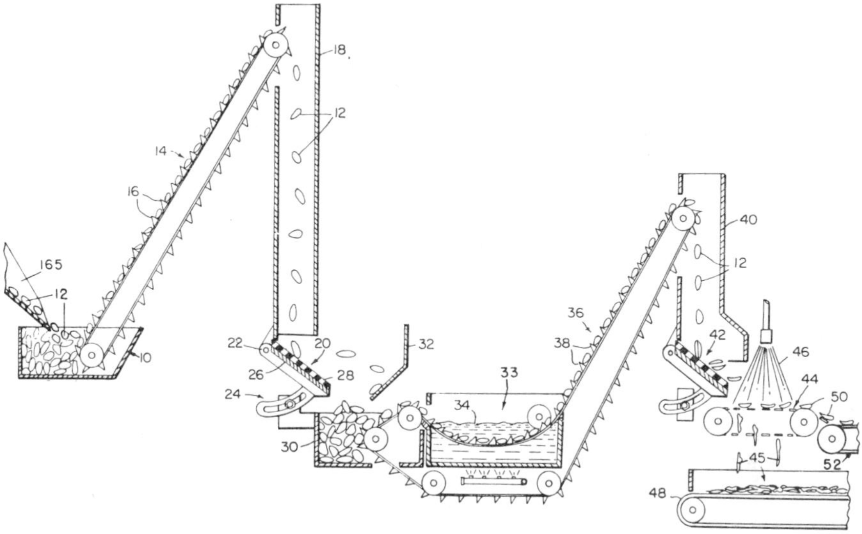

Source: R.T. Wenstrom and T.S. Gorton, Jr.; U.S. Patent 3,683,458; August 15, 1972

The next step is to pass the scallops through a tank **33** of heated water **34**. The scallops are immersed in and transported through the water by a second conveyor or carrier chain **36** also provided with spaced pockets **38** and which is arranged to pick up regulated amounts of scallops from the hopper and carry them through the hot water bath as shown. The water temperature may vary from 150° to 212°F according to variations in the time of immersion, and may also vary with different sizes and the shell condition of the scallops, the important consideration being that the time and temperature combination should be insufficient to cook the scallops but sufficient to assist in separating the meat from the shell or to prepare the scallop meats to be separated from the shell during the next step in the method.

At a temperature of 150°F it is preferred to subject the scallops to the heated water or other form of heat for 20 seconds, and at 212°F for 3 seconds. Preferably, the temperature is 180°F and the scallops subjected to the heat for 7 seconds. This may vary within reasonable limits, depending upon the condition of the scallops, i.e., whether or not they are substantially clean, covered with barnacles and whether or not the shells are thick or thin.

It will be observed that the upper run of the conveyor or carrier chain **36** extends upwardly from the tank **33** at an incline, the upper end of the conveyor being arranged to release the scallops into the upper end of a second vertical chute **40** to fall freely through the chute onto a second adjustably inclined baffle plate **42** having a rubberized fabric facing of a more or less semirigid consistency. The height of the free fall through the second chute may and preferably will be substantially less than the fall through the first chute, and preferably comprises a 20-foot drop. This may be varied within reasonable limits.

In operation, upon striking the baffle **42**, the two half shells come apart, releasing the entire inside of the scallop which is then free from the shells at this point. The shells and the meats then fall or bounce from the inclined baffle onto a relatively wide mesh flexible screen belt **44**, the meats **45** pass through the screen to fall onto a discharge conveyor **48**, and then are carried to a discharge point for further processing wherein the muscle, i.e., the meat of the scallop, is separated from the viscera. If desired, a spray of water **46** may be used to assist in washing the meats through the screen. The shells **50** do not pass through the screen and are carried on top of the screen conveyor **44** and are transferred to a second discharge conveyor **52** to be carried to a storage point.

Enzymatic Separation of Undesirable Tissue from Adductor Muscle

The process already described for enzymatic separation of undesirable tissue from clams developed by *G.B. Fehmerling; U.S. Patent 3,729,324; April 24, 1973; assigned to Ocean Research Corporation* is also useful for separating undesirable sections of scallop from its desirable adductor muscle.

Example: "Calico" scallops were shucked in a mechanical shucker. Ten pounds of scallops complete with viscera were washed with water, drained, and placed in a perforated stainless steel tub. The tub of scallops was immediately immersed in an aqueous enzyme solution containing 0.026% enzyme mixture and comprising 40% of the proteolytic enzyme, 25% of the glycoside hydrolase enzyme, and 35% of the lipolytic enzyme. Five gallons of the enzyme solution was used. Temperature of the solution was maintained at 125°F by circulating water at 125°F through the jacket of the kettle containing the enzyme solution. Agitation of the solution around and through the scallops was achieved by bubbling compressed air through the solution. The scallops were exposed to the enzyme solution for 5 minutes, removed to a revolving rod reel washer with rods spaced three-eighths inch apart.

The scallops were washed with sprays of water at 50 psi while revolving in the washer. Scallops were washed free of all undesirable edible tissue, including the viscera, the digestive system, straps (gills) and reproductive system. The clean adductor muscle, the desirable edible flesh portion of the creature, was completely free of previously attached undesirable tissue. Four pounds and fifteen ounces of cleaned adductor muscle were recovered from the original ten pounds of original material (49.3%). Quality of the enzyme treated scallops was adjudged by experts in the field and found to be identical to hand shucked and hand cleaned scallops.

Cleaning and Sterilizing Scallops in the Shell

In order to efficiently process scallops as they are brought to dockside from boats, it is necessary to provide a way of removing dirt, barnacles and the like from the outer surface of the shellfish while at the same time killing the bacteria which could lead to harmful contamination.

An apparatus for effectively cleansing and sterilizing the outer surface of shellfish has been developed by *R.C. Martin; U.S. Patent 3,811,148; May 21, 1974.*

Scallops, still in their shells, are conveyed from a measuring device or the like into a hopper which is in direct communication with the interior of the screw conveyor steamer. The screw conveyor steamer comprises an outer jacket of generally cylindrical shape, having disposed therewithin and coaxially therewith an inner jacket having symmetrically circumferentially disposed orifices in its surface. The orifices are large enough to permit the usually encountered debris to pass through but not large enough to pass a full sized scallop.

Extending coaxially with the outer jacket and inner jacket is a screw conveyor which is driven by a variable speed hydraulic motor. As the screw conveyor rotates, scallops fed to the conveyor by the hopper will be conveyed upwardly at a speed determined by the prevailing conditions, such as scallop size, relative cleanliness and the like.

In communication with the outer jacket are steam lines for forcing steam under pressure through the orifices and into contact with the outer surfaces of the scallops as they are conveyed along. The steam, which is conventionally controlled to be at a temperature between 215° and 240°F, thus is applied to the outer surface of the scallops in a substantially 360° pattern, insuring that all the dirt, barnacles and the like are removed from the surface of the scallops and the bacteria is killed.

The tumbling action induced by the screw conveyor as the scallops pass along the length of the inner jacket also insures that the cleansing and sterilizing action is thorough, since a different surface of the scallop shell is constantly being presented to the steam entering through the orifices.

The end wall of the screw conveyor steamer is closed by a plate held in place by reinforcing ribs. An exit orifice is provided in the plate to permit the debris to be washed out and away by the steam condensate on the inner walls of the outer jacket. Since the screw conveyor steamer is downwardly inclined toward the left, the debris will naturally be carried away from the system. Cleaned and sterilized scallops will exit through a chute and on to a further conveyor for additional processing.

SQUID

Skinning and Filleting U.S. Coast Squid

Squid is found in various waters of the world and in quantity on the east and west coasts of the United States. While squid is a great delicacy among certain peoples, it is virtually unknown to most Americans because known methods of preparation are not to the average American taste. Squid prepared in this country for export are generally canned or dried whole, not eviscerated, and generally not even washed or cleaned.

The body of a squid is covered with a dark, mottled membrane which is thin, film-like, sticky and gelatinous. While this skin is easily removed with the caudal fins, it has been found that the mechanical devices which have heretofore been used for this purpose do not slip or peel off the skin at a speed consistent with efficient production of the fillets. It is essential that the skin be removed from the squid because the fillet and other edible meat portions would otherwise have a dark, purplish appearance which is undesirable to our people. The skin also contains some of the ink sacs of the squid which should be removed.

B.E. Williams; U.S. Patent 3,034,906; May 15, 1962; assigned to Hodges Research and Development Company has found that the skin of the squid can be readily removed by first freezing the squid and thereafter tempering until the skin-on surface is approximately 29°F. The skin by reason of this treatment acquires almost parchment-like strength and texture, being entirely changed from the thin, film-like, sticky and gelatinous membrane. The parchment-like skin is easily removed since freezing and tempering has been found to greatly reduce the adherence of the skin to the body of the squid. The parchment-like character of the skin is inducive to quick and ready removal of the skin by both manual and mechanical means.

After the skin has been removed the resultant white meat may then be tenderized and rendered substantially noncurling upon cooking by mechanical treatment comprising scoring the meat at more or less regular intervals which prevents objectionable curling and shrinkage upon application of heat in cooking. This scoring of the ligaments also tenderizes the meat and releases the flavor of the meat more readily upon cooking. This treatment causes the meat to swell to an increased thickness when cooked.

After scoring, the fillet and other portions may be breaded or battered with conventional seasoning and then preferably frozen and packaged for sale in the frozen state. Of course, the fillets, for example, may be sold in unfrozen state. The application of batter and breading is not essential and the fillets and other portion cuts may be frozen and packaged for sale in the frozen state without breading and battering so that the ultimate user may cook the product as desired.

Fillets may be prepared for the table in many ways as by cooking in deep fat at about 350°F for about a minute and a half or they may be cooked in a skillet with a film of fat or oil for about two minutes to the side. Overcooking of the fillets is to be avoided as harmful to the delicate flavor.

Preparing Surumeika Squid

The above described process for skinning and filleting squid has proved entirely satisfactory for most varieties of squid, and is particularly useful and efficient with fillets from squid found on the west coast of the United States and particularly those squid abounding in Monterey Bay, California.

A much larger squid is found and caught in the waters off Newfoundland, around the Mediterranean area and off Japan and South Korea. The latter has the common species name Surumeika. The fillets obtained from such squid are several times larger than the fillets from the squid obtained on the west coast of the United States. These squid, also known as *Ommastrephes solani pacificus* Steenstrup, have a thin membranous outer cover which is most difficult to remove and does not come off with the polka dot skin but remains adhered to the fillet.

The large fillets from Japan and South Korea also have several strands of ligaments on the inside running the length of the fillet, that is, from head to tail of the shellfish. Upon contact with heat these several ligament strands contract causing the fillet to curl into a roll. The large fillets from these squid are too great in diameter and are too thin to be attractive in the raw frozen state and also present a packing problem. However, when cooked they curl and shrink to about one-half to two-thirds their raw size and this is quite a disappointment for the chef or the housewife and for the consumer.

B.E. Williams; U.S. Patent 3,368,908; February 13, 1968 describes a method which comprises the finding that the mantle of large squid, especially the Japanese and South Korean squid, can be simultaneously reduced in size and made plumper, tenderized and rendered noncurling upon cooking by first heat shrinking the mantle and then mechanically treating the mantle. The mantle or white meat has a membranous cover overlying the meat portion of the mantle and a series of longitudinal ligaments. Upon heating or blanching of the mantle, as in cooking, the individual ligaments and the elastic membranous covering contract causing the mantle to curl and shrink.

The large fillet produced from these mantles can be reduced in diameter and made plumper by first subjecting the fillet to boiling water for from 5 to 15 seconds and the fillet then rendered curl and shrink resistant by scoring the fillet clear through to cut not only the ligaments but also the elastic membranous covering so that cooking will influence only the immediate area surrounding each segment, if at all, with no cumulative contraction resulting in objectionable curling and shrinking. Cutting the ligaments makes the meat when cooked quite tender and scoring and filling the cuts with batter and breading material causes further swelling of the meat to produce a cooked fillet of increased thickness.

In carrying out the process, the squid are cleaned and skinned, leaving the sidewalls or mantles clean and trimmed to shape and size to produce ultimately a fillet of approximately 6 x 8 inches in size. The flat mantle is then subjected to boiling water for from 5 to 15 seconds and shrinks with some curling to a rectangular shape of approximately 4 x 5 inches in size but having increased thickness. The fillet is flattened and then subjected to a cutting or scoring treatment being cut or scored clear through the fillet, the elastic membrane and the ligaments to prevent recurling of the fillet when subjected to cooking. The flesh of the mantle is segmented by a plurality of spaced cuts extending in at least two mutually opposed or opposite perpendicular directions in such a manner that substantially all of the ligaments and the membranous covering are repeatedly severed into lengths of about one-eighth to one-half inch each.

A conventional seasoning or batter and breading is then applied to the surface of the scored fillet as is usual in the preparation of other fish products. The batter and breading fill the interstices created by the scoring. The fillet may be frozen and packaged for sale in the frozen state. Scoring the fillet clear through the meat, ligaments and membranous covering permits the steam which forms under the fillet during cooking to escape through these cuts for better, more rapid and more uniform cooking because the fillets lie flat in the pan. The batter and breading will enter these cuts for better adherence thereto providing more uniform breading and battering of the fillet.

The fillets, so prepared, can be kept chilled for reasonably long periods and can be frozen for many months. They have an excellent appearance both before and after cooking, are plump and appetizing and cook quickly without a fishy smell. These fillets may be cooked in deep fat at about 350°F for about a minute and a half or cooked in a skillet in a film of fat or oil at "browning" heat for about two minutes per side. These fillets should not be overcooked and should be served piping hot.

MOLLUSK CONSUMER PRODUCTS

Scallops Frozen in Blocks and Die-Cut to Shape

T.S. Gorton; U.S. Patent 2,938,803; May 31, 1960 describes a process for freezing scallop meat in such a way that it may be produced in any desired size or shape and in a form which enables the products to be cooked as served with advantage in restaurants and the like.

The process contemplates the production of scallops simulating in size and appearance natural scallops and other seafood products of any desired uniform size and shape. They are particularly adapted for restaurant use where "portion control" is of the utmost importance. The scallop product enables portions of uniform size either by count or weight to be rapidly determined before and after cooking and to be served with leftover waste substantially eliminated.

Uniformity of size is also of importance when scallops are cooked in a thermostatically controlled deep fat fryer in that the cooking time and temperature can be adjusted so that all of the scallops cooked will be perfectly done when they reach a golden brown color. Bulk quantities of these frozen breaded or prefried scallops may be kept in their frozen state, and only such quantities as may be needed can be easily removed from the freezer for immediate use.

In accordance with the process a number of individual scallops of various sizes are sliced or cut, preferably in a slicing machine, in random slices. The sliced pieces of scallop meat are thereafter packed in a fibrous container and frozen to thereby form a frozen compact solidified mass. The frozen mass may then be sliced to provide a number of relatively thin sheets preferably of uniform thickness. The particular thickness of the sheets may vary depending upon the size scallops desired to be produced.

Thereafter the individual frozen sheets may be die-cut in their laminated relation or may be individually die-cut to form individual scallops of substantially uniform size and thickness, each made up of a plurality of slices frozen together.

The scallops may be produced in any desired shape. The individual scallops may be round, square, hexagonal in cross-section, or may be in stick form. In practice hexagonally shaped scallops are considered to be more merchantable and are preferred. After the scallops have been die-cut from the frozen sheets formed in accordance with the process they may be dipped in or covered with batter, breaded and packed for shipment to the consumer.

In some instances it may be preferred to dip the scallops and/or scallop meat in a suitable fluid, such as a mixture of flour, egg and powdered milk, prior to packing for freezing in order to assist the scallops and/or scallop meat to firmly adhere together when frozen.

It will be apparent from the foregoing description that the surface area of the individual scallops is greatly increased by the slicing or cutting operation. Each sliced or cut piece of scallop meat is provided with relatively large, flat plane surfaces. When the pieces are packed in a container for freezing, the flat surfaces enable the corners of the container to be completely filled with meat, and the flat surfaces of each piece firmly adhere to each other to from a compact mass substantially free of moisture and air spaces. The mass is further compacted by expansion during the freezing operation, thereby producing a solid packed mass of scallop meat which when frozen may be die-cut and handled thereafter with a minimum of breakage.

The process uses scallop rims in the production of the frozen blocks described above. Heretofore, scallop rims have been of little or no commercial value. The rim of a medium size scallop is approximately equal in weight to the scallop meat itself. The rims may in some instances be used alone or mixed with scallop meat in practicing the foregoing process. The rims may be packed in the container together with whole scallops, thereby forming a closely packed mass. It is preferred to bleach the rims in water or color with a suitable dye before they are mixed with the scallop meat in order that the inside of the finished product when broken open will be more attractive in appearance.

Scallop Product Resembling Abalone

Among gourmets abalone steak is well known for its distinctively delicate flavor. However, it is also widely known that abalone is difficult even for culinary experts to prepare with optimum results, and unless properly pretenderized and cooked in just the right manner it will become tough and unpalatable. In addition to these inherent difficulties, the availability of abalone has decreased rapidly in recent years and it has become an increasingly rare and expensive product in both restaurants and markets.

J. Joaquin; U.S. Patent 3,532,512; October 6, 1970 has devised a method for producing an artificially constructed seafood product which has the same natural delicate taste, distinctive texture and nutritional value as abalone and yet one which overcomes all of the disadvantages and problems theretofore associated with it.

An important factor in the process is the discovery that marine bivalve mollusks of the genus Pecten, commonly known as scallops, could be used to simulate the more expensive meat of abalone. This method provides a manner in which the scallop fibers, which are naturally long and arranged in a coarse grainy structure, can be broken down and then reoriented to resemble not only the taste but also the grain structure or texture of abalone.

The scallop meat is broken down in essentially two forms, some of it being ground and some with its natural fibers left intact. These meat portions are then recombined with other ingredients to form a mix which closely resembles the conventional abalone steak in both taste and fiber structure.

Figure 4.10 is a schematic flow diagram illustrating the various steps used for making the seafood product. As shown, the first step is to provide a quantity of natural and seafood animal matter in its raw state.

FIGURE 4.10: SEAFOOD PRODUCT RESEMBLING ABALONE STEAK

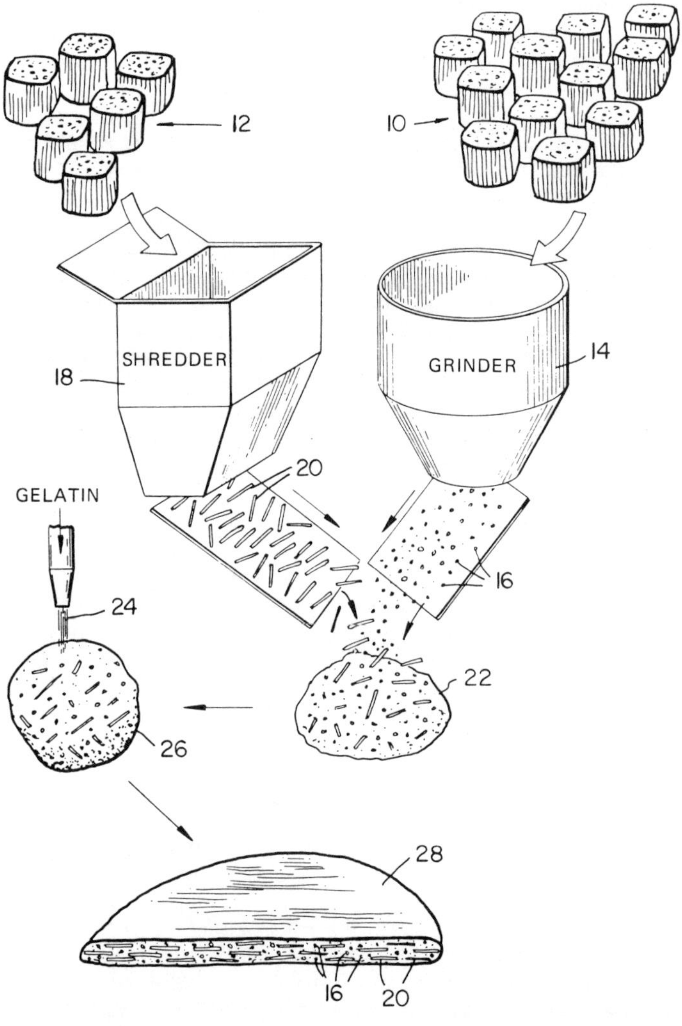

Source: J. Joaquin; U.S. Patent 3,532,512; October 6, 1970

For this matter the adductor muscles of the well known marine bivalve mollusk of the genus Pecten, commonly referred to as scallops, are used. These adductor muscles or scallops are usually tender, succulent and have a distinctive delicate taste. Their relatively long muscle fibers are utilized to help form the new product.

The raw scallop meat is divided into two portions which are indicated by the numerals **10** and **12** in the drawing. One portion of the scallops is ground up in any suitable meat grinder **14** so that it is reduced to a mass of relatively small ground bits **16** of raw scallop material. For this step, a grinder using a head having holes of approximately one-eighth inch in diameter is preferred.

The remaining portion of scallops is then treated in a manner that causes their relatively long muscle fibers to become shredded and separated from adjoining muscle fibers. This may be done in several ways, such as in a dough mixer **18** or some similar device wherein the scallops are tumbled and beaten until their grainy structure is naturally broken apart with their long muscle fibers **20** separated but intact.

In the next step the previously ground portion of scallop meat comprised of small bits and the separated portion of scallop material comprised of the elongated muscle fibers are thoroughly combined together in a suitable container to form a homogeneous mixture **22** of short bits and elongated muscle fibers of scallop material.

After the thorough mixing of the two forms of scallops a predetermined quantity of a glutinous material **24** is introduced into the mixture. This latter material is preferably pure animal gelatin which is introduced in the dry granular form. The gelatin is mixed into the scallop material by feeding it in gradually as continued mixing of the latter occurs. When the gelatin is thoroughly mixed within the combined scallop portions it dissolves and forms a binding medium in a mixture **26** which serves to hold the mass of scallop material together. Later, during storage and when the product is cooked, the gelatin helps to retain and prevent the escape of the natural juice. At this point, some taste enhancing material such as monosodium glutamate may be introduced in the granular form.

The proportions of the ingredients of the product are extremely important to its quality. For example, for best results in providing a product that closely resembles abalone, the ratio of ground scallop meat to shredded scallop meat should be approximately two to one. Also, the amount of gelatin to be used should be in the range of from 2 to 5% of the total mixture weight with 3 to 4% being preferred. In one example, the following proportions were used and produced excellent results.

	Ounces
Ground scallops	55
Shredded scallops	25
Gelatin	3

After the mixing steps have been accomplished, the finally mixed material **26** can be subdivided into smaller portions **28** of any desired size. Preferably, it is desirable to make each portion the approximate size and shape of a natural abalone steak. Thus, in a further step each subdivided portion, as shown in Figure 4.10, is rolled or flattened out to a substantially uniform thickness, (e.g., one-quarter inch). When this flattening is done by applying the necessary rolling pressure on a subdivided portion of mixed scallop material and gelatin, a grain structure of the product is formed which causes it to resemble closely in texture that of an actual abalone steak.

The elongated fibers tend to assume a generally horizontal orientation and intertwine while the interstices between them are filled by the ground up bits. Thus, with the help of the gelatin which provides an additional binding effect, a solid product is formed which will maintain its shape while being cooked and have a texture as well as a flavor closely resembling that of a natural abalone steak.

Clam Snack from Clam Foots

J. Marvin, E.E. Anderson and J.H. Tichenor; U.S. Patent 3,041,178; June 26, 1962; assigned to Saltesea Packing Company developed a clam snack food which is crisp and tender and can be eaten without any further treatment such as heating.

Shucked sea clams constitute about 50% by weight of edible material, including the rims or straps, the eyes, the foots or tongues and snouts. In this Saltesea process the clam foots are used. These are large triangular shaped muscular pieces amounting to about 50% of the edible part of the clam and they lend themselves readily to slicing.

The clam foots which are to be used are, of course, separated from the other edible portions of the clams and are washed or cleaned to any desired extent to remove sand and the like. The foots are then cut or sliced to a desired thickness, preferably to give slices which range from one-sixteenth to one-half inch in thickness.

The resulting sliced clam meats are then soaked in a suitable liquid medium which typically may be water, clam juice, or a combination of these two liquids. The soaking time may be from about a few seconds, (i.e., dipping) to several days. If soaking is to be accomplished over an extended period of time, it is preferably carried out under refrigerated conditions. This soaking aids in retaining the original size of the clam meats throughout subsequent processing operations and appears also to effect tenderization of the sliced clam meats. The soaked clam meats may be used immediately for further processing or they may be frozen and stored for subsequent processing.

Freezing at this point of the process appears to contribute somewhat to the ultimate tenderization of the clam meats, although it is not necessary. If the clam meats have been frozen after the first soaking they must, of course, be thawed by any suitable process before processing further. In any case the unfrozen, or frozen clam meats subsequent to thawing are drained to remove any liquid which will drain off the clam meats at this point.

The resulting drained clam meats are then combined with approximately an equal weight of tap water (at about 45° to 75°F) and the resulting mixture of clam meats and water is brought to a temperature of between 140° to 160°F and then cooled. In this heating step the temperature of the clam meats reaches substantially the same temperature as that of the water. The temperature range is critical.

The clam meats are then immediately drained from the processing water and are immersed in a tap water bath ranging in temperature from about 45° to 75°F. This serves both to cool the clam meats and to give them a fresh liquid surface which will facilitate the pickup or adsorption of breading or batter materials or similar types of coating which are then applied prior to final frying. The coating may be a breading or batter composition and may vary appreciably according to the desired final flavor and texture.

The coated clam meats are then introduced into hot cooking oil and cooked for a time sufficient to reduce the moisture content in the clam meats to between about ½ to 5% by weight. The moisture content is preferably reduced to less than 2% by weight. Cooking oils in this step may be maintained from about 320° to 400°F, with about 375°F being typical. The cooking time will, of course, depend on the temperature of the cooking oil and the amount of moisture contained in the clam meats at the time they are introduced into the cooking oil.

The resulting cooked clam meats are then removed from the cooking oil, drained and then may be further seasoned if desired by application of appropriate spices or seasonings. Antioxidants may be added to the cooking oil or to the post-frying seasoning mixtures.

Antioxidants may also be added to the container in which the resulting fried clam product is packed. Finally the clam product may be packaged in cans, glass jars or moisture proof containers of plastic, foil or paper materials. The air in such containers may be replaced with an inert atmosphere or a partial vacuum, but this is not required. The clam product is a crisp, tender material which has a distinctive clam flavor which may be blended with other desired flavors depending upon the type of spices or seasoning incorporated in the clam meats during processing.

Oyster Soup Mix

Generally speaking, soup making in the family kitchen is a very time consuming and troublesome operation. Furthermore, it is difficult to prepare, all year round, a soup having a good taste from raw oyster and shrimp. This is due to the fact that the taste obtained has a close relationship to the catching season and freshness of the oyster and shrimp. To overcome the difficulties previously encountered, instant soup mixtures are often used in many restaurants as well as in homes.

However, these previously known instant mixtures are not entirely satisfactory. Many concentrates or powders of broth and chicken soups prepared from meat and chicken, having relatively good taste and flavor can be found in the market, but those prepared from good quality oyster or shrimp cannot be found due to difficulties involved in the treatment of the raw materials.

S. Ohtaki; U.S. Patent 3,256,098; June 14, 1966; assigned to Nisshin Kato KK, Japan has overcome these difficulties to produce powdered oyster and shrimp soups, each with a characteristic taste, which are easily used and economical. The method of the process comprises the following steps:

(1) Raw oyster or raw or dried shrimp is sprinkled with 2 to 3% of sodium chloride based on the weight thereof and stirred. The viscous materials or other impurities on the surface thereof are washed out with waste.

(2) Primary extracting step – The resultant washed oyster or shrimp is boiled

for a short time (10 to 15 minutes) in a boiling, dilute, aqueous sodium chloride solution (2 to 3%) which is adjusted to be slightly acidic (pH 5.5 to 6.0) by addition of a phosphate buffer solution.

(3) Secondary extracting step – The boiled oyster or shrimp is ground and a composition of enzymes, consisting mainly of protease, prepared by extraction from a cultured medium of a microorganism with subsequent purification (for example, bacterial protease and mold protease: these are the proteolytic enzyme preparations which are extracted with ethanol from the cultured wheat gluten media of *Bacillus subtilis* var. *biotecus* and *Aspergillus oryzae*, respectively, and they contain amylase, lipase and hemicellulase, etc. in addition to protease) together with a 2 to 3% solution of sodium chloride having the same composition as the primary extract are added to the ground oyster or shrimp. The mixture thus obtained is decomposed by the enzymes at pH 5.5 to 6.0 and at a temperature between 45° and 60°C, for 1 to 2 hours; the action of the enzymes is then stopped by heating at 90°C for 10 minutes.

(4) The primary and secondary extracts are mixed together, and auxiliary ingredients such as soluble starch, condiments and antioxidants are added to the mixed extract and the mixture is treated in a homogenizer, dried to less than 8% water content by a spray drying, drum drying, vacuum drying or freeze drying operation.

The step wherein raw oyster or shrimp is sprinkled with sodium chloride, stirred and washed with water, is necessary to remove impurities adhering to the raw materials and viscous matter on the surfaces of the raw material. The taste and flavor of the powdered extracts are impaired in the absence of this step. In the second step, the reasons for restricting the extraction period to shorter than 15 minutes, the pH value to 5.5 to 6.0 and the concentration of sodium chloride solution to 2 to 3% are based on the fact that these conditions are necessary to obtain the most taste and flavor possible, without, at the same time, losing the original flavor of the raw materials.

More than 80% of the nitrogenous compounds which are soluble in hot water, more than 90% of all the free amino acid and more than 70% of nucleic acid relating materials are dissolved in the extracting solution. The extraction ratio of solid extracts was increased less than 5% when the boiling was continued for 60 minutes, and negative factors such as degradation of taste and flavor significantly increased.

The adjustment of pH in the secondary extraction is very important. When a nitrogenous compound such as protein is decomposed with an enzyme, the reaction solution has a strong tendency to become neutral or alkaline together with the formation of basic amino acids, basic amines, and ammonia, etc. However, under conditions in which alkali protease becomes active, the enzyme causes a bitter taste in many cases and heavily deteriorates the delicacy of the flavor. Accordingly, the pH range of 5.5 to 6.0 is the optimum condition.

The significance of employing a 2 to 3% aqueous solution of sodium chloride lies in increasing the solubility of protein in the solution in the primary extraction stage and in adjusting the enzymatic decomposition of proteins together with pH in the secondary extraction. Experimental results, comparing the extracting ratios with a 3% sodium chloride solution extract of the process with simple hot distilled water extracts of oyster, show an increase of the ratios of the extracted nitrogenous compounds by about 30% (based on the amount of total nitrogen).

When the enzymatic decomposition is carried out without removing the main portions of flavor ingredients, nucleic acid materials and free amino acids by the primary extraction, some of the useful ingredients are changed to unusable materials. This changing of the nucleic acids taste materials is especially rapid. According to results obtained a longer period of time for the enzymatic decomposition than that specified tends to decrease the amount of free amino acid, such as arginine, lysine, threonine and cystine, etc. (due to the Maillard reaction), coloration to brown, generation of hydrogen sulfide, odor of ammonia and other gases, as well as the lowering of the nutritive value.

Example: Oyster Powder – 30 kg of raw oysters was stirred gently with 0.7 kg of sodium chloride and washed well with water. The oysters are put into 30 kg of a boiling solution of 11 g of disodium hydrogen phosphate, 73 g of potassium dihydrogen phosphate, and 1 kg of sodium chloride in water (pH 6.0), and boiled for 10 minutes. The oysters are taken from the broth, chopped by means of a chopper, ground by means of a grinder, mixed with a part of the broth (about 10 kg) and 25 g of bacterial protease, and allowed to stand for 90 minutes at 55°C.

The enzymatically decomposed mixture is mixed with the remainder of the broth (the first extract) and heated at 90°C for 10 minutes to stop the enzymatic action. The mixture is filtered, 10 kg of soluble starch are added along with 1.5 kg of sodium chloride, 1 kg of monosodium glutamate, 5 g of sodium succinate, 1 g of BHA (butyl hydroxyanisole), 1 g of BHT (dibutyl hydroxytoluene), and 10 ml of vinegar.

The whole is sterilized at 80°C for 10 minutes and then passed through a homogenizer. The mass is dried by spraying the homogenized mixture from a nozzle under a pump pressure of about 2,000 lb/sq in using heated air having a temperature of 145°C and a drying chamber temperature of 90°C to yield 13 kg of a powdery product.

Dry Flavoring from Mollusks

R.D. Gray; U.S. Patent 3,529,975; September 22, 1970; assigned to Gray Chemical, Inc. has produced a storage stable flavoring material to give the flavor of mollusks. Certain types of mollusks, while having edible flesh portions and readily available at or near the sea, are not currently commercially utilized because of their lack of consumer appeal.

Mussels and ocean quahogs are two examples of such mollusks. While they have a desirable flavor, their size, texture, color and the like lack consumer appeal. However, by this process the desirable flavor of such mollusks can be obtained and utilized without the disadvantages of undesirable characteristics.

The process briefly includes the steps of cooking the edible flesh portion only either prior to or after its removal from the shell. The cooked edible flesh sections of the mollusks which are the final raw material for the food flavoring are then preferably wet ground. This grinding step can be carried out using any conventional apparatus for comminuting of the materials. Two or more mills may be used, one for pregrinding and one for fine grinding as described.

The particle size of the suspended flesh solids in the slurry formed by this grinding operation should be fine enough to allow subsequent rapid water evaporation by whatever drying method is used to give a fine free flowing powder after drying which may or may not contain small amounts of water. The concentration of the slurry formed in the grinding step for the flesh portions is not unduly critical but is at least 5% solids and preferably higher to minimize as much as possible the extent of water evaporation required in the subsequent drying step.

The slurry at this point may be treated if desired with conventional flavor stabilizers as, for example, certain starch compounds to encapsulate the flavor material and to enhance the flow properties, although such addition is not necessary. This assures retention of the flavor if the slurry is not immediately subjected to drying and processing. No other additives are necessary to give the final, dried products.

The slurry of ground edible flesh is dried in any conventional manner to produce the flavoring materials of the process. Spray drying is preferred because of the speed, ease and lack of detrimental effect on the flavor, although other forms of drying can be used. For spray drying, the slurry particle size is preferably in the range of 10 to 100 microns. The desired food flavoring material is obtained directly from the drying step.

The food flavoring powder may be packaged in any suitable and well known manner, such as in ordinary screw cap jars, for home use, or in polyethylene bags for bulk shipment to commercial food processors. No unusual precautions need be taken in packaging, since the food powders so prepared are hot hydroscopic, and do not deteriorate in flavoring power or increase in bacteria count, even during long periods of unrefrigerated storage.

Bacteria counts for food derived from seafood are generally required to have a plate count of less than 50,000. A batch of food flavoring powder manufactured by the latter described method, after a period of unrefrigerated storage, was found to have a plate count of under 5,000, a most surprising and unexpected characteristic of the material and a great advantage for such a product. This flavoring material may be utilized in the manufacture and preparation of many types of food products for human consumption such as dips, sauces, soups, crackers, and the like.

For many products, addition of the powder in the proportion of about 0.1% by weight imparts a suitable flavor thereto. The flavor may be enhanced if desired by the addition of a flavor intensifier such as monosodium glutamate. The following examples illustrate some ways in which the flavoring material may be obtained.

Example 1: For commercial preparation of the common clam strips used for fried clams only 40 to 50% of the clam flesh is used. About 100 pounds of clam waste (flesh) from a commercial process were cooked in boiling water. The cooked, edible flesh portions of the clams milled in a whirling blade comminutor to yield flesh having a particle size below one-eighth inch average diameter. This milled material was then fed to another mill and further ground to give a particle size in the range of 5 to 100 microns.

The slurry obtained was about 10 to 15% by weight solids. This slurry was then spray dried giving a product having about 5% moisture and a low bacterial count. The product was stable for long storage times without refrigeration and its flavor did not change. It was found to impart a very excellent flavor to sauces and cheese dip and was entirely fit and well adapted to human consumption.

Example 2: About 80 pounds of previously shelled and frozen sea clams consisting solely of the flesh of the clams were cooked at 240°F for 2½ hours. Approximately 47 pounds of clams (flesh) and 30 pounds of broth were obtained. The flesh of the clams and broth were then milled together in the comminutor through a one-eighth inch screen. One-half pound of starch was then added to the slurry and it was milled 3 passes through a colloid mill.

Twenty gallons of slurry were obtained. The slurry was spray dried and gave a very fine, free flowing powder entirely fit and suitable for human consumption and capable of storage without refrigeration. The bacteria count was 4,000 and the coliform was L-10. This dry powder was found to impart a very excellent clam flavor to sauces and dips.

Example 3: 1,000 pounds of liquid waste was obtained from a commercial processor, which is the water slurry and steam condensate remaining after cooking the clams (flesh). This waste water was heated to 160°F for 1 hour. The resulting slurry was spray dried resulting in 100 pounds of finely divided powder. The resulting clam flavor was excellent having a plate count of 500.

Example 4: About 100 pounds of oysters were shelled, the shells discarded and the flesh cooked. The edible flesh of the oysters was then wet milled diluting the solids content during the process to 15% by weight to a particle size of 100 microns. The resulting spray dried product was found to be stable on storage and imparted a very excellent oyster flavor to sauces and dips.

CRUSTACEANS

SHRIMP

Separating Shrimp and Fish on Board Ship

Commercial shrimp boats catch shrimp by spreading a net and pulling it through the water in the vicinity of the bottom. Needless to say, various fish are usually also caught in the net. When the net is full, it is hauled back to the surface and swung aboard the boat. The catch is dumped at a sorting station on board the boat. The shrimp and fish are then separated from one another in a manual manner. The shrimp are placed in a shrimp storage facility, while the fish are either thrown back into the sea or placed in a separate fish storage facility in the event that it is desired to use the fish for commercial purposes.

This manual separating of the shrimp and fish is cumbersome and time-consuming. Also it usually requires the use of a larger number of people than is desired. Furthermore, when performed aboard a shrimp boat as the shrimp and fish are being caught, it is usually done in such haste that as many as 40% of the shrimp are lost through failure to separate them from the fish.

J.R. Lovett; U.S. Patent 3,750,233; August 7, 1973 has developed a more rapid and efficient way of separating shrimp from fish which can be used on a shrimp boat. After a net load of shrimp and fish is pulled aboard a shrimp boat the intermingled shrimp and fish are subjected to an electric shock for purposes of electrocuting same. The catch is then dumped into a flotation tank forming part of a sorting machine located on the shrimp boat. The lower portion of the flotation tank is filled with water and the shocked shrimp sink to the bottom of the tank, while the shocked fish float to the surface of the water therein.

In sinking to the bottom, the shrimp fall through an inclined screening tray and onto a conveyor which removes them from the tank. After the shrimp have had time to settle to the bottom, the lower portion of the screening tray is elevated for purposes of dumping the larger fish onto a conveyor which carries them to a fish meal machine. A continuous flow of water is supplied to the tank and this serves to flow the smaller fish out of an outlet at the rear of the tank, from whence they are dumped back into the sea or into the fish meal machine. The apparatus is generally illustrated in Figure 5.1

FIGURE 5.1: SHRIMP AND FISH SORTER

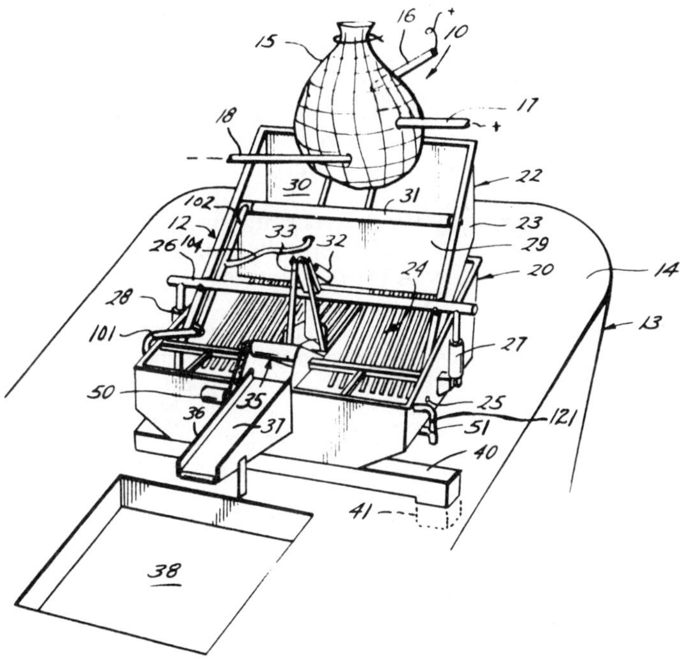

J.R. Lovett; U.S. Patent 3,750,233; August 7, 1973

The separating apparatus includes an electrical shocking apparatus, part of which is indicated at **10**, and a sorting machine which is generally indicated at **12**. The sorting machine is mounted on the rear portion of a shrimp boat **13** with parts of the machine being located below the deck **14** of the shrimp boat. A typical net load of intermingled shrimp and fish are indicated at **15**. For simplicity of illustration, the hoisting mechanism and lines and tackle for pulling the net out of the sea and swinging it aboard the boat have been omitted.

The electrical shocking apparatus includes metal probes or electrodes which are inserted into the net load of shrimp and fish after it is swung aboard the boat. Probes **16** and **17** are connected to positive terminals of a high voltage source, while the probe **18** is connected to the negative terminal of such source. The electrical shock applied to the shrimp and fish is preferably of sufficient magnitude to electrocute substantially all of such shrimp and fish. After the shrimp and fish are thoroughly shocked, the high voltage is turned off and the probes are removed from the net. The sorting machine includes a flotation tank **20** for holding a supply of water and for receiving the intermingled bunch of shrimp and fish which are dumped therein from the net following the electrical shocking operation. Within the flotation tank is a tiltable fish-removing mechanism **22**, which includes a three-sided vertically extending frame structure **23** having a screening tray **24** attached to the bottom thereof. The fish-removing mechanism is pivotally coupled to the flotation tank at point **25**. A controllable actuator mechanism for raising and lowering the rear portion of the fish-removing mechanism includes a lifting bar **26** and a pair of hydraulic rams **27** and **28** which enable the raising and lowering of the lifting bar and, hence, the rear portion of the fish-removing mechanism.

A transverse gate **29** may be opened and closed and, when closed, provides a shrimp and fish-receiving compartment **30** into which the shrimp and fish in the net are dumped. The lower portion of the gate **29** can be swung upward or downward by means of a hydraulic control cylinder **32**.

A basic concept employed by the process is the fact that dead crustaceans, such as shrimp, will sink to the bottom, while dead fish will float to the surface of a body of water. Thus, when the fish-removing mechanism **22** is in its lowered position such that it is mostly inside of the water-filled flotation tank **20** and the electrocuted shrimp and fish are dumped into the receiving compartment **30**, the electrocuted shrimp fall through the screening tray **24** and into the lower region of the flotation tank **20**.

The shrimp are then removed from the lower portion of the flotation tank by means of an inclined conveyor **35** which dumps the shrimp onto a culling table **36**, which has a moving conveyor belt **37** for conveying the shrimp to a shrimp storage facility. A human operator is stationed alongside of the culling table for removing any remaining trash (small fish, sea creatures, plant life, etc.) which may have reached the conveyor.

The bigger fish are periodically removed from the flotation tank by raising the fish-removing mechanism to its elevated position. At this point, the screening tray is completely out of the water and the bigger fish slide down the screening tray and fall into a lateral conveyor mechanism **40**, which may simply dump the fish over the side of the boat or it may instead be used to convey the fish to a fish meal machine located on board the boat. In the drawing, the conveyor mechanism is shown as dumping the fish into a chute **41** which runs to a fish meal machine located below the deck. Details of the sorting machine are illustrated by further figures in the patent, and are described with reference to those figures.

Soaking Shrimp in a Phosphate Solution Before Peeling

P.W. Bynagte; U.S. Patent 3,705,040; December 5, 1972; assigned to Westgate-California Foods, Inc. has found that dipping fresh shrimp into a solution of water, sodium acid pyrophosphate or calcium acid pyrophosphate and sodium tripolyphosphate or sodium metaphosphate or sodium hexametaphosphate or sodium trimetaphosphate or sodium orthophosphate for a period of at least two minutes during which time the solution is agitated produces greatly improved results in subsequent peeling operations.

The solution reduces the strength of the underskin of the shrimp, allowing the crust and shell to be easily and quickly removed making the peeling easier. Thus not so many small pieces of shrimp meat are broken in the shell removal. Further it has been found that in this process the red colored portions of the shrimp between the shell and the shrimp meat stays on the meat better and longer. In the practice of this process, the shrimp are taken from the water and placed in a solution comprising one part of water, a second part of particular acid pyrophosphates and a third part of particular sodium phosphates. The particular acid phosphates include sodium acid pyrophosphate or calcium acid pyrophosphate and the particular sodium phosphates of the third part comprise any of sodium tripolyphosphate, sodium metaphosphate, sodium hexametaphosphate, sodium trimetaphosphate, and sodium orthophosphate.

The percentage of the parts of the solution range from 2% to 30% by weight of the second part, 2% to 15% by weight of the third part with the remainder comprising fresh water or salt water. The salt water may have a sodium chloride content of about 3% by weight of the water. The shrimp are soaked in the solution for at least two minutes.

Normally the shrimp are soaked in the solution from two to ten minutes depending upon the size of the particular shrimp. The larger the shrimp, the longer time the shrimp would be soaked in the solution. While the shrimp may be soaked for long periods of time in the solution without damaging the shrimp meat, there is normally no economic justification for leaving the shrimp in the solution for excessively long periods of time. The action of the solution on the skin of the shrimp is normally increased by agitating the solution during the soaking.

While there are many ways in which the solution may be agitated, it has been found particularly advantageous to bubble air through the solution. The shrimp are then removed from the solution and are cooked in a fluid such as water at a temperature near boiling for at least two minutes. The temperature of the water should be at least 200°F. The shrimp are then removed from the cooking solution, are cooled and then peeled. The shell or crust is removed by hand in the normal manner of peeling shrimp.

Example 1: 25 lb of shrimp were placed in an agitating solution, which solution comprised by weight, 96% fresh water, 2% sodium acid pyrophosphate and 2% sodium tripolyphosphate. The shrimp were left in the solution about 2 minutes and then removed. The shrimp were then cooked for about 2 minutes at a temperature of about 200°F. After cooling for 1 hour, 6½ pounds of whole shrimp were recovered with about 5% broken.

As a contrast with mere boiling, 25 lb from the same catch were boiled with no solution treatment for about 2 to 3 minutes at the same temperature as above. The shrimp were then cooled and peeled. 5.3 lb of shrimp were recovered and 25% of the shrimp were broken.

Example 2: 1,000 lb of shrimp were placed in a solution of 2% sodium acid pyrophosphate and 2% of sodium tripolyphosphate with the remainder comprising fresh water. The shrimp were left in the solution about 8 minutes. The shrimp were then removed from the solution and were cooked for a period of about 3 minutes at a temperature of 212°F. The shrimp were then removed from the cooking solution and cooled for 1½ hours. The shrimp were peeled in 1 hour and a 26% yield by weight of whole shrimp meat was recovered.

By contrast, 1,000 lb of shrimp were cooked for about 5 minutes. The shrimp were cooled for 1½ hours and were then peeled. A 23.5% by weight yield of shrimp meat was recovered; 20% of the shrimp were broken and it took 1½ hours to peel the shrimp.

Process and Apparatus for Shrimp Peeling

Shrimp are sometimes referred to as prawns and are a type of crayfish. Shrimp is not only a savory and delicious seafood and in high demand but also is often used to decorate and thus enhance the value of seafood dishes. Thus, shelled whole shrimp meat which is unbroken and nonmutilated often has twice the market value of broken or mutilated shrimp meat. This originally led to a widespread employment of hand peeling of shrimp.

As labor costs have relatively increased, machine peeling of shrimp has become more and more desirable. However, mechanical shrimp peelers have often produced too great an amount of broken and mutilated shrimp meat so that hand-peeled shrimp are still sold on the market.

J.C. Willis and O.B. Sundberg; U.S. Patent 3,466,699; September 16, 1969; assigned to Alaska Peelers, Inc. have produced a shrimp peeling or shrimp shelling process and mechanical means for carrying out the process which will provide shelled shrimp of a quality comparable with hand-peeled shrimp, and characterized by a relative absence of broken and mutilated shelled shrimp. The process will be more clearly understood by reference to Figures 5.2a and 5.2b.

FIGURE 5.2: APPARATUS FOR SHELLING COOKED SHRIMP

a.

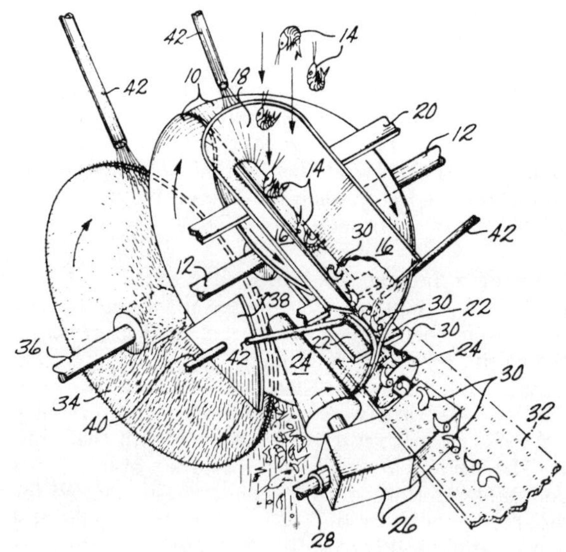

(continued)

FIGURE 5.2: (continued)

b.

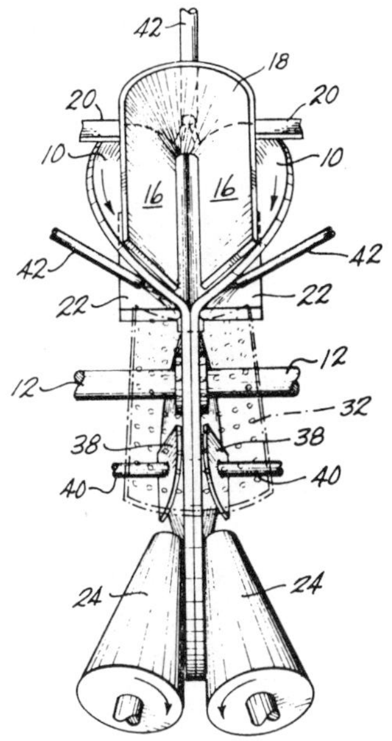

Source: J.C. Willis and O.B. Sundberg; U.S. Patent 3,466,699; September 16, 1969

Figure 5.2a is a perspective view, with parts broken away, showing the essential operative parts of a device for carrying out the process. Figure 5.2b is an elevational view of a portion of the structure of Figure 5.2a. In the apparatus shown in the figures, two discs **10** are employed. These discs are formed of natural or synthetic rubber or a combination of both to provide for desired flexibility and desired frictional contact with the shrimp.

These discs are mounted for rotary movement on a shaft **12**. Shrimp **14** have been previously cooked, as by boiling, and it is preferable that the process operate upon them soon after they have been boiled. The apparatus has had great success in the shelling of pink shrimp, which are generally characterized as being small and do not exceed a length of approximately 4" from head tip to tail when measured stretched out.

An infeed hopper may comprise side plates **16** and an end plate **18**. The discs are disposed substantially vertically and the infeed hopper is disposed at an upper elevation on the discs and may be disposed to function as a spreader of the discs. By any suitable means, the shrimp with their shells are fed consecutively, one after the other, into the infeed hopper and thus between the two rotary driven discs. The side plates of the infeed hopper extend only partially down in a radial direction between the discs **10**. The infeed hopper is supported in the general frame of the machine (not shown) by a hopper support **20**. Stationary bars **22** are supported by the general frame of the machine and extend from above the centers of discs radially outwardly and downwardly so that the lineal travel of a disc surface increases as the shrimp move down the throat under the influence of the forces of gravity and centrifugal forces of the rotating discs. The stationary bars are provided with an angle of inclination to the horizontal of approximately 20°.

A greater or lesser angle may be used and the greater the angle, the lesser the time the shrimp will remain in the shrimp shelling throat. This provides a desired combination of tumbling and turning the shrimp for frictional engagement with the surfaces of the discs. Such action and mode of operation are further enhanced in shrimp shelling action by the extended planar surfaces of the discs.

The stationary bars are spaced far enough apart so that the discs can rotate between them and at the same time the bars are mounted to urge the discs into contact with each other. Preferably the bars are mounted so that at least one is movable toward and away from the other so as to provide for an adjustable shrimp shelling throat. When the surfaces of the discs are contacting the shrimp to shell them, it is preferred that the material forming the discs be stretched, as this appears to enhance the shelling characteristics of the device. This may be accomplished by driven cones **24** in frictional-driving position against the outside faces of the discs.

The speed of rotation of the driven cones is such as to impart a rotary speed of the discs in the order of 50 to 150 rpm. The driven shaft is preferably connected to a variable speed motor which permits the use of selected speeds of the discs in accordance with the resulting shelling. In other words, if the shelling is not as complete as desired, the speed of rotation may be decreased providing for a longer time of the shrimp in the shelling throat.

As the shrimp are being shelled, shell portions are engaged in the nip in the shrimp shelling throat and are carried by the inner contacting surfaces of the discs. The shrimp tumble and turn in their travel down the shrimp shelling throat, until they are shelled. The shelled shrimp **30** leave the shrimp shelling throat and are delivered onto a discharge shrimp meat conveyor **32** which is shown in phantom in Figure 5.2a of the drawings.

To remove shrimp shell fragments from the inner contacting surfaces of discs **10**, a rotary driven brush **34** is mounted on a driven shaft **36**, and rotates counter to the direction of the discs. To remove shrimp shells at location desired, the various parts of the apparatus are continuously washed down with running water. This may be accomplished by the use of supply water pipes **42** directed to suitable locations.

Also, with such water being available, it is desirable that the discharge conveyor be perforated so that the water on the shrimp will be removed as they are delivered from this device to suitable packing or storing containers. Thereafter, the shrimp are subjected to the usual further processing of all shrimp, whether machine or hand-peeled.

In a separate patent, *J.C. Willis and O.B. Sundberg; U.S. Patent 3,465,383; September 9, 1969; assigned to Alaska Peelers, Inc.* describe suitable mounting assemblies for their mechanical shrimp shellers and apparatus for feeding shrimp to such shelling apparatus so that during travel, the shrimp are oriented in a given direction, with the shrimp traveling on their backsides and with their heads trailing. A way of conveying cooked shrimp to the shrimp peeler includes the use of a relatively wide conveyor belt which may be endless and formed of a wire screen mesh so that shrimp will be supported thereby and moisture will drain through the belt. Also, the belt is of a width to feed the various units of the shrimp peeler such as a commercial unit having eight sets or pairs of shrimp peeling discs. Shrimp will be initially spread relatively uniformly crosswise of the belt and then a plurality of delivery chutes are provided with funnel-shaped side guides so that shrimp on conveyor belt will be divided into the same number as there are shrimp peeling discs.

Air under pressure is blown on each chute and, due to the contour of the shrimp and the fact that it is descending by gravity down the chute, substantially all the shrimp descend downward on their backs, with their heads trailing. By having the shrimp oriented to a particular manner of travel down the chutes, substantially all of the shrimp are subjected to the same treatment by the shrimp peeling discs. The feeding of shrimp in a uniform manner to the shrimp peeling discs is an important feature of this process as it provides for uniform peeling of the shrimp and the elimination of most hand labor on shrimp.

When the shrimp with their heads trailing are fed to the throat between the two shrimp peeling discs and the discs are urged together to provide a nip at the bottom of the throat, the head portions, including the whiskers, of the shrimp are firmly, mechanically engaged but other shrimp portions are not. This action of engaging and pulling on the head portions of shrimp generally causes the shrimp to turn and to subject the tail portions of the shrimp to the mechanical gripping and pulling action of the discs and at the nip thereof.

The combined action of the discs at the head and tail portions appears to eject the shrimp meat from the shell portions and where water is present this shrimp meat is slippery and tends to float on the water and from the nip between the discs. The result is that shrimp is consistently recovered from the shell, is not damaged or bruised by the discs and is readily washed out of the throat between the discs and thus readily discharged by way of a discharge chute.

An improvement on the described process was made by *J.C. Willis; U.S. Patent 3,576,047; April 27, 1971; assigned to Alaska Peelers, Inc.* in which a shaker table is added to the shrimp shelling apparatus to maintain the heads-trailing orientation of the shrimp during their travel down the air chute and to space the shrimp so they travel in sequence one after the other.

Peeling Shrimp by Counterrotating Knives

E.H. Matthiesen; U.S. Patent 3,634,909; January 18, 1972 has developed a machine for peeling cooked shrimp in which shrimp are advanced in a channel formed by two counterrotating rollers. A shrimp is aligned so that its ventral side faces downwards in the channel, and while the shrimp is pressed down into the channel its dorsal shell is loosened from the meat by means of knives rotating with the rollers.

Due to the aligning or orienting of the shrimps in the dorsal-to-ventral direction and the subsequent cutting action of the rotating knife, the dorsal shell of the shrimps, i.e., the front part of the shell behind the so-called head, which actually also include the body portion of the shrimp with the viscera, is forcibly cut free of the meat of the shrimp and thereafter the head and the tail shield can easily be removed from the meat, e.g., by pulling forces exerted in opposite longitudinal directions upon these parts of the shrimp.

Dependent upon the circumstances, the dorsal shell may either be pulled off in the same operation as the head, or the head may be removed separately and the loosened dorsal shell may subsequently be removed simultaneously with the tail shell by means of a device operating with oscillating rollers. In all cases the complete separation of the meat from the shell may be effected faster than by previously employed machines and methods, so that the harmful effects on the quality and appearance of the shrimp meat which are due to a relatively slow and long-lasting processing are avoided.

The separate removal of the head has the further effect that the undesired parts of the shrimps, in particular eyes and stomachs, which have a soft consistency similar to that of the shrimp meat, and which therefore follow the meat through most peeling machines, may now be removed in one operation from the shrimp meat, so that the previously necessary and very time-consuming manual after-peeling or cleaning of the shrimp meat may be dispensed with.

The means for orienting or aligning the shrimp in the dorsal-to-ventral direction comprises at least one nozzle located at that roller section which, during the rotation, moves downwards into the feeder channel, and is arranged to discharge a jet of water substantially tangential in relation to the roller section.

It has been found that in this way a more reliable turning of those shrimps may be achieved, which lie incorrectly in the feeder channel, i.e., with the dorsal side downwards, than with a turning means in the form of a stationary guide rail having a helical edge which, by mechanical action on the shrimp presses it downwards to effect the turning. It is a particular advantage that the action of the jet of water is independent of the size of the shrimp, so that a less accurate sorting or grading of the shrimp supplied to the machine does not influence its function.

FIGURE 5.3: SHRIMP PEELED BY COUNTERROTATING KNIVES

a.

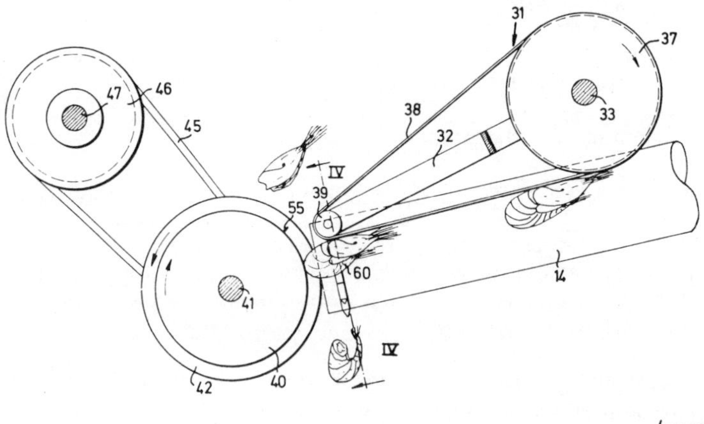

(continued)

FIGURE 5.3: (continued)

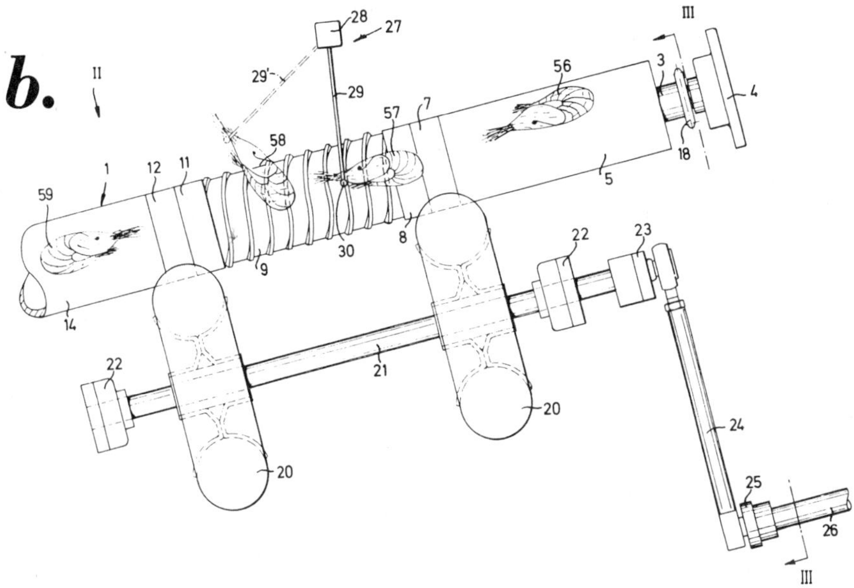

Source: E.H. Matthiesen; U.S. Patent 3,634,909; January 18, 1972

Three other nozzles for discharging water jets against the shrimp may be used, if desired, to insure that the correct alignment of the shrimp is achieved before they arrive at the knives. Figures 5.3a and 5.3b, showing a vertical section of one form of the apparatus, provide a better understanding of the process. Detailed drawings and descriptions of two forms of the apparatus are given in the patent.

Shrimp Deveiner, Sheller and Grader

John M. Jones, Sr. devised a process for deveining shrimp which is described in U.S. Patent 2,684,500. John M. Jones, Jr. developed a more sophisticated apparatus for processing shrimp which was patented on September 15, 1970 as U.S. Patent 3,528,125. The apparatus described in that patent receives the shrimp in individual troughs, with the shrimp heads projecting beyond the trough ends and conveys them to a head-removing knife and then to a transfer mechanism.

The troughs travel longitudinally of the machine at one level, and are transferred after head removal to a lower level where they move along a reverse direction, during which the shell is slit, the vein removed, and the shell separated from the meat. The meat, during and after the shell-separating operation, is transferred to still lower level for travel along a third path in the direction of the first, during which the meat is graded for size and deposited according to size in chutes for delivery from the machine. The machine is quite practical, but its speed is not as great as might be desired. In addition, some of the specific operating mechanisms were not as efficient as could be hoped for.

Improvements in the apparatus were provided by *J.M. Jones, Jr.; U.S. Patent 3,698,038; October 17, 1972.* They include the following:

(a) An increase in speed of carrying out the process;

(b) The provision of a transversely moving shrimp feed to the head-removing station, eliminating the orienting disk formerly required to arrange the shellfish bodies longitudinally of the path of travel after head removal;

(c) Improved troughs for the shellfish to transport them to and beyond the head-removing station;

(d) More positive means for moving the shellfish from the trough to the transfer mechanism; and

(e) Improved means for removing the slit shells from the meat during transfer of the meat from the slitting and vein-removing level of the operation to the size-grading level.

Detailed drawings, accompanied by exact descriptions of each step of the process for feeding the shrimp to the apparatus, deheading and deveining them, slitting and removing their shells and grading them by size, are included in the patent.

A somewhat less complicated apparatus, designed for use on board a fishing trawler, is described by *J.M. Jones, Jr.; U.S. Patent 3,703,746; November 28, 1972.* Figure 5.4 is a view in side elevation of the apparatus, portions thereof being cut away to reveal the basic conveyor and related deveining and deshelling components.

Referring to Figure 5.4, the device **100** comprises essentially three levels **200**, **300** and **400**, the mechanisms of each level being generally represented by the series indicated. In practice, the shellfish are individually dropped on to the delivery plate **132** (head to left), and transported clockwise via the troughs **212**, whereupon the respective shellfish are positioned headfirst by the action of the sprays **220** against the horizontally graduated slide plate **214**.

The slide plate is disposed at an angle and has shellfish-head-engaging grooves **216**, adapted to effect correct transverse positioning of shellfish of varying size with respect to the deheader **230**. The sprays, being directed to the base of the respective troughs, are discontinued at a position which is substantially adjacent the axis of the deheader. In the vicinity of the delivery chute **240**, the action of the deheaded shell fish in transfer is augmented by the water spray **250**, directed adjustable angularly downwardly with respect to the chute.

The level **300** includes the turntable housing **310** having the turntable **312** disposed centrally thereof. The turntable is preferably radially grooved to provide a friction surface for the shellfish. An arcuately coaxial spray element **314** comprises in the right-hand-most portion at least five sprays which are adapted to jet against the circular walls of the housing, keeping the shellfish on their right sides and preventing the backs thereof from sliding toward the center of the turntable or dropping on their left sides.

FIGURE 5.4: SHRIMP PROCESSOR FOR USE ON TRAWLER

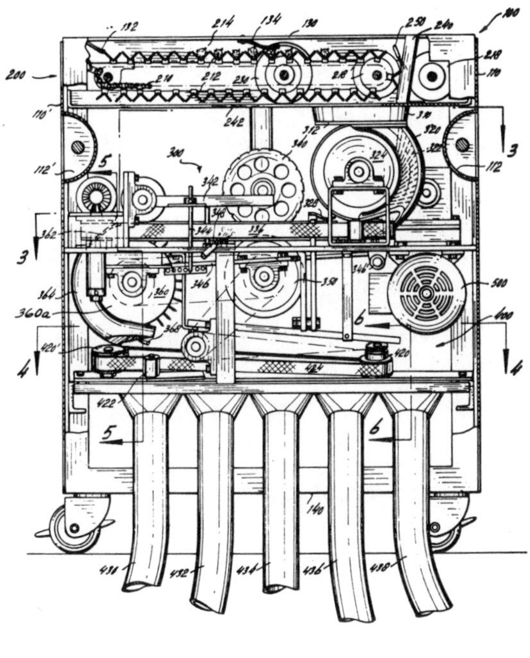

Source: J.M. Jones, Jr.; U.S. Patent 3,703,746; November 28, 1972

Moving by the turntable in a counterclockwise direction, as shown, the shrimp are engaged by other plural sprays **316** which are adapted to force the shrimp to stand on their feet and to enter the feeding chute **320** headless end first, whereby the spring elements **322** engage the back of the shrimp, forcing the legs thereof into the v-shaped cavity of the transfer wheel **324.**

Thus the respective shellfish are urged against the transfer wheel feet in, back out, to be received by the conveyor **336** feet up, back down, and secured therein for compression conveyance beyond that point. Then the shellfish are depressed downwardly thereafter by the compressor spring **328,** prior to further adjustable compression by the wheel **340.** As the wheel rides up over a large shellfish, the deveiner wheel **350,** moves upward proportionately.

Following the cleaning station, the shrimp are transported to the transfer station generally indicated at **360** including tined wheel **360a** and deflector **362.** This tined wheel is adapted for tine engagement with the shrimp to pull them from the abraded compression action of the conveyor belts **336,** whereby the shells are removed from the body and the meat transferred arcuately downwardly through the chute **364** to be forcibly ejected by the fixed plow prong **368** and thereafter to be engaged by the grader conveyor **400.** The respective shrimp are deposited size for size in the grading chutes **430, 432, 434, 436** and

438 respectively for freezing and storage thereafter. The chutes are foraminous at **440** to relieve water spray residue.

Improving the Color of Raw Shrimp with Acid

Shrimp may be classified according to color. Generally speaking, various colors of shrimp are found in nature: pink, brown, and white. The pink shrimp are found in deep water; the brown and white shrimp are found in shallow water. All types of shrimp have in their shells, epidermis and hypodermis the same pigmentary carotenoid substance known as astaxanthin.

This astaxanthin is developed by natural process to the best degree in the pink shrimp. Brown shrimp develop this astaxanthin better than the white shrimp, but not as well as the pink shrimp. White shrimp have developed this substance to the minimum degree, and not as well as the brown and pink shrimp. Pink shrimp have the greatest consumer appeal. Pink shrimp therefore are the most desired of the three colors.

J.V. Vilutis; U.S. Patent 3,365,302; January 23, 1968; assigned to Consolidated Foods Corporation found that brown and white shrimp could be changed to pink shrimp, the appearance of pink shrimp could be improved, the shrimp could be sterilized by killing the virus and bacteria on its surface, and spoilage could be reduced by treating shrimp with an acid solution for a short period of time. During this period, a significant change occurs. The brown and white shrimp develop a pink appearance, while the pink shrimp take on a fuller color.

The acid may be employed in a pH range from 0.1 to 3.0. The acid treatment may be at ambient temperatures, although for some purposes lower temperatures may be preferred. Following the acid treatment, the shrimp may be washed to remove any acid residue. Alternatively, and preferably, the shrimp may be treated with a base such as an alkali, for example a solution of sodium carbonate or sodium bicarbonate, in order to neutralize traces of acid, desirably to pH of 6.8 to 6.9, which is the natural pH of shrimp.

The inorganic mineral acids which have been found to bring out the natural color of shrimp include hydrochloric, sulfuric and phosphoric acid and special preference is given to hydrochloric acid. Ascorbic acid admixed with hydrochloric acid is especially effective in the process. The ascorbic acid acts as a reaction stimulator and a stabilizer of the pigment. Generally speaking, the inorganic acid may be present in the solution from about 0.1% to 3.0%, preferably from about 0.2% to about 0.5% by weight, and the ascorbic acid admixture therewith may be present from about 0.005% to about 2.5% by weight, preferably about 0.01% by weight.

A mixture of hydrochloric and ascorbic acid in a ratio of 10:1 is preferred. A solution of 0.1 N hydrochloric acid and 0.01 N ascorbic acid has been found satisfactory. For most purposes, the acid solution is maintained at a pH from about 0.7 to about 3.0, and most desirably from about 1.0 to about 1.5.

The process is conducted at temperatures preferably in the range from ambient temperature and lower, and to as low as freezing at about 32°F, and most desirably from about 35°F to about 40°F. The time required for treatment is from about one second to about sixty seconds, and usually about fifteen to thirty seconds. In most cases it will be found desirable to keep the time of treatment to as short a period as possible, in order to prevent changes in the quality or texture of the product. The shrimp to be treated by the process may include any of the three types of shrimp: pink, brown, or white. The shrimp is first shelled and deveined. The method is applicable to raw, uncooked shrimp. The color is produced by merely contacting the shrimp with an acid solution.

Increasing Color of Shrimp Using Canthaxanthin

It is well-known that shrimp die soon after being removed from their natural habitat and that deterioration of the shrimp sets in immediately and proceeds at a rapid rate. This deterioration not only takes the form of spoilage but also of discoloration. This discoloration normally occurs as unsightly brown blotches on the outer surfaces of the shrimp.

Simultaneously therewith there also occurs a deterioration in the natural pink color of the shrimp, i.e., they become dull in appearance. The cause of this deterioration in color is not precisely known; however, it is believed that it is due to oxidative reactions and the deleterious effect of various bacteria normally found in the natural coating of shrimp and on their shells.

Moreover, shrimp have different natural colorings, dependent upon the whereabouts of their natural habitat. Shrimp are generally classified as white, brown and pink. The pink shrimp are generally found in deep water whereas the brown and white shrimp are found in more shallow waters. The same pigmentary carotenoid substance controls the coloring of all types of shrimp. The carotenoid is developed to its fullest degree by natural processes in pink shrimp; to a lesser degree in brown shrimp and to a minimum degree in white shrimp.

By present standards, in the United States, pink shrimp have the greatest consumer appeal. Notwithstanding their original color, however, it is desirable that the shrimp have a bright appearance and not display the dull appearance commonly associated with deteriorated shrimp.

Although other methods have previously been suggested for providing shrimp and other crustaceans with improved color characterizations, they all suffer from various disadvantages.

R. Blomstrom; U.S. Patent 3,846,566; November 5, 1974 has therefore developed a method for the enhancement of the color of shrimp and other crustaceans which is simple, effective and does not require the introduction into the shrimp of foreign substances. This method comprises treating seafood with a solution of canthaxanthin or apocarotenal at an elevated temperature.

This treatment operates to enhance the intensity of the natural color of seafood, particularly, crustaceans, e.g., shrimp, king-crab meat, langostinos, etc. Canthaxanthin operates to enhance the pink coloring of the seafood whereas the apocarotenal enhances the brownish coloration of the brown crustaceans.

It is well-known that most seafoods, particularly crustaceans, e.g., shrimp, depend for their color upon the presence therein of certain carotenoid substances. Canthaxanthin and apocarotenal being carotenoids and closely related to the color-producing carotenoids present in seafood, particularly shrimp, apparently activate the natural carotenoids and stabilize them against deterioration and subsequent discoloration. In addition, the canthaxanthin and apocarotenal impart their own characteristic colors to the treated material.

The treatment may be carried out in any convenient manner, i.e., by dipping, brushing, etc. It is preferred, however, to spray the shrimp with a solution of the material. It has been found that the pick-up of carotenoid may be more closely controlled where the shrimp are sprayed. Preferably, the shrimp are spaced on an oscillating or vibrating conveyor belt or screen which passes under a series of spray nozzles. The shrimp are sprayed with a fine fog or mist of the solution and shaken on the vibrating belt until the sprayed solution has dried thereon. The colored shrimp are then removed and packaged for consumption or conveyed for further treatment.

In order to increase the solubility of the carotenoids in water and render the application of the aqueous solutions to the shrimp more efficient, it is preferred to include therein varying amounts of a volatile organic liquid which is miscible with water and is a solvent for the particular carotenoid. The solvent operates to increase the solubility of the carotenoid in water thereby enabling the introduction of a greater amount of the carotenoid into the shrimp in a shorter period of time.

Any volatile, organic liquid which is miscible with water and is a solvent for the carotenoid may be employed. The preferred solvent, however, is ethanol due to its relative low cost and availability. Generally, the aqueous treating solution should contain from about 5 to about 20%, by weight, of carotenoid. When employed, the content of volatile organic solvent, may range from about 15 to about 50% by weight.

Generally, the shrimp should be exposed to the treating solution for a time sufficient to ensure a pick-up of from about 0.001 to about 0.003 parts of carotenoid per part of shrimp. The shrimp are then dried to remove water and volatile organic solvent. The carotenoid penetrates into and is firmly bound up in the fleshy or fatty portions of the shrimp.

In order to ensure a complete penetration of the carotenoid into the shrimp and the enhancement of the natural color thereof, it is necessary to conduct the treatment at an elevated temperature, i.e., from about 55° to about 65°C. This is generally accomplished by heating both the shrimp and the treating solution.

Example 1: A quantity of peeled raw shrimp were heated to 60°C, spread on an oscillating conveyor screen, passed beneath a series of spray nozzles and sprayed with a fine mist of an aqueous solution containing 10%, by weight, of canthaxanthin and 20%, by weight, of ethanol, which has been previously heated to 60°C. The treatment was conducted in such a manner that the shrimp were in contact with the sprayed mist for approximately one-half minute.

The conveyor screen was continuously oscillated thereafter until the shrimp had completely dried. They were found to have an enhanced natural pink color and remained color stable for several days thereafter.

Example 2: Freshly peeled raw shrimps were placed on a conveyor belt under heat transfer lamps long enough to obtain a surface temperature of 50° to 80°C. A canthaxanthin solution identical to that of Example 1 was sprayed onto the shrimp after the latter were placed on a vibrating screen until all were completely wetted. The shrimps were conveyed under the heat transfer lamps again until dry. The shrimps were frozen and stored for several days during which no change in the pink coloration of the shrimp was noticed.

Shrimp Cooking Process

Raw shrimp may be cooked under a variety of conditions. In general, it is conventional to cook shrimp in water or in brine solution at a temperature of from about 170° to about 212°F for periods of the order of 5 to 20 minutes. When cooking in an aqueous solution, a substantial weight loss is noted. This weight loss generally runs from about 30 to 40% so that 100 pounds of shrimp in the cooking process results in only 60 to 70 pounds of shrimp in the final product. Since shrimp is a quite expensive food commodity, this loss of weight in the final product increases its price per pound to the ultimate consumer.

P.L. Veltman; U.S. Patent 3,501,317; March 17, 1970; assigned to W.R. Grace & Co. describes a process for cooking shrimp which results in a minimum weight loss, is simple and inexpensive and can be accomplished in a minimum amount of time. It comprises heating the shrimp with steam under pressure at a temperature of from about 220°F to about 260°F for a period of from about 1 to about 12 minutes and thereafter cooling the hot shrimp under pressure with water at a temperature below about 125°F.

Preferably shrimp are cooked with saturated steam at a temperature of from about 225°F to about 260°F and a pressure of from about 4 to about 21 psig for a period of from about 3 to about 6 minutes and are then cooled by water at a temperature of from about 32°F to about 50°F. The preferred cooling medium is ice water but water at a temperature of about 100° or even 125°F, can be employed.

Generally, saturated steam is employed for the cooking process and typical temperature-pressure relationships are as follows: 227.1°F at 5 psig, 239.4°F at 10 psig, 249.8°F at 15 psig and 258.8°F at 20 psig. Some water is always present in the pressure vessel due to condensation of the steam on the cooler shrimp.

In some instances, it is desirable to have water in the pressure cooker. The fastest cook, however, is obtained with the shrimp above the water level. Condensate from the cooking of the shrimp may be cooled and utilized as the cooling liquid for subsequent batches of shrimp. Salts and other materials may be added to it.

The raw shrimp employed may be peeled and deveined or they may be cooked directly without peeling. In general, a higher percentage yield will be obtained when utilizing unpeeled shrimp but such shrimp generally requires a slightly longer cooking time. The cooking time depends on the size of the shrimp and temperature employed and varies from 1 to about 12 minutes. With shrimp of about 20 to 30 count size, a cooking time of 3 to 6 minutes is generally employed with temperature of 240° to 250°F. Obviously, with a lower temperature a slightly longer cooking time may be required. Larger shrimp may require a slightly longer cooking time and smaller shrimp a shorter cooking time.

The process is generally carried out with unfrozen raw shrimp. If frozen are employed, they should be either thawed before use or an additional period added to the cooking time so that they will thaw and be cooked in the process.

In cooling the cooked shrimp, it is essential that they be cooled under pressure to a temperature below 212°F prior to the opening of the cooker. In general, ice water, or at least water below 125°F will be used for the cooling process. To prevent the condensation of the steam (under pressure) in the cooker when the cold water is admitted, air pressure, or other gas pressure, is applied to the cooker.

A convenient method of cooling is to build up about 5 pounds air pressure while turning off the steam and then constantly supplying air to maintain the pressure as the cold water is admitted to the hot cooker. Simple pressure regulators are used to accomplish this purpose. After cooling, the air is merely vented through a top drain.

In the following example a 3-gallon pressure cooker is modified as follows. A petcock is attached to the central portion of the cover with a tube running to a sink. A ½"-steam line is attached to the bottom of the pressure cooker to act as a source of supply of saturated steam at a preregulated pressure. The cooker contains a 1"-drain (with valve) at both the top and bottom portions of the cooker.

A ½"-air line (with valve) is attached to the top portion of the cooker so that air at 20 lb psig is supplied to the cooker. The cooker also contains a safety valve set at 23 to 25 psig. A thermocouple is placed in the cooker so as to read the temperature (°F) in the space above the top of the basket of shrimp when it is positioned in the cooker.

The cooker also contains a second small pointed thermocouple which is inserted into a shrimp in the middle of the basket so as to read the shrimp temperature. Tap water at 55° to 65°F or ice water is supplied to the cooker at 25 to 30 psig by means of a ½"-water line. The cooker is also equipped with a pressure gauge and a sight gauge so that the pressure and the level of water in the cooker can be easily read.

Example: Two gallons of 25 to 30 count raw shrimp are peeled and deveined. A 1-gallon basket is filled with shrimp (after weighing the shrimp) and placed in the above-described 3-gallon cooker which has been preheated by bleeding steam into the open cooker for about 5 minutes. A metal plate is placed on top of the shrimp to hold them in the basket.

The basket takes up the bottom half of the cooker due to the extra space around the basket. The cover is closed and all valves are turned off with the exception of the top petcock and top drain. The steam valve is then turned on supplying steam at about 15 psig and approximately 250°F. As soon as the steam is turned on it flushes the air from the cooker and the top drain is closed.

The thermocouple shows a rise in temperature from room temperature to 250°F with a corresponding rise in pressure (15 psig) in approximately 1 minute. The shrimp are cooked at this temperature for 4 minutes. The thermocouple inserted into the shrimp shows a slower rise in temperature to the cooking temperature.

The shrimp temperature during most of the cooking is actually in the range of 230° to 240°F with the higher temperature at the end of the cooking period. After cooking 5 minutes the steam is shut off while simultaneously opening the air valve to build up the pressure to 20 psig. The air line is left open to maintain 20 psig in the cooker as the water valve is opened to admit water for cooling the shrimp.

When the water has filled over half of the cooker and covers the shrimp, the thermocouple in the shrimp indicates that they have been cooled to a temperature of below 100°F. The air supply is shut off and the pressure vented through the top drain. By the time complete venting has taken place, the pressure cooker has filled with water and overflows in the top drain. The water is turned off and the bottom drain opened. The cover of the cooker is removed and the shrimp taken from the cooker and weighed. The ratio of wet weight of the cooked shrimp to the wet weight of the uncooked shrimp is about 0.85 showing about 15% loss in weight from the cooking process.

Cooking Shelled Shrimp to Prevent Curling

J.A. Chauvin; U.S. Patent 3,581,652; June 1, 1971 describes a method and apparatus for cooking shelled shrimp one of the objects of which is to maintain the shrimp in a generally straight or slightly curled condition as opposed to the well-known curled condition which previously has resulted with conventional methods and apparatus.

The apparatus includes two overlying conveyor belts which receive shelled shrimp between them and move them through a steam zone for cooking. The upper conveyor belt is provided with weights which urge it downwardly to engage and press the shrimp against the underlying conveyor to insure that the shrimp will remain in substantially straight condition during the steaming operation. The steam for cooking is introduced in coils located in a tank underlying the lower conveyor belt.

The steam rises and passes through apertures in the lower conveyor belt to cook the shrimp without excessive removal of moisture and while maintaining the proper pigmentation of the shrimp. The conveyor belts travel in synchronism and preferably their speed is made variable depending on the temperature at the cooking zone. To this end, a suitable temperature sensing device is placed at the cooking zone to control a variable speed drive gear assembly associated with the drive mechanism of the conveyor belt.

When the shrimp are cooked and reach the discharge ends of the conveyor belts, they fall through a chute on to a second conveyor which carries the shrimp through a cooling zone. Upon emergence from the cooling zone, the shrimp fall through another chute which deposits them on another conveyor for grading or further processing.

Uncurling and Shelling Cooked Shrimp

H.G. Muller; U.S. Patent 3,600,744; August 24, 1971; assigned to N.V. Machinefabrick B & S Bedrijven v.d. Woerdt, Netherlands has developed a machine for shelling cooked shrimp operating in the following sequence. Cooked shrimp which are curled by the cooking are shelled by accurately orienting them and transferring them to a shelling station where the body portion of the shell is gripped and the tail is straightened out and the shrimp is beheaded. The body and tail portions of the shell are then separated and the shrimp is pushed part way out of the gripped body portion of the shell. The tail portion of the shell is then stripped off and the shrimp is then pushed all the way out of the body portion of the shell. The patent contains 12 drawings showing the apparatus and includes a detailed description of the steps of the process.

Removal of Nonedible Shrimp Parts

In the shrimp-processing industry, processors are confronted with the problem of egg and shell removal from shrimp. They are further confronted with the problem of removing heads and tails from shrimp. Female shrimp carry their eggs on their abdominal legs during the autumn and winter months, hatching these eggs in the spring. It is the autumn and winter months which are the harvesting time for shrimp and all female shrimp harvested during this period still have eggs attached to their body.

Removal of these eggs are necessary in order to market these shrimp in the United States since U.S. consumers will not purchase the shrimp with the eggs attached. Further, shrimp processors process shrimp both with and without their shell. Whenever shrimp are processed for marketing without their shell, the processors are confronted with the problem of peeling or removing the shell. In conjunction with the removal of the shells, shrimp processors also remove the heads and the tails.

Many problems have arisen in utilizing of the commonly used mechanical means for removal of the nonedible parts. These problems arise from the fact that shrimp vary in size depending upon their species and the geographical area where they are harvested, cost of mechanical means is high in comparison to the product yield, and rate of production is limited to the production rate of the mechanical means.

The object, then, of the process described by *O.T. Aepli and R.W. Schultz; U.S. Patent 3,622,347; November 23, 1971; assigned to BASF Wyandotte Corporation* is the removal of the inedible parts of shrimp by chemical means, thus reducing loss of shrimp and reducing processing costs. The process resides in the discovery that an aqueous solution containing hypochlorite ions or sodium hydroxide on contacting the fiber between the nonedible and the edible parts of shrimp breaks down the fiber, thereby enhancing the removal of these nonedible parts.

Three variables in the removal process, namely, period of immersion of shrimp in the solution, concentration of the hypochlorite ions or sodium hydroxide in the solution, and solution temperature, can be varied to any set of conditions in a particular operation. For example, high concentrations of active ingredients break down rapidly the fibers holding the nonedible parts to the body of the shrimp. However, this high-active concentrate also attacks the outer layers of the edible part of the shrimp, thereby decreasing the quality and the yield.

The result is the same with increased temperature. Low concentration and low temperature increase the time for sufficiently breaking down the aforementioned fiber, but the resulting product is of better quality, higher yield, and requires less washing time to rid the edible product of active ingredients. It has also been found that the addition of fatty acids of sodium hydroxide solutions enhances the removal of nonedible parts of shrimp.

The composition comprises an aqueous solution containing from about 1 to 10% by weight of a hypochlorite-ion-containing compound or from about 1 to 15% by weight of sodium hydroxide, water making up the balance of the solution. In a preferred embodiment, from about 0.1 to about 1.0% by weight of fatty acids defined by the formula $C_nH_{2n}O$ is included in the solution containing sodium hydroxide. The solution may also include small amounts of wetting agents and/or defoaming agents.

The hypochlorite-ion-containing compounds that may be used include alkali metal hypochlorites, such as, for example, sodium, potassium, lithium, and calcium hypochlorites, chlorinated triazines, such as, for example, potassium and sodium dichloro-S-triazine trione, 1,3,5-trichloro-2,4,6-trioxyhexahydro-S-triazine, [mono (trichloro)-tetra-(monopotassiumdichloro)]-penta-S-triazine trione, and chlorinated hydantoins, such as, for example, 1,3-dichloro-5,5-dimethylhydantoin, and the like.

The fatty acids that may be used have the general formula $C_nH_{2n}O_2$ wherein n is from 5 to 10 and even higher. These acids include, for example, capric, caprylic, caproic, valeric, 2-ethylheptanoic acids and the isomers thereof and the like.

If desired, a small quantity of any suitable wetting agent and/or defoamer may be added to the treating solution. Compounds suitable for use as wetting agents of defoamers include nonionic surfactants such as a polyoxyethylene-polyoxypropylene block polymer. Other suitable wetting agents or defoamers are the alkyl aryl polyethers. Polyethoxylated alcohols as well as amine polyglycol condensates are examples of still other nonionic surfactants that can be employed. The amount of nonionic surfactant actually added is that which is sufficient to give the desired wetting and/or defoaming action. It is generally in the range of 5 to 20% by weight based on the total additives or in the range of 0.01 to 0.04% by weight based on total solution.

Raw or cooked shrimp are treated by immersing the shrimp in an aqueous solution containing hypochlorite ions or sodium hydroxide under conditions of concentration, time, agitation, and temperature sufficient to partially disintegrate fiber between the nonedible and edible parts of the shrimp without any substantial damage to the edible parts of the shrimp; removing the shrimp from contact with the solution; and washing the shrimp to provide peeled edible shrimp products.

The concentration of the active ingredients, e.g., sodium hydroxide or hypochlorite ions, in the solution may be varied over a wide range. For example, the concentrate of the active ingredient may be as low as about 1% by weight or as high as about 15% by weight for sodium hydroxide and from about 1% by weight to about 10% by weight of hypochlorite ion, based on the total solution. When a fatty acid is used in combination with sodium hydroxide, the concentration of the fatty acids may be varied from about 0.1% by weight to about 1% by weight, based on the total solution.

The temperature employed in the process can be varied over a relatively wide range. Usually, the temperature ranges from about 75°F to about 160°F. The time of immersion of the shrimp products can also be varied over a wide range. For an egg removal process, the immersion should be in the range of from about 2 to 5 minutes. However, when the process is for peeling of shells or shells, heads, and tails of the shrimp, the period of immersion depends upon the condition of the shrimp.

For example, if the shrimp are fresh and have only been out of the water for a few minutes, the shell is extremely soft and may be removed in as short a time as 5 minutes. If the shrimp have been out of the water and stored for an extended period such as 1 to 6 months, an immersion period of as long as 1 hour may be required. Also, if the shrimp have been cooked prior to peeling, the shell is again tougher than raw shrimp and will require longer immersion times. However, for most situations the preferred immersion period ranges from about 3 to about 10 minutes.

Example: The following example demonstrates the procedure that was followed in removing shrimp shells, tails, and heads. An aqueous solution was prepared by adding 5 parts by weight of flake sodium hydroxide to 94.5 parts by weight of water. The solution was agitated for about 2 minutes and 0.5 part by weight of a mixture of fatty acids consisting of 24% pelargonic, 11% caprylic, 38% caproic and 3% valeric acid was added.

With continued agitation, the temperature of the solution was increased from ambient temperature to 120°F. To this solution was added 1 pound of pink shrimp (*Pandalus borealis*). Vigorous agitation was applied and the temperature was maintained at 120°F for about 5 minutes. At the end of this period, it was found that the shells, tails, and heads had become dislodged from the meat.

CRABS

Removing Meat from Crabs

J.M. Lapeyre; U.S. Patent 2,978,334; April 4, 1961; assigned to The Peelers Company has developed a method for extracting meat from crustaceans without damage to the meat, and allowing it to be recovered in the commercially valuable lump form. The process is described as applied to crabs although it will be understood that it is equally useful for extracting the meats from lobsters, crawfish, shrimp, etc. It consists essentially in solidifying the meat of the crab in its natural state, as by freezing, and thereafter exploding the crab so as to deliver the meat in the frozen condition undamaged from the shell. The following steps, in the sequence set out, will be found most convenient for carrying out the process.

(1) Cooking the crab.
(2) Solidifying the crab including the meat, as by freezing.
(3) Exploding the crab.
(4) Separating the recovered frozen meat from the shell and trash.

(1) This first step is carried out by parboiling or cooking the whole crab. This cooking of the crab serves to drive out a portion of the moisture contained in

the meat of the crab, thereby shrinking such meat within its particular cavity. Where it is desirable to extract the meat already cooked, this first step of cooking will be carried out but it is not absolutely necessary in order to effect separation.

(2) The second step includes solidifying the meat as by freezing the entire crab or that portion of the crab which contains the meat. This step solidifies the cooked or partially cooked meat into a hardened condition, rendering it substantially impervious to damage from mechanical shock or abrasion.

(3) The exploding step may be accomplished in two ways: (a) By simply subjecting the crustaceans to a certain negative pressure, such as a very high vacuum in which the crustacean would explode due to its interior pressure suddenly becoming greater than its exterior pressure. (b) By subjecting the crustaceans to high pressure, and thereafter releasing same after it has accumulated within the interior of the shells. The entire crab or that portion containing the desired meat is subjected to a gaseous pressure for a sufficient length of time to allow such pressure to enter and build up within the various cavities of the crab shell which contain the hardened edible meat which has been shrunken within its cavities due to the preliminary cooking operation.

Pressures will build up within these cavities since there are small cracks or openings through the shell walls into the cavities which will allow seepage of the gas under pressure. This pressurizing of the crab will conveniently be conducted in a closed chamber having controlled access to a source of air or other gas under pressure. After this pressure has been built up internally in the crab structure, the gaseous pressure is suddenly released on the exterior of the crab which will permit the crab shell to explode due to its cavities containing gas under pressure which will expand rapidly when the exterior pressures are removed suddenly. This may be accomplished by suddenly opening the closed chamber to atmosphere after a pressure above atmospheric pressure is built up internally in the crab.

(4) Once the shells of the crabs have been exploded and the meat in its still frozen condition has become released from the shell and the shell structure of the crab, separation of the meat and the shells or trash can be made in a number of ways, for instance by flotation in chilled brine; or the lightweight shells may be blown off or vacuumed off a conveyor which would receive both the heavy meat and the lightweight shells from the exploding apparatus. By the use of this process, crab meat is retained perfectly in its lump form and can be sold at premium prices, either frozen or canned. The patent contains drawings and full descriptions of apparatus suitable for carrying out the process.

In an improvement to his previously described process, *J.M. Lapeyre; U.S. Patent 3,812,271; May 21, 1974* subjects the crustaceans to centrifuging prior to, or during freezing. The crabs or lobsters, either cooked or raw, are thereafter frozen to rigidify the edible meat to protect it from damage during subsequent steps of the process. The crustacea may be subjected to the increase in G-force of a centrifuge to compact the meat at one end of the exoskeletal shell during the freezing process to define a positive chamber within the exoskeletal shell structure which may first be subjected to a vacuum to remove air and moisture followed by charging the chamber when the vacuum is destroyed in the presence of an explosive gas/air mixture, with the explosive mixture which is ignited as by passing the crustacea through an open flame which will ignite the gaseous mixture causing an explosion of the exoskeletal shell away from the edible meat at which time, the meat still being frozen and rigid will not be injured by flying shell segments. An operative explosive gaseous/air mixture is a mixture of natural gas and air.

After the explosion of the exoskeletal structure the meat and shell segments fall into a refrigerated brine tank where the meats float and the shells sink. The meats are then skimmed off the top of the brine solution.

Crab Meat Extraction Apparatus

The main problem in the machine-extraction of crab meat lies in the machine's inability to separate the meat cleanly from the shell fragments and, perhaps most importantly, the inability of the machine to remove the large lump sections of crab meat in unbroken form.

A relatively simple, compact machine for this purpose and a procedure which assures maximum extraction of the crab meat in forms calculated to have the highest market value has been developed by *W.L. Lockerby; U.S. Patent 3,597,792; August 10, 1971; assigned to Avico, Inc.* The machine uses centrifugal force, and is characterized by structures and procedures permitting operation at two different speeds to separately and successively extract the more valuable, relatively large lump meat sections in substantially unbroken condition; followed by extraction of the smaller, less valuable, flake meat sections. The patent contains a number of drawings of the apparatus and a detailed description of its operation.

Removal of Haemocyanin from Crustaceans

I. Osakabe; U.S. Patent 3,007,802; November 7, 1961; assigned to Nippon Kanikanzume Yushutsu Suisangyo Kumiai, Japan has developed a method of removing haemocyanin from the meat of such Crustacea as crabs, lobsters or the like. According to the process, a crab or a lobster that is to be processed for food and in which the body fluid (the blood) contains blue-colored hematin (haemocyanin), has the viscera removed and parts of its shell and the like and is washed, as the first step, in the usual way.

It is then cooked or heated under such conditions that both the body fluid and the haemocyanin which is the principal ingredient of the fluid are not coagulated by heat but the muscles are coagulated by heat. For example, a crab is cooked at a temperature from 40° to 70°C for a period of from several to over 30 minutes, whereby not only the myosin occupying the greater parts of the muscles but also the myogen which is relatively small in amount are perfectly denatured by thermal coagulation while the body fluid and the haemocyanin contained therein remain as a liquid.

The muscles and the shell are then in an easily separable state owing to the thermal shrinkage of the muscles, so that the meat may be easily taken out of the shell and tendon. Further, when it is cooked at a temperature of from 95° to 100°C for a period of from 20 to 60 seconds, the temperature thereof rises only to a lower temperature than the thermal coagulating point of haemocyanin owing to the shortness of the period, and so only the muscles are thermally coagulated and the meat may be readily taken out of the shell. The meat thus taken out is accompanied by the body fluid which remains liquid and this fluid may be easily removed by washing.

It has been experimentally clarified that there is a difference of about 20°C between the thermal coagulating points of the body fluid of a crab and of its muscles. The myosin which makes up the greater part of the crab's muscles coagulates from 43° to 50°C, while its body fluid and the haemocyanin contained therein are thermally coagulated at 70° to 71°C. By putting this difference of the coagulating temperatures to work, the muscles may be readily separated from the body fluid and the haemocyanin.

Extracting Meat from Crabtails

Crabtails and the crab meat therein have in the past, normally been discarded as waste by processers because the difficulty in removing the meat makes its recovery uneconomical. Yet there is approximately 2 to 3 ounces of usable crab meat between the skin and the shell in crabtails that is recoverable as a unit. Therefore it is advantageous to have a process for economically recovering the crab meat in crabtails.

P.W. Bynagte; U.S. Patent 3,577,243; May 4, 1971; assigned to Westgate-California Foods, Inc. has developed a process in which the crabtails that are normally removed by breaking from the body of the crab, are placed in a solution comprising salt water, sodium acid pyrophosphate or calcium acid pyrophosphate and sodium tripolyphosphate. The percentage of the parts of the solution range from 2 to 30% by weight of sodium acid pyrophosphate or calcium acid pyrophosphate, 2 to 15% by weight of sodium tripolyphosphate with the remainder comprising salt water or fresh water having a sodium chloride approximately 3% by weight of the water.

For best results the bath is in moderate motion during immersion of the crabtails which may be accomplished by a circulating pump system or by other known agitating means. The crabtails are immersed in the solution for a period of time that is in excess of two minutes. The solution loosens the crab meat from the shell and skin. Upon removal from the solution, the crab meat is removed by breaking through the film of the intestine projection into the crabtail and grasping and pulling the crab meat out of the shell in substantially one piece. The crab meat is then precooked in boiling water for approximately 12 minutes, which removes the solution. The following example will serve to further illustrate the process.

Example: Sixty-five pounds of crabtails were placed in a circulating solution comprising by weight about 95% seawater, 3% sodium acid pyrophosphate and 2% sodium tripolyphosphate for about 5 minutes. The crabtails were then removed from the solution and the crab meat was removed from the shell and skin in about 1 hour. The recovered raw meat weighed 54.6 pounds. The meat was then precooked in fresh water with about 3% sodium chloride by weight added at 200°F for 12 minutes yielding 35.7 pounds or recovered crabtail meat.

Chemical Treatment for Producing Soft-Shell Crabs

E.L. D'Aquin; U.S. Patent 3,222,186; December 7, 1965 describes a process for in situ softening the relatively hard shell portions of common crustaceans such as crabs, crayfish, shrimp, lobsters, and the like belonging to the class Crustacea. More particularly it provides an economically feasible chemical process for treating natural hard-shell crustaceans to enhance their quality and value by converting them into relatively soft-shell crustaceans without substantially altering their physical composition, size, appearance and palatability.

The process provided here is particularly applicable to the blue crab *Callinectes sapidus* Rathbun, one of the most widely consumed crustaceans that grow in size by the natural process of molting or shedding. These crabs are most prevalent in the waters of the Chesapeake Bay and the Gulf Coastal waters of the United States.

Practically all crustaceans are captured in the live hardshell state. They must be cooked to render the meat palatable. The meat is separated by hand-picking it from the exoskeleton. The blue crab, however, is an exception in that a sizable proportion of the catch is converted by a segment of the seafood industry into soft-shell crabs, the shells of which are soft to the touch, and paper-shell crabs, the shells of which are somewhat less soft. Both are well established food delicacies, and command a market price of 4 to 6 times, on an equal size basis, of that obtainable for crabs in the hard-shell state.

The principal reasons for this are that soft-shell crabs have a more appealing flavor, and when fried or broiled, practically the entire crab can be eaten. In contrast, to prepare hard-shell crabs for eating, they are first boiled; and after cooling, the edible meat fragments must be hand-picked from the shell cavities, which is a tedious procedure. Further, the inedible shell or skeleton portion constitutes about two-thirds or more by weight of the whole crab, and must be discarded as waste.

The method employed by commercial fishermen for the harvesting of hard crabs, and for the production of soft crabs therefrom is as follows. Briefly, the catch is first classified to select only those crabs which are judged by experience to be ready to shed in a period of about 1 to 15 days. These are then further segregated, and are placed in submerged wooden or wire cages in accordance with the number of days in which they are expected to shed.

Different fishermen employ different procedures, but in all cases, the boxes or cages containing the crabs that have reached the stage where they are nearest to shedding their hard shells are carefully examined every few hours in order to remove them from the water after they have completed shedding for it is at this point they they are most desirably soft. The above procedure is traditional and is employed throughout the world for the production of soft crabs.

As far as is known, there are no other methods of techniques that have proved successful in reducing the time and labor required; and moreover, none have proved successful for producing soft crabs from hard crabs which are not full (fat), i.e., not ready for early shedding, and none have proved successful in producing soft crabs from paper-shell crabs.

Hard-shell shellfish belonging to the class Crustacea can be converted into substantially soft-shell shellfish by a simple and inexpensive chemical treatment, followed by a simple soaking or washing in water to remove the excess of the treating chemicals. This process produces soft-shell products of acceptable appearance and edibility, and in the case of crabs, products which are comparable to those attainable by natural molting.

In carrying out the process with a crustacean such as for example, the hard-shell blue crab, live specimens of legal size and preferably full (fat), are held submerged in an excess of an aqueous solution of an organic or a mineral acid which kills them within 10 to about 30 minutes. They are then allowed to continue soaking in the solution for a prolonged and sufficient period of time until the outer and inner shells have softened to an acceptable degree as determined by the feel or touch.

The soaking is most important as it is the step which accomplishes the softening of the shells. After the crabs are softened, they are removed from the acid solution bath and are subjected to a sequence of soakings or washes in cool or heated tap water to remove substantially all of the residual entrained and absorbed acid.

If desired, the last traces of residual acid can be effectively neutralized by employing a final soaking in a weak solution of a mild alkaline agent such as sodium bicarbonate. The substantially deacidified softened crustaceans are the final end products, and in this form are suitable for cooking for edible use. However, it is customary prior to cooking, to lift the ends of the large top shell with the finers, and to remove any undigested food particles that were originally contained in the craw of the crab when caught. At the same time, the lungs and eyes may also be removed.

It has been discovered that there exists a limit in acid concentration and also a limit in temperature of the treating of the bath and of the combination of the two, beyond which the shells of the crustaceans develop an objectionable mottled brownish-red discoloration.

As will be shown in the specific examples that follow, it is preferred to soften the shells of raw crabs by soaking them in water that contains organic acid in an amount of about 3 to 5% by weight. Concentrated acetic acid which assays about 90 to 100% by weight is the preferred organic acid. Temperatures should not exceed about 80°F, because higher temperatures tend to unevenly discolor the shells of the crabs. The following example is presented to illustrate the process.

Example: Live hard crabs of medium and large sizes were subjected to a series of separate soaking treatments in which the acid concentration and the temperature of the treating solution were varied within the limits given in the table. The acid used was reagent-grade acetic acid, assaying from 95 to 100% of acetic acid by weight. Range of acid concentration in the treating bath was 0 to 10% by weight. Bath temperatures employed were 40°, 80°, and 100°F. Temperatures were thermostatically controlled.

In each experiment, the raw crabs were held submerged in the bath until the shells of the crabs were judged by tactile examination to be acceptably soft. At this point, the presence or absence of shell discoloration was recorded, and also the elapsed treatment time. Experiment 2 was discontinued after 35 hours, as the crabs showed little or no signs of softness, due probably to the low acid concentration employed. Experiment 1 was discontinued after 72 hours as the crabs showed little or no signs of softness, probably because of the low bath temperature employed. The results in the table show clearly that the best or optimum results were obtained in Experiments 3 and 4, in which was employed a combination of 80°F bath temperature and acid concentrations of the treating bath of 3 and 5% by weight. These conditions yield final crab products of acceptable softness, having no discernible shell discoloration.

The data given below show that acetic acid is definitely a suitable acid for the softening of raw hard crabs by the subject process. The softened crabs resulting from the treatment in Experiment 4 at 80°F with a 5% concentration acid were soaked in running tap water for 2 to 4 hours to substantially free them of residual acid.

They were then dipped in a seasoning batter preparation and were deep-fried in cooking oil. The fried crabs compared favorably in palatability, flavor, etc. with fried soft and paper-shell crabs that were purchased live in the local market.

Exper. No.	Degrees Temp. Fahren.	Acetic Acid Concentrat., Wt. Percent	Soaking Time, Hours	Properties of Softened Crabs	
				Acceptable Softness	Discoloration of Shells
1	40	0-8	72	No	No.
2	80	2	35	No	No.
3	80	3	25-35	Yes	No.
4	80	5	15-30	Yes	No.
5	80	6-10	15-10	Yes	Yes.
6	100	3	25	Yes	Yes.
7	100	5	12-22	Yes	Yes.

In another patent by *E.L. D'Aquin; U.S. Patent 3,346,395; October 10, 1967* the process previously described is utilized for softening shrimp shells. Shrimp in their raw state as harvested, or after boiling, can be converted into substantially soft-shell shrimp. The deacidified shrimp in their whole state, i.e., including head and claws can suitably be shredded or chopped up, seasoned, and frozen or hermetically canned for use directly as animal food, or in animal food formulations.

Chemically softened shrimp could also find use as human food in areas where severe protein food shortages exist, since the head of the shrimp (normally discarded) contains considerable amounts of fats, meat, vitamins, minerals, etc. which are of high nutritive value.

There is also the possibility that chemically softened shrimp could find use under circumstances where the only shrimp that are avilable for harvest are too small in size for hand-picking to recover the meat. The process would also be applicable to the softening of krill (family Euphasids), a relatively small shrimp-like crustacean that inhabits Antarctic waters in quantities estimated in billions of tons. In connection with the highly imminent problems related to the world population explosion, scientists are already considering krill as a potentially valuable food resource and are exploring means for harvesting krill and means whereby they can be suitably prepared, deshelled, etc. for human consumption. For the softening of raw or boiled shrimp by this process when using hydrochloric acid and a soaking bath temperature of about from 80° to 145°F, it is preferred to use an acid concentration of about 0.30 to 7.9% of hydrogen chloride by weight, and a soaking time of about from 30 hours to 40 minutes; and when using a soaking temperature of about from 145° to 212°F an acid concentration of about from 0.30 to 7.9% by weight of hydrogen chloride is preferred, and a soaking time of about from 10 hours to 10 minutes, the longer times being required for the lower acid concentration and the lower temperature.

Pasteurized Shellfish Meat

(This process has been placed under the heading of Crabs, but can be used for various varieties of shellfish such as shrimp, oysters, clams, or lobsters.)

E.L. Walker and A.R. Ferrandini; U.S. Patent 3,852,486; December 3, 1974 developed a process for providing a packaged shellfish product which may be stored over extended periods of time under mild refrigeration without growth of the spores of *Cl. botulinum*, which may be safely frozen and then thawed and stored under refrigeration, and which may be processed rapidly and economically in large volume and can be safely distributed to the consuming public under normal marketing conditions.

In accordance with the process, shellfish meat (e.g., crabmeat, shrimp meat, oyster meat, clam meat and lobster meat) is first cooked in the usual way, as by submerging fresh shellfish in vats or tanks of cooking water for appropriate periods of time in accordance with well-known practice. This cooking makes the meat edible and it destroys harmful surface bacteria, shrinks meat away from the shell and facilitates removal of the meat from the shell by hand or by mechanical means; also it removes certain of the moisture from the meat which facilitates absorption of the treating solution described below.

This cooked but otherwise fresh shellfish meat may then be typically treated as follows: The meat is placed in stainless steel wire mesh containers and dipped for about two minutes in a cold water bath containing about 100 ppm of chlorine. This step prevents or diminishes the transfer of harmful bacteria from the shells (which are more likely than the meat to be contaminated by bacteria including *Cl. botulinum*) to the meat.

Then the meat, while still in the mesh and after drainage of the chlorine bath, is submerged in a treating solution which is an aqueous solution of an inorganic chloride (preferably sodium chloride), an antibacterial agent and a strong organic acid. Since these ingredients penetrate and remain in the meat, they should be nontoxic and approved food additives accepted by the U.S. Food and Drug Administration. The NaCl is preferably employed in an amount of about 2½ to 3% by weight of the treating solution.

The preferred antibacterial agents are sodium benzoate, sodium nitrite and nisin. The concentration of antibacterial agent in treating solution should be sufficient for the purpose, that is, to inhibit bacterial growth under conditions of mild refrigeration over a substantial period of time (e.g., to reduce bacterial growth to an insubstantial degree for a period of three to four months when the product is stored at 32° to 40°F.) Examples of suitable strong, nontoxic organic acids are citric, ascorbic, fumaric, malic and chloracetic acids. Enough of this acid should be used to lower the pH of the treating solution to about 5 or less.

Examples of suitable formulations of treating solutions are as follows:

Formula 1: In a 240-gallon dip tank add enough sodium chloride salt to obtain a 3% solution (60 lb). Then, prepare a mixture of 50% sodium benzoate and 50% fumaric acid and add 16 oz in liquid measure of this compound to the 240-gallon solution.

Formula 2: To a tank of cold water add enough salt to obtain a 3% solution. Then prepare a mixture of 50% of sodium nitrite and 50% citric acid, and add 8 oz by liquid measure to the water tank.

Contact between the cooked meat and the treating solution described above is brought about by any suitable means, e.g., by placing or retaining the cooked meat in a strainer basket, submerging the meat in the treating solution, holding the meat submerged in the solution long enough to insure a uniform penetration of the meat by the treating solution, and then removing the basket and allowing the meat to drain. Usually a time of about 5 to 10 minutes is adequate.

Penetration is accelerated by having the meat in small pieces, e.g., not exceeding about 1 inch in thickness. This treating step is preferably carried out at about room temperature. As noted above, the initial cooking step removes some moisture from the meat and facilitates penetration of the treating solution into the meat.

Next the treated meat is subjected to pasteurization, i.e., to heat sufficient to bring the temperature of the meat uniformly to about 150° to 210°F, preferably 180° to 210°F. This pasteurization step may be carried out in various ways and heat may be applied in various ways. It is important that the treating solution which has been caused to penetrate the meat not be leached out of the meat during or after pasteurization. Therefore, the pasteurizing step should be carried out in such a way as to avoid washing or leaching the meat and removing all or a substantial part of the treating solution which impregnates the meat.

Thus the meat, preferably immediately after treating with the treating solution and draining, is placed in a suitable container which is effective to seal the meat against entry of contaminants such as water that is used as the heating medium, etc. For example, the meat recovered from the step of impregnating with treating solution may be placed in plastic bags which are then sealed as by heat sealing. The bags may be made of opaque or transparent (preferably transparent) plastic material which is resistant to penetration by moisture and is resistant to the degree of heat employed, such bags being of a type that are readily heat-sealed and which act to seal the contents from the atmosphere, from water, etc.

Pasteurization can be carried out in this manner by available microwave techniques and apparatus. Exposure of shellfish meat to microwave energy at power levels recommended by the manufacturer of the equipment for short periods of time, e.g., 2 to 3 minutes will bring the temperature of the entire body of meat to the requisite temperature, e.g., 180°F, very quickly and will accomplish pasteurization. Cooked shellfish meat, preferably given a chlorine treatment as described above and impregnated with the treating solution as described above, is placed in packages, for example plastic bags or envelopes or in glass containers, but the packages or containers are left unsealed.

The reason for leaving the containers unsealed is to prevent the rapid expansion of air (which is due to the very rapid rate of heating) from bursting the containers or creating leaks in the containers. Exposure of the containers to such treatment results in a very rapid heating of the meat including interior portions of the meat, even when present in relatively large bodies as in conventional glass containers. Pasteurization can be accomplished in approximately three to four minutes.

The containers are sealed immediately after pasteurization, for example by heat sealing of plastic envelopes or applying the lids to glass containers. Such sealing is preferably carried out in automated sterile sealing equipment, for example automatic heat-sealing equipment in the case of plastic packages and automatic capping equipment in the case of glass containers. The product is chilled to 40°F or below, preferably within 1 to 1½ hours after pasteurization. Whatever the means of pasteurization, the meat is held at the pasteurization temperature for a time sufficient to effect pasteurization, which will vary with the species.

Example: Two lots of Dungeness crab meat were employed. Both had been given preliminary processing including chlorine treatment as described above. Lot 1 was impregnated with an aqueous solution of 3.2% sodium chloride and 0.1% citric acid (pH of the solution was 3.5).

Lot 2 was impregnated with an identical solution which, however, contained 0.1% of sodium benzoate. That is to say, the meat of Lot 1 was treated with a solution containing only sodium chloride and citric acid, whereas Lot 2 was treated with a solution containing sodium chloride, an antibacterial agent and acid. The meat of both lots was treated in the same manner (and as described above) with the respective treating solutions and was then pasteurized as described above, packaged in sealed plastic containers and immersed in water at 190°F for 14 minutes.

Both lots of meat were delivered to a laboratory in sealed, pasteurized, impregnated condition. Each lot was then divided into two sublots (sublots 1A, 1B, 2A and 2B). These sublots were then treated as follows: Lots 1A and 2A were washed with two ounces of sterile water; Lots 1B and 2B were left unwashed. Then each lot was inoculated with 50,000 spores each of Type E, Boluga and Saratoga strains of *Cl. botulinum* and incubated at 74°F for 11 days. Each lot was sampled at 4, 7 and 11 days, with the results as set forth in the table on the following page.

At the end of 11 days, Lots 1A, 1B and 2A were all gassy and had a cheesy odor characteristic of *Cl. botulinum,* Type E spoilage, and under the microscope were observed to have very many vegetative cells resembling *Cl. botulinum.* Lot 2B showed no gas after 11 days and had very few vegetative cells. Mouse inoculations proved the absence of toxin development. All of the lots (including 2B) produced growth of *Cl. botulinum* after washing with distilled water (which removed the treating solution).

Lot No.	Days Incubated	Presence of Toxin
1A	4	Negative
1B	4	Negative
2A	4	Negative
2B	4	Negative
1A	7	Positive
1B	7	Negative
2A	7	Negative
2B	7	Negative
1A	11	Positive
1B	11	Positive
2A	11	Positive
2B	11	Negative

These results illustrate the superior effectiveness of the three-component solution (salt, sodium benzoate and citric acid) and the fact that the continued presence of this solution in the meat maintains its nontoxic character and prevents the growth of *Cl. botulinum.* Thus only Lot 2B (which had been treated with the three-component solution and which had not been subjected to the leaching action of water) remained free of *Cl. botulinum* and its toxin and the gross indications of the same. As noted, even Lot 2B developed toxin after washing it and extracting the treating solution.

LOBSTERS

Preparing Frozen Lobsters

It is well-known that lobsters are generally cooked alive for the reason that they quickly lose their optimum flavor and tenderness when they are not so cooked. Therefore, great expense is incurred in keeping lobsters alive in the proper environment and transporting them alive. No methods have been found to date for successfully storing whole lobsters after they have been killed.

The result of this is that lobsters are relatively expensive and unavailable in many places. Many attempts have been made to quick-freeze lobsters alive, but they have not been successful due to the fact that the meat sticks to the outer skin or shell in a very tenacious manner so that it cannot be easily removed without shredding. Also, under such treatment the lobster loses its flavor and the meat decomposes and loses its firm texture.

Attempts have also been made to preserve cooked whole lobsters. But these attempts have not been successful since the lobsters lose their flavor and the meat becomes tasteless and tough. Some success has been attained in freezing lobster tails which are generally imported from South Africa. However, the American lobster, which is generally procured from the northeastern coast of the United States, is eaten whole since there is good meat in the body portion of the lobster and in the claws.

R.W. Jung; U.S. Patent 3,261,693; July 19, 1966 has developed an improved method for preparing whole lobsters for freezing. Steps are as follows:

(1) Quickly paralyzing the nervous system of the lobster by electric shock by inserting a pair of electrodes into the body portion.

(2) Applying the electricity until the lobster becomes immobile and then quickly removing the electrode and placing the lobster into a quick-freezing compartment. It is not even necessary to wrap the lobster since the shell and body provide a natural seal.

It is believed that slow killing of a lobster such as by placing a live lobster into a freezing compartment, causes the lobster to secrete certain fluids or acids which tend to decompose the meat and to make the meat spongy and tasteless. This method paralyzes or kills the lobster practically instantaneously, before such action can take place and then the lobster is quick frozen with the meat in good condition. By the use of this process the lobster may be kept for long periods of time and the meat does not freeze onto the shell so that it cannot be easily removed upon normal thawing.

The process is effective by either killing or paralyzing the nervous system with the first shock. In one mode of this process, two long needle-type electrodes are inserted under the shell forward of the tail and forward into the body portion at least half or two-thirds into the forward portion of the shell. The vital organs of the lobster are located in this locality mainly the heart and brain.

The two electrodes are inserted along the top underside of the shell spaced apart so that the electricity will pass between them and paralyze these portions of the nervous system or kill the lobster. In one successful process 110 volts AC were applied to the electrodes with a resistance of about 20 ohms, in one lead and the lobster of about 1¼ pounds becomes relaxed and immobile in about 20 seconds.

The electrodes were then removed and the lobster placed in a quick-freezing compartment. The frozen lobster may then be kept in a freezer compartment for a long period of time of at least several months. Lobster prepared by this process has been kept for as long as six months with good results. After the lobster is quickly frozen it may be desirable to put it in a paper or plastic bag so it does not stick together. Suitable moistureproof and vaporproof coatings or protective wrappings may be used to effect interminable frozen storage.

After the lobster has been frozen, it may be thawed out in a conventional manner by letting it rest at room temperature for a period of a few hours or it may be thawed in hot water, depending upon the size. If the lobster is thawed out at room temperature, then it may be cooked in conventional manner by boiling or broiling. If the lobster is cooked while it is still frozen, then it is necessary to cook it slightly longer.

Frozen Cooked Whole Lobster

Although lobsters are caught in the coastal waters in many regions, Maine lobsters have a reputation for succulency that has resulted in increasingly widespread consumption of this crustacean. Traditionally, consumption was limited to the northern New England coast summer resort areas and surrounding areas which could be served by trucks having brine tanks, ice or artificial refrigeration facilities. With the growth of the air freight business the market for live Maine lobsters increased. However, the market remains risky and has troublesome aspects, because some lobsters shipped live arrive dead and are unsalable, because the immediate attention which must be given lobster shipments often routs consignees from bed at odd hours to meet planes bearing their shipments, etc.

The lobsters which may be used in this process are properly designated cold North Atlantic lobster and are taken from the North Atlantic ocean in the waters off Cape Elizabeth, Maine running in an easterly direction along the entire coast of Maine to New Brunswick, Canada, including the entire Bay of Fundy, continuing in an easterly direction along the entire coast of Nova Scotia ending at Sydney, Cape Breton Island, Nova Scotia, up to more than twenty miles out into the ocean. The family is Homaridae.

R. Moore; U.S. Patent 3,692,545; September 19, 1972 describes a process for boiling whole lobsters and then freezing them for shipping. The process is conducted in the following way: Fifty pounds of live cold North Atlantic lobsters are put into two fifty-pound open-wire, noncorrosive baskets so that one hundred pounds are processed at a time. The two baskets are then lowered into two stainless steel cookers which each contain forty gallons of fresh boiling water and fifteen pounds of granulated salt. When the water returns to a boil the cooking time begins for a period of 9 minutes. Then the baskets are removed and immediately rinsed and cooled with 45°F fresh water.

The precooking and rinsing procedure accomplishes the following: elimination of bacteria and solidification of the internal substances of the lobster which is of particular importance when lobsters are shedding their shells and the roe is more liquid in the late spring and summer months. This procedure releases the meat from adhering to the inside of the lobster shell. After each one thousand pounds is processed the cookers are emptied, cleaned and refilled with new water and salt.

Boiling for a shorter period of time than about 8 minutes, for instance blanching the lobster for 30 seconds to 5 minutes is definitely insufficient to prevent lobster blood, roe (before spawning), and other internal nonmeaty substances of the lobsters from flowing into the tail and contaminating the tail meat so it is unsightly and repulsive to some consumers.

If the lobster is cooked much longer than 9 minutes, for instance for 10 minutes or more, its meat suddenly toughens irreversibly. In order to be legally kept, a caught cold North Atlantic lobster must measure at least 3 3⁄16 inches long along the back from the eyes to the beginning of the tail. This process is applicable to such lobsters, weighing at least 12 ounces up to not more than about 22 ounces. Preferably, processing is begun as soon as possible after the lobsters are caught, most advantageously within 24 hours.

Throughout the entire process until fully frozen, the lobsters are placed on their backs so that the lobster blood will remain in and around the center of the lobster body and the internal substances will have less tendency to run into the tail.

The precooked and rinsed lobsters are immediately removed from the baskets and placed on a stainless steel table upside down on their backs. Then the lobsters are sleeved into flexible plastic film bags of low vapor permeability. The dimensions inside the bags are fifteen inches long by six inches wide. Before use the bags are heated to a temperature of about 90°F to make them more flexible and less subject to puncture by sharp parts of the lobster. The lobster is sleeved into the bag by the use of a light gauge steel sleever which is curved and sixteen inches long with a four-inch extension at the end for protrusion into the end of the bag. The lobster slides into the bag tail first.

Each bag containing a lobster is then placed with the lobster upside down on its back for fifteen minutes before the next step. This allowance of time is for the remaining heat from the lobster itself to transfer to the warm bag. This results in the bag becoming more pliable for the next step.

One bag which has been found to be acceptable is made of 6 mil polyester film but it is expensive relative to a bag made of 6 mil polyethylene nylon film laminate. Pliability, resistance to puncture, ability to hold a vacuum and safeness in contact with food are the more important qualities of the film packaging. Since the package is preferably heat-sealed, the capability of being heat-sealed is also important. Other films than the two just mentioned will fit the above criteria.

In the next step the air is drawn out of each bag through the use of a vacuum sealing machine so effectively that the bag collapses around the lobster so that it is form-fitting with very little air content. Also, this is where the pliability of the bag is important, because the sharp points located over the lobster would puncture the bag if the bag were brittle allowing air to reenter the bag. As soon as the vacuum is drawn on the bag, the bag is heat sealed across the width at the end. Then, as each bag is sealed, it is laid upside down on an open mesh wire tray until each tray contains twenty-four lobsters.

When each tray is filled it goes into a cooling room for four hours where temperature is maintained at 35°F. Then at the end of four hours, the tray is transferred to a freezer for twelve hours where the temperature is maintained at -20°F. Upon completion of the freezing process the lobsters are ready for packing which merely consists of weighing the film-packaged frozen cooked lobsters, putting them into retail boxes, marking the net weight on each retail box and then packing twelve retail boxes to a shipping case. The shipping case is sealed and put into a freezer until shipment by refrigerated trucks.

Rapid freezing of the packaged cooked lobster, e.g., by liquid nitrogen blast-freezing, dries out the meat, because the tiny ice crystals which form can evaporate or sublime so easily during storage. Accordingly, freezing in a conventional, e.g., circulated brine, indirect, refrigeration system is preferred with complete freezing taking at least 2 to 4 hours.

The frozen lobster is prepared by the consumer as follows: The bagged lobster is removed from the retail box and without opening the bag or defrosting the bagged lobster is submerged in a pot of water. The water is brought to a boil and allowed to boil for about seven minutes. Then the lobster is removed from the bag ready for eating.

If the lobster is to be broiled, it may be placed on its back and the entire length of its body and tail cut with a knife. The black intestinal vein and stomach (the hard sac near the head) are removed. The coral roe and tomalley or liver of the lobster, which turns green when it is cooked should not be discarded, as many people consider these to be the best eating parts of the lobster. The claws should be cracked and the lobster placed as flat as possible, brushed with butter, seasoned and broiled until lightly browned. Many people like a sauce of melted butter and lemon juice on their broiled lobster.

Processed and prepared as suggested above, the frozen whole cooked cold North Atlantic lobsters are as tasty as those freshly caught and cooked. In particular, the blood, roe and intestinal contents which would be inconsolidated and runny if the lobster were blanched and/or positioned right-side-up during processing, are consolidated, easy to deal with and have not run from where they belong. Furthermore, the bacterial count of the lobster as purchased by the retail consumer is very acceptably low.

Frozen, Cooked, Eviscerated Lobsters

R.D. Trelease, S. Malinow and J. Baranauskas; U.S. Patent 3,773,962; November 20, 1973; assigned to Swift & Company have developed a method of preparing cooked and frozen whole lobsters wherein the internal organs of a whole lobster are first cleaned or removed, and the lobster is subsequently subjected to a two-step cooking sequence.

Whole live lobsters are selected for processing according to their weight. Although lobsters weighing from about 1 to 2 pounds are preferred, the method is applicable to lobsters of any size, and generally may be employed to process lobsters weighing from about ½ pound up to about 5 pounds and more. While the process has been devised for Maine lobsters, it is also applicable to all other varieties such as spiney lobsters, African lobsters and Bahama lobsters, as well as to other crustaceans.

It has been discovered that off-flavor and off-odor development in frozen, cooked whole lobsters can be traced directly to degradation of the contents of the internal organs of the lobster, such as the stomach, intestinal tract, and cloaca. Therefore, in order to minimize development of off-odors and off-flavors, the lobsters' internal organs (stomach, intestinal tract and cloaca) are eviscerated or otherwise cleaned. For convenience, the lobster's cloaca and intestinal tract (vein) will be collectively referred to as the hindgut in the description of this process.

The description of the process, taken in conjunction with the drawings on the next page, will clarify the procedure. Figure 5.5a is a top view of the whole lobster, while Figure 5.5b is a side view in half section. One technique of cleaning out the lobster's internal system is by flushing with a suitable solution. Referring to Figures 5.5a and 5.5b, flushing of a lobster **2** can be carried out by inserting a spray nozzle into the lobster's mouth **3** through mandibles **4**. Preferably, seawater or salt water, which is at room temperature or colder, is sprayed through the nozzle under a pressure of from about 5 psig to about 20 psig.

FIGURE 5.5: FROZEN COOKED LOBSTER

a.

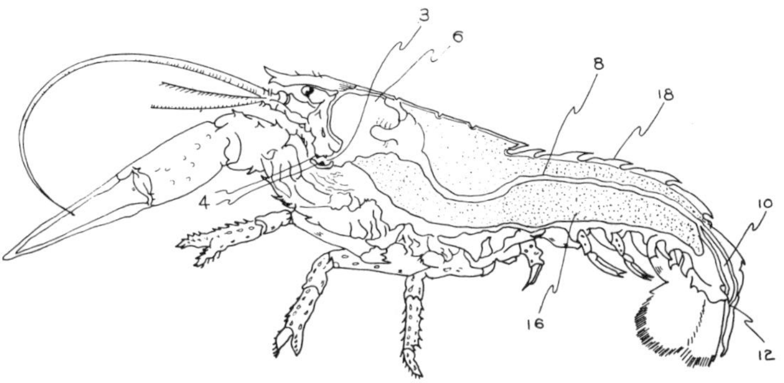

b.

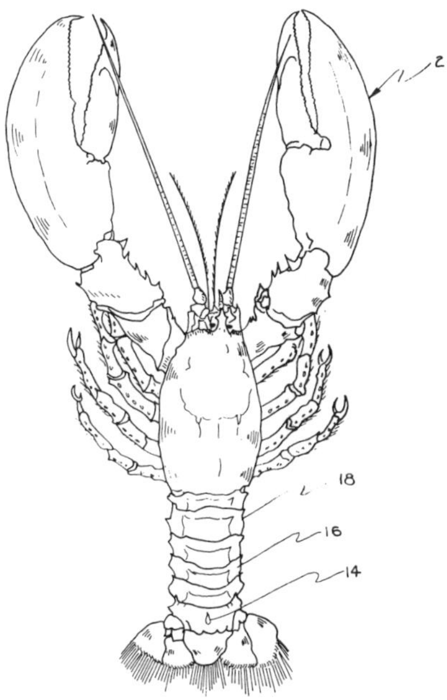

Source: R.D. Trelease, S. Malinow and J. Baranauskas; U.S. Patent 3,773,962; November 20, 1973

Water pressures of less than about 5 psig are generally ineffective to flush out the contents of the lobster's internal organs. On the other hand, pressures of greater than about 20 psig tend to cause undesired rupturing of the internal organs. Most lobsters can be flushed out in about 3 to 20 seconds using this technique. However, when the water pressure is within the most preferred range of from about 10 psig to about 15 psig, flushing can often be accomplished in as little as from 3 to 5 seconds. The stomach contents are generally flushed back out through the mouth, thus cleaning out the stomach **6**.

Moreover, the pressurized water is effective to flush out the undesirable contents of the intestinal track **8** and cloaca **10** through the anus **12**. In this manner organic and inorganic debris is eliminated from the gastrointestinal system of the lobster. A preferred spray nozzle device for carrying out the flushing step would comprise a thin, hollow tube having about a 1⁄16-inch orifice at the end, and about 4 additional 1⁄16-inch orifices spaced at 90° intervals about the side walls of the tube, and located about ⅛ to about ½ inch from the end of the tube. Most effective flushing is achieved when the spray nozzle is inserted through the mouth and mandibles, and is thereafter rotated about by hand within the lobster.

Although the above-described flushing technique for removing the contents of the lobster's internal organs is effective, the preferred technique for cleaning out the lobster's hindgut involves suction evisceration. This technique comprises making a small slit **14** in the abdomen **16** (lobster tail) in the area of the lobster's cloaca. Although this slit may be made on the top or the underside of the lobster, it is preferred to make the slit through the posterior part of the abdominal shell **18** on the top side of the lobster. A vacuum tube is then inserted through slit **14** and is used to suck out the whole hindgut comprising the intestinal tract **8** (vein) and cloaca **10**. Suction evisceration of the hindgut is generally always carried out prior to cooking.

The suction nozzle will generally have an inside diameter of about ¼ inch in order to effectively remove the hindgut of the lobster. Also, the nozzle will generally be operated under a vacuum of from about 15 to 25 inches Hg, and removal of the hindgut can generally be accomplished in from about 3 to 5 seconds.

The stomach of the lobster can subsequently be removed by insertion of the suction nozzle through the mouth between the mandibles. Suction removal of the stomach and stomach contents generally requires a longer period of time than suction removal of the hindgut, and usually will require from about 5 to 15 seconds.

Because of the longer period of time needed to remove the lobster's stomach and stomach contents by suction, it is preferred to remove the whole stomach by grappling it with a hooked-end rod and pulling it out through the mouth. Thus the stomach can be completely removed by inserting through the lobster's mouth a thin tube or rod having a plurality of hooks formed on the end thereof to engage the stomach lining. Preferably these hooks are formed so as to point slightly outward and back toward the mouth of the lobster. Although it is feasible to remove the lobster's stomach either before or after cooking, it is preferable to remove the stomach prior to cooking.

When cleaning or evisceration of the lobster has been completed, either by flushing, suction evisceration, or grappling, or a combination thereof, the lobster is ready for cooking. The lobster is first placed in boiling water (about 210° to 212°F) until an internal temperature of from about 110° to about 120°F is achieved. The water preferably contains up to about 3% salt. This first heating step acts to terminate enzyme action in the liver (tomalley) of the lobster, and also is effective in minimizing future adhesion of the meat to the lobster's exoskeleton.

For most lobsters, this first heating step will require from about 3 to 12 minutes in boiling water. Thus for a 1½-pound lobster, a first heating step of about 5 minutes in boiling water is sufficient. On the other hand, heating for up to about 10 to 12 minutes may be required for a 2½-pound lobster. It is important to remember, however, that the necessary criteria to be met in this first heating step is the internal temperature of the lobster. The internal temperature achieved during the heating step should not exceed about 120°F with a temperature of from about 110°F to about 115°F being preferred.

When the required internal temperature is obtained, the lobster is transferred to a cooking vessel containing water (preferably up to a 3% salt solution) held at from about 180°F to about 190°F. Again, the internal temperature achieved by the lobster during the cooking step is the important criteria. An internal temperature of greater than about 165°F is required to heat-denature the proteins.

Once an internal temperature of from about 165° to about 170°F is effected, the lobster is removed from the cook vessel. If internal temperatures of greater than about 170°F are effected, the meat of the final lobster product tends to be tough and dry. On the other hand, if internal temperatures of less than about 165°F are achieved, there will usually be insufficient denaturization of the lobster meat protein. For most lobsters, a cook period of from about 8 to about 20 minutes is sufficient to achieve the desired internal temperature of about 165°F. It should be clear that the optimum heating time and cook time for any given size lobster can be readily determined by utilizing the described internal temperature guidelines.

Once the two-step cooking sequence has been carried out, the cooked lobsters are removed and are immediately cooled to stop the cooking process and to prevent bacteriological incubation. The cooling step should bring the internal temperature of the lobsters down to within the range of from about 32° to about 50°F, and preferably to about 40°F. This chilling step may be accomplished in cold water or cold air as by cooling for 1 to 2 hours at 40°F or placing in a freezer for 15 to 30 minutes.

The lobsters are then placed in individual bags and are frozen, or in the alternative, they may be first frozen and then bagged. Preferably, however, the lobsters are first placed in individual plastic bags or pouches which are moisture impermeable, puncture-resistant, and which have low-temperature flexibility. The bag is then sealed and the product is ready for freezing. If desired, in order to achieve an even greater shelf life, a vacuum may be drawn on the bags and they can be backfilled with a nitrogen atmosphere prior to sealing.

Finally, the cooked whole lobsters are frozen, preferably to an internal temperature of less than 0°F. Placing the lobsters in sub-0°F freezer for from about 6 to 12 hours with chilled air circulating through the freezer at an air velocity of about 500 feet per minute would be sufficient. When the lobsters are frozen, they may be packaged in suitable insulated boxes for future storage and shipment.

The cooked and frozen, whole lobster product prepared in accordance with this process may be kept for up to 6 to 9 months or longer when stored in a sub-0°F freezer, without development of off-flavors and off-odors, and without loss of desirable color and pigmentation. Moreover, the lobster meat is tender and juicy, and may be easily removed from the exoskeleton without shredding. To prepare the frozen product, the consumer need merely reheat the product as by broiling or placing in boiling water.

CRAWFISH

Large quantities of frozen meat obtained from crustaceans, such as crawfish, lobster, shrimp, and the like are marketed annually for human consumption. Such meat, especially crawfish, when preserved and distributed in the frozen state is regarded as decidedly inferior to the fresh product. When stored for even short periods, the frozen meat loses the characteristic color of the fresh product, becomes tough and corky, and has an inferior taste as opposed to the fresh material.

A.C. Heumann, Jr., W.L. Denton and J.D. Fitzpatrick; U.S. Patent 3,852,500; December 3, 1974; assigned to Mr. Crawfish, Inc. have found that by heating the hepato-pancreatic tissue of the crawfish to a temperature and pressure sufficient to cause continuous distillation from the tissue material, distilling and collecting from about 50 to about 70% by weight of the charge of the hepato-pancreatic tissue and subjecting the edible crawfish to a sufficient amount of the distillate material, the meat will be preserved in both taste and quality.

In carrying out the process, the liver and pancreas of the crawfish are first removed and the hepato-pancreatic distillate prepared from them. The hepato-pancreatic distillate which is isolated is obtained as a complex aqueous solution of substantially zero bacterial count which does not spoil under freezer storage conditions. This is contrasted with the frontal sections of the crawfish from which the liver and pancreas tissue materials are obtained which are detrimental to crawfish tails upon storage.

The hepato-pancreatic tissue of the crawfish is distilled under conditions of temperature and pressure which will cause continuous distillation of the material, inactivates the proteolytic enzymes of the distillate material, destroys bacteria, and, therefore, produces a substantially sterilized distillate. Such conditions of temperature are about 60°C to about 100°C. The optimum distillation temperature is about 80°C. The distillate obtained at this temperature under reduced pressure with agitation imparts superior flavor qualities. At these temperature conditions, the enzymes will be inactivated and bacteria destroyed during the period of distillation.

The distillate collected should be from about 50 to about 70% by weight based on the weight of the initial charge of the hepato-pancreatic tissue material, the optimum quantity being about 60%. The material thus collected may be kept under any reduced temperature condition range which is suitable. It has been found that such temperature should be preferably from about 10° to about -10°C. Lower temperatures may be used. The thus collected distillate material which is both of a preservative quality and a flavoring agent to the crawfish meat material may be stored at reduced temperatures for a prolonged period of time prior to usage in conjunction with the crawfish meat to be preserved.

The crawfish meat (from the tail), often separated, peeled and cleaned, is normally sorted according to size as well as to other factors which are conventional to the industry. The crawfish tails are then placed in containers, such as plastic bags, and subjected to at least 5 ml of distillate with preferably from about 5 ml to about 20 ml of concentrated distillate material per pound of crawfish meat. Larger amounts of the distillate may be used to increase the flavor of the resultant meat. The addition of the distillate material may be done in various ways, such as by spraying the distillate material into the bag containing the crawfish meat or injecting into the bag the required amount of distillate material and then tumbling the bag and its contents to allow for an even distribution of the distillate with the crawfish tail meat.

The distillate is allowed to saturate the processed tails in the bag for several minutes. The meat is then frozen in any conventional manner, such as by air-blasting, immersion of the closed container in a reduced temperature solution, and the like. It has been found preferable that the material should be quick frozen, such as being air-blasted, at subzero temperatures, as this method has been found to cause less mechanical destruction of the intact cells of the meat and a more prompt prevention of the microbial growth.

Example: Scalded fresh crawfish were placed on a stainless steel table and the heads and tail shells were removed. During this peeling process, the hepato-pancreatic tissue found in the heads of the crawfish was removed and saved.

100 parts of the removed hepato-pancreatic tissue was subjected to simple batch distillation at a temperature of 80°C under a reduced pressure of 0.47 atm while the tissue material was continuously agitated during the distillation. The distillate was collected through a condenser (temperature 0° to -10°C) into a receiver maintained at an ice-brine bath temperature of about -5°C. The distillation was continued until 60 parts of distillate was collected.

One pound of crawfish tails, previously peeled, cleaned of fatty material, sorted by sizes, and placed in plastic bag containers, was sprayed with a 15-ml portion of the distillate. The distillate containing crawfish tails was allowed to remain at ambient conditions for several minutes. The bag was then sealed closed and air-blasted frozen at -30°F. When the processed tails were defrosted after being frozen at least four weeks, they showed no signs of putrefaction and were rated substantially indistinguishable from freshly prepared crawfish tails.

CONSUMER PRODUCTS FROM CRUSTACEANS

Dry Flavorings

It would be desirable to provide a stable flavoring material for food products of various types which will impart thereto a flavor of certain edible crustaceans, for example, the flavor of lobster, shrimp, and crabs. Such a flavoring material should be capable of prolonged storage without refrigeration and also without deterioration or loss of flavoring power.

R.D. Gray; U.S. Patent 3,264,116; August 2, 1966 has described a method of manufacturing a flavoring material of the type described which utilizes only certain portions of the crustacean which are normally discarded in commercial processing. In the commercial preparation of edible crustaceans such as lobster, crabs, and shrimp for food products, it is customary to first cook the whole crustacean in boiling water, steam, or hot air, after which the edible meat portions are separated from the shell for canning, freezing, or other utilization in appropriate food products. The shells as well as other inedible parts such as viscera, heads, etc., are discarded as having no commercial value, and must be promptly disposed of, since they putrefy rapidly causing sanitation problems.

It has been discovered that the relatively useless by-product shells, unremoved flesh and fiber can be utilized to provide a food-flavoring material by a suitable series of processing steps. This food-flavoring material has been found to be entirely useful for human consumption as indicated by storage characteristics, stability, flavor and bacterial count both immediately after preparation and following storage.

The process briefly includes the steps of preparing the raw material from the cooked shell and fibrous portions from which the meat has previously been removed for preparation as foodstuffs for human consumption. As soon as possible after the meat is removed from the crustacean shells, and before sufficient time has passed to permit any deterioration, the cooked shells and fibrous portions are washed with water, either fresh or salt, to remove the viscera and like wastes from the shell and fibrous portions.

The shells and fibrous sections which are the final raw material for the food flavoring are then preferably wet milled. This milling step can be carried out using conventional apparatus for wet milling. Additional water may be introduced to aid milling, but this is not necessary. Two or more mills may be used, one for pregrinding and one for fine grinding.

A whirling blade commutator has been found very useful in pregrinding while a colloid mill has been found very useful in fine grinding although other apparatus can be used. The particle size of the suspended solids in the slurry formed by milling should be fine enough to allow subsequent drying by whatever drying method is used. The concentration of the slurry formed in the milling step is not critical but is at least 5% solids and preferably higher to minimize as much as possible the water evaporation required in the subsequent drying step.

After preparation, the slurry is preferably acidified to stabilize the liquid thus minimizing bacterial growth both during processing and in the final product. This acidification step while not necessary is desirable in that it simplified production of a bacterially stable food-flavoring material. Any food-grade acid either organic or inorganic may be used in this stabilization step. Preferred acids are citric, hydrochloric, phosphoric, lactic and the like. The final pH of the slurry is preferably in the range of 4.5 to 8.0.

The slurry is dried in any conventional manner to produce the desired flavoring material. Spray drying is preferred because of the speed, ease and lack of detrimental effect on the flavor, although other forms of drying can be used. For spray drying, the slurry particle size is preferably in the range 10 to 100μ. The desired food-flavoring material is obtained directly from the drying step.

The food-flavoring powder may be packaged in any suitable manner, such as in ordinary screw-cap jars, for home use, or in polyethylene bags for bulk shipment to commercial food processors. It has been found that no unusual precautions need be taken in packaging, since the powder is not hygroscopic, and does not deteriorate in flavoring power or increase in bacterial count, even during long periods of unrefrigerated storage. Bacteria counts for foods derived from fish are generally required to have a plate count of less than 50,000.

A batch of flavoring powder manufactured by the described method, after six months unrefrigerated storage, was found to have a plate count of only 2,000. The flavoring material disclosed herein may be utilized in the manufacture and preparation of many types of food products, such as sauces, cheese, crackers, and the like.

For many products, addition of the powder in the proportion of about 0.1% by weight imparts a suitable flavor thereto. The flavor may be enhanced if desired by the addition of a flavor intensifier such as monosodium glutamate, and for special flavors, powders prepared from different types of crustaceans may be mixed. The following examples illustrate the process.

Example 1: About 40 lb of shell and fibrous portions obtained from lobsters which had been previously cooked in boiling salt waters were washed to remove the viscera. The shells were then milled wet in a whirling blade commutator to a particle size below ⅛-inch average diameter. This milled material was then fed to a colloid mill and further milled to give a particle size in the range of 5 to 100μ.

The slurry obtained was 9% by weight solids. The slurry was acidified to a pH of 7.0 by addition of citric acid. The acidified slurry was then spray-dried giving a product having about 5% moisture and a low bacterial count. The product was stable in long storage without refrigeration and its flavor did not change. It was found to impart a very excellent flavor to crackers and cheese dip and was entirely fit and well adapted to human consumption.

Example 2: About 44 lb of shell and fibrous portions obtained from lobsters which had been previously cooked were washed to remove the viscera. The shells were then put in a tray dryer operating under 15" vacuum at about 210°F and dried overnight. The dried lobster was then milled in a whirling blade commutator to a particle size ⅛ to ¼ inch.

The preground material was then micropulverized to a particle size of 100% 100-mesh. The pH of the powder was adjusted to about 8.0 by addition of citric acid and was ready for use as a flavoring agent. The product was found to be stable on storage and imparted a very excellent flavor to crackers and cheese dips.

Shrimp Soup Mix

The process by *S. Ohtaki; U.S. Patent 3,256,098; June 14, 1966; assigned to Nisshin Kato KK, Japan* for making oyster soup mix has been reviewed in Chapter 4. This process is useful for making shrimp soup mix also, as illustrated by the following example.

Example: Shrimp Extract Powder – 40 kg of dried shrimp are further dried to 32 kg. The mass is then ground, and placed in a jacketed kettle. 30 kg of a solution of 22 g of disodium hydrogen phosphate, 146 g of potassium dihydrogen phosphate and 0.5 kg of sodium chloride in water is added, and the mixture is boiled for 30 minutes, followed by filtration. The residue is mixed with 10 kg of the same aqueous saline solution as that mentioned above and 200 g of mold protease, and the mixture is maintained at 45°C for 2 hours, allowing enzymatic decomposition to take place.

Then the mixture is boiled for 10 minutes for decomposition of the enzyme, filtered, and mixed with the first broth to give 20 kg of the total broth. To the mixture there is added 10 kg of soluble starch, 150 g of sodium glutamate, and 15 g of sodium succinate. The mixture is sterilized at 80°C for 10 minutes, and dried on a drum of 1-meter diameter having a surface temperature of 120°C and rotating at 2 rpm, to yield 11.5 kg of the powder product. A similar product is obtained when 200 kg of raw shrimp are stirred gently with 4 kg of sodium chloride, washed well with water, and the mass ground and treated in the same manner as the dried shrimp mass.

Frozen Controlled-Size Shrimp Patties

A popular frozen food product is a portion-controlled patty prepared from shrimp, sometimes referred to as a shrimp steak. The frozen product is prepared from small (Titi) sized imported shrimp. Titi-size shrimp is the commercial term for tiny shrimp which have a count of 90 to 130 shrimp per pound. The patties are prepared by combining Titi-size shrimp with suitable flavorings and thereafter printing the shrimp into portion-controlled patties.

The use of imported shrimp, however, presents a considerable problem to the industry in that imported shrimp received from any available foreign source is often found to be infected with extremely high bacterial count. This is attributed to the fact that handling of shellfish at ambient temperatures is extremely difficult without contributing bacterial contamination. Even though the foreign shrimp are frozen after harvesting, they incur sufficient handling at ambient temperatures to incur an extremely high bacterial count.

While it is true that cooking of the shrimp reduces the counts of bacteria found in the shrimp, it would be desirable to provide a more careful control of microbiological contamination of shrimp prior to the incorporation of the shrimp into a product, such as a shrimp steak. Better control of microbiological contamination could be maintained if domestic shrimp were used to prepare the shrimp steaks. However, domestic shrimp are generally too large to be used in preparing shrimp steaks and a grinding step increases the possibility of microbiological contamination. It would therefore be desirable to provide a method for grinding shrimp or other shellfish that limits the possibility of microbiological contamination.

Also, known methods for preparing frozen portion-controlled patties from shrimp are not entirely suitable. In accordance with known methods, a processor receives shrimp as frozen blocks. The blocks of shrimp are then permitted to thaw while immersed in cool water by the processor for a period of 5 to 7 hours until the shrimp have become completely unfrozen. The shrimp are then mixed with suitable flavoring materials. It is then necessary to spread the shrimp onto pans and to chill the shrimp to a temperature of about 35°F to provide a suitable consistency for printing of the shrimp mixture into portion-controlled patties.

After chilling, the shrimp are transferred from the pans to the hopper of a printing machine and printed or molded into portion-controlled patties which are then breaded. The printing operation is sometimes difficult to control due to a tendency of the shrimp mix to stick to the printing apparatus.

Each of the steps indicated above are potential sites for bacterial contamination. It would be desirable to provide a method for preparing portion-controlled patties of shellfish wherein the potential for bacterial contamination is reduced. It would also be desirable to provide a method for preparing shellfish patties from sections or segments of shellfish which are considered too large to be directly incorporated into a patty.

Accordingly, a method for preparing frozen shrimp patties has been developed by *A.S. Partyka; U.S. Patent 3,607,315; September 21, 1971; assigned to Kraftco Corporation* in which bacterial contamination is reduced and in which shrimp, previously considered too large for use, can be incorporated into the patty.

Commercially available blocks of frozen shrimp are used in the process. The shrimp may be any size or from any source, but it is preferred to use shrimp which are harvested off the coast of Maine. Such shrimp are harvested during a season which extends from about the first of January to the last of March each year. By utilizing shrimp which are harvested during seasonally cold months, it is possible to provide reduced bacteriological counts on the shrimp. Also, by using domestically harvested shrimp, it is possible to insure that the shrimp are processed and frozen within a reduced period of time and to thus reduce the bacteriological count even further.

More particularly, frozen blocks of shrimp are ground to the desired particle size in suitable apparatus, such as high-energy impact-grinding apparatus, or an auger-type meat grinder. When high-energy type grinding apparatus, such as a hammer mill, is used to effect a reduction in particle size of the shrimp, it is preferred to maintain the temperature of the shrimp below at least about 28°F. At higher temperatures, the work energy imparted to the shrimp during grinding may increase the temperature of the shrimp to thawing conditions and the possibility of contamination is increased. The operation of high-energy compact-type grinding apparatus is such that efficiency of grinding of the frozen shrimp is increased as the temperature of the shrimp is decreased below freezing.

As the temperature of the frozen shrimp which are to be fed to the impact-type grinding apparatus is lowered below freezing they become more friable and easier to grind. Consequently, in a particularly preferred embodiment, frozen shrimp are removed from freezing apparatus and are ground to the desired particle size in high-energy impact-type grinding apparatus before the temperature of the shrimp has risen substantially above that temperature maintained in the freezing apparatus. For this embodiment the temperature of the shrimp is not increased above about 10°F during the grinding step.

When auger-type meat grinding apparatus is used to effect grinding of the shrimp it is necessary that the shrimp be at a temperature of from about 15° to about 28°F during grinding. At lower temperatures, the meat grinder will not usually operate. Therefore, when a meat grinder is to be used, frozen shrimp which are removed from commercial freezers at a temperature of –10°F are permitted to warm for a period of about 2 to 3 hours to permit the temperature of the shrimp to increase to within the indicated range. It is sometimes desired to provide at least a portion of the particles of shrimp produced by grinding of a size that is comparable to the Titi size of shrimp. This size particle is most easily produced with high-energy impact-type grinding apparatus, such as a hammer mill. In a preferred embodiment the conditions of grinding frozen shrimp in a hammer mill are adjusted to provide ground shrimp having about 20% by weight of particles which are comparable to Titi-size shrimp with the remaining particles being smaller.

The shrimp particles are then transferred to suitable mixing apparatus where they are combined with other ingredients for preparing the portion-controlled patties. Because of the low temperature of the shrimp particles, the ingredients which are introduced to the mixer to be mixed with the shrimp particles tend to lump and to prohibit uniform dispersion of the ingredients with the shrimp particles. Single and double-arm dough mixers are found to be most suitable, and sigma-type mixers are also suitable.

During the mixing step the particles of shrimp are maintained at a temperature below freezing. It will be readily understood that when the other ingredients to be mixed with the shrimp particles are at ambient temperature that the level of the other ingredients that can be added to the shrimp particles is limited to that level which will not raise the temperature of the mix above freezing. For example, when the temperature of the shrimp particles is about 28°F, the shrimp particles should comprise at least 80% by weight of the mix. Lower levels of shrimp particles may be used if the temperature of the shrimp particles is reduced or if the temperature of the other ingredients is reduced prior to mixing.

It is desirable to incorporate a binding agent into the mix which is to be made into the patties. A preferred binding agent is an alginate, either sodium or calcium alginate. The use of a binding agent, however, is not essential. When used the binding agent is added at a level of from about 0.2 to about 0.5% by weight of the mix.

The mix of ingredients at a temperature below about 28°F is then transferred to suitable printing or molding apparatus. The printing apparatus preferably includes means for breading the printed patties with a light coating of bread or cracker crumbs. After printing, the patties of shrimp, at a temperature still below freezing, are packaged and the packages are immediately transferred to freezing apparatus where the temperature is reduced to a level suitable for cold storage, such as below about 0°F.

Example: The process is used to prepare 100 pounds of a shrimp mix suitable for making portion-controlled patties. The shrimp mix has the following ingredients:

Ingredients	Weight Percent
Frozen shrimp portion-controlled	87.44
Cracker meal	9.62
Spices	2.97
Calcium alginate	0.44
Water	0.44

The shrimp are removed from a freezer which is maintained at a temperature of -10°F. The frozen shrimp are allowed to stand for a period of 2 hours until they have reached a temperature of 15°F. The shrimp are then removed from the polyethylene bags in which they are packaged and are loaded into the hopper of an auger-type grinding machine. The grinding machine is equipped with a plate provided with a plurality of ¾-inch holes.

The ground shrimp at a temperature of 22°F are collected in a sterilized stainless steel container and are transferred to a double paddle mixer. The cracker meal, spices, calcium alginate and water are preblended and are then gradually added to the ground shrimp, as the mixer is operated to provide a shrimp mix. After blending, the shrimp mix at a temperature of 25°F is transferred to a printing machine. The shrimp mix is then printed into portion-controlled patties which are subsequently battered and breaded.

The portion-controlled shrimp patties are then packaged and transferred to a freezer for storage. After storage for a period of 6 weeks the shrimp patties are examined to determine the level of microbiological contaminants. The total microbiological count is less than 4,000. The shrimp patties are then prepared for eating by deep frying at a temperature of 350°F for 3 minutes. A delicious product is obtained with a mild shrimp flavor.

Molded Shrimp Product

L. Rubin; U.S. Patent 3,852,505; December 3, 1974; assigned to Atlantic Corporation has developed a process for preparing a molded shrimp product for market from very small decapitated and deshelled refrigerated shrimp mixed with leftover fragments of larger shrimp without the use of any extraneous binding material.

In general this is done by tempering the prepared very small shrimp with or without the admixed fragments of larger shrimp from this refrigerated state within the range of -5° to 0°F by then bringing them relatively gradually up to a temperature of +20°F and ultimately to 30° to 33°F. The tempered shrimp is then subjected to braying through a one-eighth inch aperture following which the brayed material is agitated at 85 to 125 rpm for precisely 5 minutes. Thereafter the tempered, brayed and agitated material is discomposed by tituration and then fluttered for 2 to 3 minutes only at a temperature of between +40° and +55°F, the material at the end of this period usually having approximately the latter temperature.

The treated material is then molded by introducing it into single or multiple molds of aluminum, stainless steel, synthetic plastic, synthetic rubber or other suitable inert material at low temperature and high pressure under such conditions that no undesired discoloration or off-taste is imparted to the molded shrimp which when removed from the shrimp-shaped molds, which are ordinarily coated with an approved mold release agent such as Teflon, have unusual self-cohesive properties so that the molded shrimp can be handled, stored, shipped and ultimately used in restaurants and by the housewife in all the regular ways, such as boiling and breading and frying, in which natural commercial size shrimp are used without disintegrating, crumbling or otherwise losing their molded shape. No heating or helical extruding is used during the processing of the shrimp and shrimp fragments.

The shrimp are treated in the tempering room until they reach a more or less ambient temperature, the best and correct temperature being 33°F. It has been found that by doing so the shrimp, which are, for instance, piled on a sheet of aluminum or on a wire grill, are thereby conditioned to withstand the friction and incidental heat created during the subsequent step of braying. Braying means essentially forcing the shrimp or shrimp mixture through equipment in the nature of a meat grinder having a plate provided with a one-eighth inch hole or opening.

This places the shrimp or shrimp mixture in smooth, moist condition for subsequent treatment and exposes the right amount of surface area so that when the shrimp are next agitated at 85 to 125 rpm in a suitable mixer or agitating machine, the proteins in the shrimp attain a special degree of trituration so that exceptional bonding becomes possible. It has been found in practice that agitation should be carried out for precisely 5 minutes and while some minor variation is permissible, the full benefits of the process are not obtained if the agitation time appreciably departs from the 5-minute period.

After agitation the shrimp are discomposed to make them deranged, i.e., disrupted physically as in random orientation. While this can be carried out by hand, for commercial reasons it is done mechanically. The discomposing is a strong or almost violent churning action as by a mechanical rotating and/or vibrating element which is carried out until by observation and experience the protein present reaches the correct amount of derangement and the desired viscosity and consistency.

Thereafter the shrimp is fluttered (in the same machine if desired) by rapid, frequent and irregular or eccentric vibrations of a vibrator for a period of 2 to 3 minutes, but no longer and at a temperature in the range of +40° to +55°F. While it is possible to carry out the fluttering action by hand, such is only practicable on a commercial scale by using suitable commercial equipment and both the fluttering time and temperature have been found to be critical and should not be appreciably varied.

At this point the shrimp are ready for molding which is the final step in the process. Due to the sequence and character of processing steps above-described it has been found that the shrimp when molded at 125 psi and then subjected to freezing below 32°F and down to -300°F as in the presence of liquified nitrogen are not only in a form closely resembling or duplicating natural shrimp, but they do not disintegrate or break up into small pieces either when they are removed from the mold and handled or when they are packed and shipped, or when they are cooked as by being boiled or breaded and fried. They thus have exceptional cohesiveness.

ANIMAL FEEDS

FEED FOR POULTRY AND LIVESTOCK

Fish Broth plus Milling By-Products

Attempts have been made to use fish or fish products as an animal feed, because fish or fish products are not only abundant and inexpensive, but also have a particularly high degree of nutritive and generally wholesome value, especially in regard to their contents of albumin and vitamins. The fish products used, for instance, in mixtures with comminuted straw, potato mash, or similar carbohydrate-containing products and fermented at slightly elevated temperatures, are mainly so-called fish offal comprising the heads, the spines and similar fishbones, ventricular portions, the skins and sometimes the intestines of fish.

Fish offal may also be treated with flour mill by-products such as bran, chaff, residual flour or the like by fermentation with suitable microorganisms to obtain an animal feed.

Processes utilizing these raw materials suffer from the drawback of requiring the handling of large volumes of liquid. Furthermore, all of the known animal feeds obtained from these fish or fish products have a particularly pronounced taste and odor which is generally repulsive to animals and hence decreases their appetite.

The animals fed with them, therefore, are prone to consume less food than is desirable and necessary in order to fatten mammals and fowls for marketing and slaughtering, or they soon refuse to consume such feed at all. In addition, the meat of animals fed with fish or fish products assumes a fishy flavor which makes this meat taste less good than is generally exacted by the modern consumer.

J. Krüss; U.S. Patent 2,986,469; May 30, 1961 has developed a feed for poultry and livestock which avoids these drawbacks. According to the method, fish broth obtained from fish waste or by-products, is de-oiled and concentrated. The concentrated fish broth is then subjected to a fermentation brought about by first thoroughly and intimately mixing the fish broth with carbohydrates suitable as a fodder for animals such as milling products and the like, and then exposing the resulting mixture to the action of bacteria belonging to the group of aerobic or anerobic bacteria, for example, diplococci, which is facultatively anaerobic. The following example illustrates the process.

Example: 100 kilograms of de-oiled, concentrated fish solubles having a water content of 40%, and 150 kilograms of cereal grist containing about 15% of water are given into a stirring apparatus and are intimately mixed. The mixture obtained is subjected to a process of fermentation by filling the mixture into suitable containers and exposing the same for five days in a fish processing plant to the influence of the air having a room temperature approximately from 16° to 25°C.

The fermentation products by the aerobic bacterial flora always present in the plant bring about a biochemical reaction changing the components of the fish solubles. As a result, odorous and flavoring substances are produced which create pleasant and appetite stimulating sensations to fowls and mammals fed therewith. Upon completion of the fermentation, approximately 240 kilograms of animal feed are obtained.

Livestock Feed from Fermented Fish Solubles

In the scientific feeding of livestock, a variety of synthetic feeds have been used, particularly to increase the nitrogen intake of the animals. Nitrogen is utilized by the animals in the synthesis of protein. Grains and other plant foods contain protein and are the traditional feed for, say, cattle. However, supplements have been used which are more rich in nitrogen or supply it in a more readily utilized form. These supplements are capable of increasing the rate of growth of the animals, with consequent economic advantages.

Certain waste products which contain natural protein have been used which are available at a sufficiently low price. In particular, work has been done in feeding to cattle, as a supplement, a product known as fish solubles. This is obtained as a waste product in canning fish such as tuna and salmon.

In a cannery, fish is first cooked in water or, more commonly, steam. The steam extracts natural oils, proteins and other nutrients, and is condensed and separated from the fish meat. The liquid thus obtained is referred to as cooker water. The heads and other waste portions of the fish are removed and pressed to extract water, prior to converting to fish flour, and the liquid separated is known as press water. Then the two liquids are added to each other, and the liquid obtained is known as stickwater.

It will be understood that this is a kind of fish soup. It usually contains from 4 to 7%, most commonly about 5%, solids. The stickwater is fairly quickly concentrated in an evaporator, operating at elevated temperature and under partial vacuum, to 50% solids; the liquid obtained is known as fish solubles. It is particularly significant that in carrying out these steps, where necessary, precautions are taken to avoid any spoiling or fermentation, which would make the material unpalatable and possibly unsafe.

A feed for livestock has been produced by *M. Senyi; U.S. Patent 3,329,504; July 4, 1967; assigned to Liquidos Diversificados, SA* by combining stickwater, molasses and brewer's yeast and fermenting it. A material is obtained which is best described as a beer and this is distilled in a manner similar to that known in the making of whiskey. A high proof distillate or beverage spirits is obtained which might be referred to herein as a kind of rum. It will be appreciated that the distillate is not the ordinary kind of rum obtained by fermentation and distillation of molasses and other cane sugar extracts, since it also contains constituents derived from the stickwater.

The undistilled slops are collected and then concentrated, for example, by evaporation in equipment of the type used in making fish solubles. Concentration is continued until the solids content is increased to about 50 to 60%. Then, the distilled spirits is recombined with the concentrate, and additional molasses is added. The result is a feed supplement which is palatable to all livestock and also provides a balance of nutrition. The process is illustrated by the following example.

Example: In a large open vat 700 gallons of tuna stickwater having a solids content of 7% was mixed with 75 gallons of mill run grade molasses. Sulfuric acid also was added until the pH was 4.8, followed by 40 lb of brewer's yeast. The mixture was agitated with air as necessary for 48 hours while its temperature was kept below 90°F by cooling coils in contact with the liquid, and then passed into a vertical still column. The reflux ratio was adjusted so that about 30 gallons of rum distilled over having a proof of 180.

The slops recovered at the base of the still was passed through an evaporative concentrator and the solids content was raised to about 50% to give 100 gallons of concentrate. Then the concentrate and rum were recombined and 370 lb of molasses was added.

One of the principal advantages of this process is that it converts fish waste into a form which cattle like to eat. The material known as fish solubles, when mixed with molasses, is not palatable to cattle. However, when the stickwater, from which the fish solubles are made, is subjected to fermentation with molasses as described above, there appear to be changes in the protein and other constituents of the stickwater. The change in constitution also changes the flavor and, when it is mixed with molasses, cattle find the flavor very satisfying. Thus, the usefulness of fish solubles as a feed for cattle is considerably increased.

In undergoing the changes which take place in fermentation, the nutritional value of the fish constituents is not reduced but actually increases. The protein is believed to undergo certain changes which may be more readily utilized by the animal and certain unknown nutrient factors are formed. Futhermore, the alcohol produced has a tendency to relax the animal and thus improve his appetite and digestive processes.

Chick Growth Factor from Fish Solubles

E.L. Stephenson; U.S. Patent 3,421,900; January 14, 1969 has developed a process which relates to two solid crystalline substances of low molecular weight having growth promoting activity when incorporated into the diet of chickens, and to a process for obtaining these substances from extracts of condensed fish solubles. Two methods of operation of the process are best illustrated by the following examples.

Example 1: Two pounds of condensed fish solubles of sardine origin are employed in this experiment. Condensed fish solubles is the product obtained by condensing the water resulting from the hydrolytic extraction of oil from fish. Two pounds of the condensed fish solubles equal to a volume of approximately 400 ml are mixed with five volumes of methanol for a period of several hours to extract the organic material from the fish solubles and precipitate impurities including inorganic salts. The precipitated impurities are removed from the methanol filtrate concentrated in vacuo to a thick syrup which is slurried in approximately 500 ml of water.

The water-insoluble material is removed by filtration and the aqueous filtrate containing the extracted material is subjected to dialysis against distilled water at room temperature. The dialysis is continued until the solids content of the water extract drops to a constant value. The resulting large volume of dialystate is concentrated in vacuo to yield an aqueous concentrate of dialyzed material having a volume of approximately 500 ml. The dialysate of condensed fish solubles is then subjected to continuous elctrophoretic separation into fractions by the use of a commercially available electrochromatography or electrophoresis apparatus.

The electrophoretic fractions derived from the anode of volume of approximately 100 ml were evaporated under reduced pressure to produce a residue of approximately 10 ml. To this aqueous concentrate is added approximately 35 ml of methanol which precipitates a white crystalline solid.

The solid is removed by centrifugation and dried in vacuo and identified as the methanol-insoluble fraction. This methanol-insoluble fraction is characterized by the following properties: a low molecular weight; the presence of sulfur; slight solubility in water; slight solubility in hydrochloric acid and in aqueous sodium hydroxide solution; insoluble in ether, methanol, acetone and aqueous sodium bicarbonate; aqueous solutions of the methanol-insoluble fraction are stable on treatment with acid or on prolonged heating; decomposes without melting when heated to temperature in excess of 100°C.

The supernatant methanolic liquid is then cooled to approximately 0°C and acetone added dropwise with constant stirring until the solution becomes turbid. After the first appearance of turbidity, a few additional drops of acetone are added to effect precipitation of a solid crystalline material which is removed by centrifugation and dried in vacuo. This crystalline material is identified as the acetone-insoluble fraction. This acetone-insoluble fraction is characterized by the following properties: a low molecular weight; the presence of sulfur; slight solubility in water; soluble in hydrochloric acid and aqueous sodium hydroxide; insoluble in ether and aqueous sodium bicarbonate solution.

Example 2: The procedure of Example 1 is repeated omitting the initial methanol extraction step. Thus, two pounds of fish solubles equal to a volume of about 500 ml are dialyzed against distilled water until the solids content of the fish solubles drops to a constant value. The remaining steps of the procedure of Example 1 are then repeated including the electrophoresis and the methanol and acetone crystallization. The fractions obtained by this alternate method were substantially identical to the corresponding fractions obtained by the procedure of Example 1.

The activity of the two fractions obtained was determined by measuring the effect on the growth of chickens fed an otherwise nutritionally adequate diet. Three chick growth trials were conducted using the fractions isolated by the above procedure. The composition of the basal diet used in the experiments was adequate in all known growth essentials.

Day-old chicks of mixed sex were weighed and placed into groups which had a weight range of no more than 2 grams. The group of chicks were distributed at random into pens of 10 birds each. The diets were mixed according to the experimental outline and assigned to pens at random. There were four replications per diet. The chicks were housed in electrically heated battery brooders with raised wire floors, and feed and additives supplied ad libitum throughout the experimental periods.

An initial pen weight was taken at the beginning of the experiment and the chicks weighed weekly. The weights were recorded to the nearest gram. The diets fed to the chickens were prepared to test the ability of certain materials to stimulate the growth of newly hatched chicks. Thus, in addition to the basal diet, diets were tested which contained added condensed fish solubles and added solid crystalline fractions including the methanol-insoluble fraction and the acetone-insoluble fraction. Fractions were added to the basal diet as follows: 2% by weight of condensed fish solubles; 1 gram of crystals of methanol-insoluble fraction per 100 lb of basal diet; 10 grams of crystals of acetone-insoluble fraction per 100 pounds of basal diet. Results were as follows:

Treatment	Avg. Wt. Gain in Grams at 14 Days	Feed/Gain
Basal diet	98.40	1.28
Basal diet + condensed fish sol.	113.40*	1.29
Basal diet + methanol-insol. fraction	113.75*	1.36
Basal diet + acetone-insol. fraction	117.60*	1.32

*Differences are considered significant at wt. gain < 0.05 g.

The results of the experiments clearly show that stimulation in growth rate of newly hatched chicks occurs when condensed fish solubles, the methanol-insoluble fraction or the acetone-insoluble fraction is added to the basal diet.

Expanded Vermiculite Impregnated with Fish Solubles

Extensive research work conducted within recent years has established the existence of growth-stimulating nutrients in such natural animal feed ingredients as fish solubles, whey and alfalfa meal which are referred to as "unidentified growth factors." Fish solubles are not only distinctive in their growth stimulating properties, but are known to be troublesome in feed formulation because of their inherent hygroscopic properties. Alfalfa meal while also recognized as a source of unidentified growth factors, is normally not hygroscopic but as a carrier of fish solubles has rather specific limitations in arresting or correcting its inherent hygroscopic properties.

Nevertheless, these natural products because of their indigenous growth-stimulating properties have enjoyed wide acceptance among animal nutritionists and are particularly employed in so-called "starter rations" in the feeding of young poultry and swine with demonstrable economic advantages from the standpoint of effective and efficient utilization of their total feed consumption. However, of the various feed supplements, fish solubles still enjoys a relatively high degree of commercial acceptance.

Fish solubles have never been successfully dried commercially to a solids content substantially higher than 50% by weight due to the fact that the proteinaceous constituents possess such a high degree of hygroscopicity as to preclude their effective drying and particularly their subsequent dry storage.

In instances where fish solubles have been dried experimentally to a very high solids content the dried product has been found to absorb moisture so quickly and tenaciously, and thus become so tacky and sticky that handling and bagging the resulting product is commercially impractical. Furthermore, the product, even when it has been packaged, cakes so badly on storage that utilizing the cake material poses many troublesome problems in feed formulating operations.

M.E. Parker; U.S. Patent 3,130,054; April 21, 1964; assigned to W.R. Grace & Co. is able to produce a unique dry, relatively nonhygroscopic product of greatly increased fish solubles by mixing increased quantities of condensed fish solubles with relatively low concentrations of exfoliated vermiculite. Due to the enormous drying surface areas per unit of volume of exfoiliated vermiculite the moisture in the condensed fish solubles is much more readily removed without resort to the relatively high drying temperatures that are necessary in attempting to dry condensed fish solubles on a drum or rotary drier in admixture with leguminous carriers.

With the exfoliated vermiculite carrier, no recycling of the condensed fish solubles-vermiculite mixture is necessary when using a drum drier and without the operational difficulties encountered with the leguminous carriers in comparably low concentrations as constrasted with that of the condensed fish solubles even when its solids contents are as high as 80 to 85% of the dry weight of the finished product. Furthermore, the product is characterized by its extremely high, surprising and unexpected resistance to any subsequent absorption of moisture vapor even when the vermiculite concentration is in proportionately lesser amounts such as 10 to 20% by weight, for example.

Normally, exfoliated vermiculite, 95% by volume of which would be retained on a 30 mesh screen with less than 10% by volume retained on an 8 mesh screen, will enable the drying of a nonhygroscopic fish solubles product containing up to 65% fish solubles when using a conventional rotary dryer. However, to increase the fish solubles solids to 80 to 85% by weight, it will be necessary to recycle adding sufficient condensed fish solubles to the predried 50 to 65% fish solubles solids product so as to result in an ultimate final product containing 80 to 85% by weight of fish solubles solids.

"Trash" Fish from Shrimp Trawling

It is estimated that an average of six pounds of "trash fish" (fish not normally sold) are taken with every one pound of shrimp caught along the Texas Gulf

Coast. These fish are usually simply discarded at sea.

C.W. Keys and W.W. Meinke; U.S. Patent 3,249,442; May 3, 1966 describe a procedure for handling such trash fish during operation of a shrimp trawler so that high quality animal food together with various useful by-products can be derived from them. The process will be more readily understood by reference to Figure 6.1, which is a flow diagram illustrating the methods of handling fish according to this process and indicating the various fish products obtained by use of the methods.

FIGURE 6.1: PROCESSING OF FRESHLY CAUGHT "TRASH" FISH

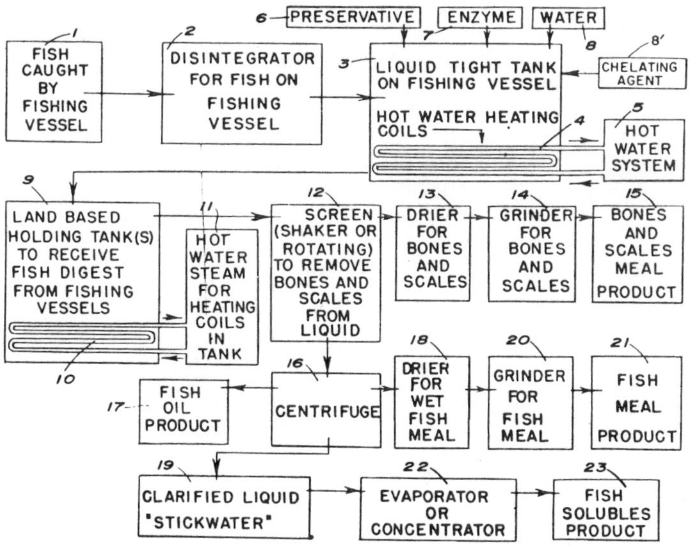

Source: C.W. Keyes and W.W. Meinke; U.S. Patent 3,249,442; May 3, 1966

Referring to the figure, the process follows immediately after fish are caught aboard the fishing vessel as indicated by block **1**. As the fish are caught they are fed to a disintegrator, chopper or grinder, indicated at **2** aboard the fishing vessel. The disintegrated fish are then fed into a liquid tank **3** aboard the fishing vessel which may, however, have an open top. Tank **3** may be fitted with heating or cooling coils **4** in the bottom of the tank.

The heat for the hot water system to supply the coils **4** in tank **3** may be utilized either from the waste heat of the engine of the fishing vessel, or from an independent fuel-fired hot water heater **5**. In many cases, no heat needs to be added, and alternatively, cooling provided by pumping sea water through the coils at sea water temperature may be utilized. The tank **3** is insulated to conserve heating or cooling and to prevent transfer of heat to or from other parts of the fishing vessel.

In most if not all instances, both heating or cooling of the various vessels and tanks may be eliminated completely, and the entire method, at least on the fishing vessel and possibly during further processing at a land installation, be carried out at ambient temperatures. Of course, temperatures below freezing may require some moderate heating simply to maintain the processing materials in a fluid state, and some heating may at times be desirable to accelerate the enzymatic process.

In general, the preservation and storage of fish material aboard the fishing boat and the transfer of digested fish to land-based holding tanks may be accomplished at ambient temperatures. Desirably, however, elevated temperatures are employed in the land-based tanks to aid in the separation of fish oil from the other components and to provide for cleaner separation of bones and scales from the liquid digest. Also, elevated temperatures are essential to the drying of the recovered fish solids and for concentrating the stickwater to fish solubles.

Additives introduced into the tank **3** are indicated as comprising the preservative **6**, enzyme **7**, and water **8**, and as noted hereinafter, a chelating agent is added, as indicated at **8**. When the fishing vessel reaches port, the contents of the tank **3** on the vessel are pumped to land-based tanks represented at **9** for storage and further processing, it being noted that additional water for pumping purposes need not be added.

These tanks **9** are equipped with hot water or low pressure steam coils **10** heated from the heater **11**, which will permit the attainment and holding of a temperature of 70° to 80°C for a period of thirty to sixty minutes. This heating is provided to inactivate enzyme activity and to permit easier removal of bones and scales from oil and liquid digest.

The land-based tanks **9** serve as auxiliary tanks whereby settling of bones and scales to the bottom of the tank and a top layer of oil is realized. This operation permits decanting of the oil layer, within limitation, from the aqueous liquid layer. Also decantation of aqueous liquid layer from the bones and scales can be realized. By this process hot rinse water at approximately 70°C. can be added to the drained bones and scales and thus remove more of the liquid digest from the bones and scales when they are conveyed to the screen separator **12**.

When the process is carried out on a factory ship at sea, a series of tanks to receive the disintegrated fish are utilized both as receiving tanks and settling tanks. The capacity of the factory ship in handling a certain number of tons of fish determines the number of tanks to be employed. On a factory ship at sea the operation of the process is the same. The only change in the flow sheet is that the land-based holding tanks are located on the factory ship as discussed hereinafter. The separation is accomplished by a shaker or rotating screen **12**.

The liquid separated from the bones and scales goes to a centrifuge **16** and a three-fold separation is realized, i.e., fish oil product **17**; a wet fish meal **18**;

and a clarified liquid **19** known as "stickwater" are produced. The wet fish meal, delivered from the centrifuge **16**, is dried in the drier **18** and ground at **20** to produce a fish meal product **21**. The clarified liquid **19** is concentrated by the removal of water at **22** by an evaporator or concentrator **22** to produce a fish solubles product **23**. This final fish solubles product **23** is acidified to approximately pH 4.5. The process as outlined yields four products:

(a) Bones and scales meal—protein, 28.2%; moisture, 8.0%; ash, 54.6%; oil, 4.6%
(b) Fish meal—protein, 60.7%; moisture, 8.0%; ash, 11.8%; oil, 4.8%
(c) Fish solubles—protein, 32%, total solids, 42 to 50%
(d) Fish oil—assay values cited are representative of experimental samples prepared from mixed rough or trash fish of the shrimper's trawl.

The two meals (a) and (b) and the fish solubles (c) are designed for use in mixed feed formulations, specifically for poultry feed formulations. Fish oil finds use as a human food and as an industrial raw material either for direct or further processing. The following example describes the process as related to a shrimp boat operation.

Example: The catch, representing a mixture of shrimp and fish, is emptied from the trawl onto the deck of the boat. Shrimp are sorted from fish, deheaded and iced in the hold of the boat. Fish and shrimp heads are fed into a grinder. Ground fish are discharged into a watertight, air vented tank in the hold of the boat. Water equal to 10% of the weight of the fish containing 10 to 400 ppm bacterial activity soluble Terramycin and 180 to 720 ppm bacterial activity of nf 180 (ppm based on fish weight) along with citric acid at the rate of 1 gram per pound of fish (0.2% by weight based on weight of fish, or 0.1 to 4.0%) is added or was present in the tank before ground fish was added.

This procedure is repeated with each trawl catch. In practice water may be charged into the tank before putting to sea. The charge of water is determined by anticipated catch of trash fish, 10 tons, 20 tons, etc. In this sitution only preservative, chelating agent and enzyme are added as sea operations. Under these conditions the boat can be at sea for a period of eight days.

The boat returns from sea and ties to the dock of the land-based operations. Here the liquid digest and undigested portions of the fish are pumped into holding tanks on land. Boat tanks are rinsed with small volumes of water and the rinse solution also pumped to land holding tanks. Finally, the storage tanks on the boat are sterilized by steam and/or disinfectants. This eliminates the possibility of development of a strain of bacteria resistant to the preservatives used in the process.

Land-based operations involve further digestion at 45° to 50°C, or at higher temperatures such as 70°C, decantation, dewatering of solids by shaker screens, centrifugation, concentration of liquid digest in an evaporator, drying of fish bones and scales for grinding to a meal, drying of undigested fish (sludge from the centrifuge) for grinding to a meal, and oil recovery. Products produced are fish solubles (32% protein), bones and scales meal (appoximately 30% protein), fish meal (approximately 60% protein), and fish oil; based on one ton of rough or trash fish caught in shrimper's trawl (Texas area), approximately 740 lb of fish solubles,

137 pounds of bone and scales meal, 58 pounds of fish meal, and 79 pounds of fish oil are derived as products from the preservative-enzyme fish handling process. These yields are from fish assaying 16% protein, 73.5% moisture and 4.5% oil.

The superior food product obtainable by the method of this process results from the fact that the fish are disintegrated as they are caught on the fishing vessel and the fact that the treatment process is commenced at once and occupies the time normally required for transportation of the fish to shore or for the catching of more fish. Thus the fish are disintegrated and treated in the nascent state and reduced to an agglomerate which may be pumped through a pipeline to the land-based holding tanks from which the agglomerate is treated for deriving the fish products as heretofore explained. No such superior results would be obtained using existing knowledge where the fish are transported to shore and thereafter subjected to treatment.

Method for Grinding Dogfish

It is recognized that the class of fish commonly known as "dogfish" have considerable commercial value for use as animal feed, fertilizers, etc. It is generally required, as a part of the processing of the fish for these and allied uses, that the fish be ground into small particles. One of the difficulties in grinding the fish is that the skin portion of a dogfish is composed of a very hard substance, consisting mainly of keratin and having a sandpaper-like texture, and being in consequence very damaging to the grinding equipment.

E.M. Buerman; U.S. Patent 3,383,216; May 14, 1968; assigned to Buerman Fish Products Co. describes a method for so conditioning the skin of dogfish that it is practical to grind the fish in a commercial process. The process also properly conditions the skin of the dogfish in a manner to remove the urea therefrom without damaging the flesh of the fish.

The process comprises: conveying the fish without interruption along a travel path which includes within its length a water bath maintained at a temperature close to 180°F and having a depth exceeding the thickness of the fish; subjecting the fish to a gentle downward pressure during its traversal of the bath to cause the fish to be completely submerged; and establishing for the conveyed fish as it traverses said bath a length of travel so correlated to the speed that the conveyed fish is immersed for a period of approximately eleven seconds.

Under these conditions, the skin of the fish becomes sufficiently softened so that the fish can, with relative ease, be then processed through a following grinding operation. Should the fish be held in contact with the 180°F water for unduly long periods, say a half minute, skin shedding takes place and the flesh beneath the skin becomes "cooked" to a degree making the same unpalatable as a feed.

A characteristic of dogfish is that they excrete urea through the skin, with a good portion of the urea contained in the fish being carried in the skin or along the skin line. The skin conditioning process described herein has been found to remove the urea in significant quantities, the urea passing into the water of the tank. Such a lowering of the urea content of the fish had been found highly desirable in conditioning the fish for some of the commercial uses to which dogfish are suited.

It has been found that even a very high concentration of urea in the water of the tank does not impede the skin conditioning procedure of this method so that the process can be carried on continuously for extended periods without requiring the water in the tank to be drained and replaced. Should it be desired, a continual "bleed" introduction of fresh water can be provided, while continually draining a corresponding volume from the tank, or small-quantity withdrawal and fill operations can be performed intermittently at moderately short time intervals so that a constant or near-constant water level is maintained.

Pelleted Feeds

Pelleting fish scrap or meal is desirable because it reduces the volume of the meal by about one-fifth to one-fourth thereby permitting greater loads of meat to be shipped in a carrier. The reduction in volume of the meal reduces the labor of "turning" the meal in open air bulk storage. Turning comminuted meal particles is necessary to prevent burning or rapid oxidation of the meal and subsequent charring, resulting in loss of protein and palatability.

Furthermore, fish meal, in an unpelleted condition, often becomes airborne due to high winds when the meal is transferred from one carrier to another or to a storage vessel in open air. On the other hand, pellets of fish meal are considerably larger and heaver than comminutes of fish meal and thus are not subject to being airborne as readily as particles of fish meal. Accordingly, transferring pelleted fish meal from one carrier to another or to storage in open air results in lower losses than occur when comminuted particules of fish meal are transferred in the same manner.

G.T. Lanz; U.S. Patent 3,359,115; December 19, 1967; assigned to Ralston Purina Company has developed a method in which the addition of condensed fish solubles to fresh fish scrap or meal produces a mixture which can be formed into durable pellets. Condensed fish solubles have about 50% dry matter and 50% water. They also have about 29% protein and a high content of certain B complex vitamins. They may contain about as much riboflavin as dried skim milk and they are rich in pantothenic acid and especially high in niacin. The process may be described as follows:

From a supply bin of fish scrap or meal, particles of fish scrap are fed by a conveyor to a press where the fish meal is pressed. The pressed fish scrap or meal, upon leaving the press may be fed to a de-watering grinder, although such a grinder is optional. The fish scrap is then fed into a rotary dryer. As the fish scrap enters the dryer, condensed fish solubles are delivered from a storage tank and sprayed on the fish scrap. An antioxidant, such as ethoxyquin, may also be added at this point.

The amount of condensed fish solubles added to the fish scrap or meal may be about three parts by weight or less of condensed fish solubles for every seven parts by weight of fish scrap or meal. The temperature within the dryer is usually in the range of 500° to 600°F, but may be higher.

A substantial amount of water is evaporated from the mixture while it is in the dryer so that the mixture has about 6 to 8% moisture content upon leaving the dryer. Rather than adding the antioxidant to the fish scrap or meal just as it enters the dryer, the antioxidant may be added to the mixture after it is discharged from the dryer.

The mixture of fish scrap or meal, condensed fish solubles and antioxidant is then fed into a grinder where it is ground, reduced in size and further mixed. Upon being fed out of the grinder the mixture usually includes one-sixth to one-fourth pound of condensed fish solubles for every one pound of fish scrap or meal, but these proportions may be varied. Considerable moisture was driven off from the condensed fish solubles in the dryer.

The mixture of fish scrap or meal, condensed fish solubles and antioxidant is then delivered to a surge bin from which it is fed by a feeder to a conditioner of a pellet mill. Additional condensed fish solubles may be added to the mixture just as it enters the conditioner. In the conditioner the mixture is heated and conditioned for pelletizing. The additional condensed fish solubles added to the mixture increase the lubricity of the product and act as a binder when the pellets are cooled and dried.

A rotating mixer in the conditioner thoroughly mixes the condensed fish solubles by the feeder. Upon leaving the conditioner the mixture is formed into pellets by the pellet mill. The pellet mill is conventional and under pressure produces short cylindrical pellets. Other shapes of pellets, such as rectangular or oval may be produced if desired depending upon the die in the pellet mill.

During the pelleting process particles are squeezed through dies which compress and heat the pellets. Strands or lengths of compressed particles are automatically cut to pellet length. As the pellets are ejected or discharged from the pellet mill they are in the form of tacky, slightly moist, plastic material. The pellets are in a heated condition, such as in the range of 140° to 180°F when ejected from the pellet mill and then passed through a cooler where a blast of air cools the pellets to ambient temperature. The moisture content in the pellets is between about 10 to 12% upon leaving the cooler. Upon leaving the cooler the pellets may be delivered to a storage unit.

PET FOODS

From Clam Waste

With the increasing demand for processed and canned foods for animals and pets, it is necessary to seek out new sources of economical food to be processed, food which at the same time is high in protein and nutrient value, and which is also appealing and appetizing to the animals or pets which are to consume it.

So-called "sea clams" (Mactra solidissma) have been used for human foodstuffs for some 20 years or more. However, certain portions of these sea clams have always been discarded in the shucking plants during the processing of the sea clams for human consumption. Sometimes the necessity to dispose of these portions has resulted in considerable expense in the clam processing. The portions of the sea clams which are discarded are generally referred to as clam "bellies" and they include the stomachs, livers and other organs. There may be, of course, small amounts of muscle attached which are left in the preparation of the clams as foodstuffs. These parts of the clams, hereinafter referred to as "clam waste," have never before been used.

J. Marvin and E.E. Anderson; U.S. Patent 3,017,273; January 16, 1962; assigned to Pet Kitchens, Inc. have found that by processing these clam wastes with a thickener, a composition may be made which has the desired consistency, texture and taste appeal for a canned animal food. In preparing the composition of this process, the clam wastes are cooled with the thickener or binder until the desired consistency is attained. The resulting uniformly textured and consistent material is then packed into cans, the cans sealed in a manner to create a vacuum therein, and the material sterilized. In as much as it is necessary to prevent any bacterial or enzymatic degradation of the clam wastes, it is necessary to process this material soon after its extraction from the meat.

In the processing of the clam waste to form the animal food it may, of course, be desirable to wash the clam waste to remove sand, shells and any excess salt. It may also be found desirable, although it is not necessary, to grind or otherwise macerate the material before mixing with the thickener.

The final thickened clam waste which comprises the composition of the process may be described as a soft, gel-like material which is cohesive, but which may be easily spooned out or otherwise divided into discrete portions. It may be further described as having a consistency of or similar to existing pet foods. The final material has an even and uniform consistency and texture, and there are no bones or other hard or undigestible pieces in the material.

The binder or thickener used with the clam waste is added to absorb the excess liquid contained in the waste and to give the final composition the desired gel-like consistency defined above. The binder or thickener is present in a concentration equivalent from 1 to 10% by weight of the clam material, the amount used depending upon the character of the binder and upon that quantity required to give the desired consistency.

The binder may be either a naturally occurring material or a synthetic material, or a combination of both. Suitable binders or thickeners may include cereals, meats, various starches and flours, gums such as gum arabic, locust bean gum, gum karaya and tragacanth, seaweed extractives, agar, pectins and the like. Among representative synthetic products which may serve as binders may be listed methyl cellulose, sodium carboxymethyl cellulose, propylene glycol esters of alginic acid and the like. The following example, is given to further describe the product process.

Example: Clam wastes were heated to a temperature of 160°F until the material had been converted into a relatively free-flowing liquid. To this was slowly added a quantity of low methoxy pectin, equivalent to 2% by weight of the clam wastes used. The mixture was maintained at the 160°F temperature and slowly stirred until the pectin had been thoroughly mixed therein and the composition resulting was a thick liquid of uniform texture and consistency. This thick liquid was then introduced at a temperature of 160°F into a size 300 by 407 can (15 to 16 oz), and the can sealed while the material was maintained at that temperature. Subsequently the material in the can was sterilized by exposing the can to superheated steam, maintained at 240°F. Sterilization was carried out for about 70 minutes.

When the can was subsequently opened the material contained therein was a dark gray gel-like, finely textured material which could be spooned out or otherwise removed in discrete protions without any crumbling, but also without any adhering of the material to the spoon or to other utensils.

From Frozen Fish

Notwithstanding the well-known fondness of cats for fish, the pet food industry has found from experience that whereas a given cat food prepared from fresh fish may be readily accepted by cats, a cat food identically prepared and formulated except for having been prepared from thawed frozen fish is so unacceptable to cats as to be an undesirable commercial item. The attendant disadvantage to manufacturers of pet foods is a serious one which is aggravated by the fact that the availability of fresh fish is seasonal and great economies could be had if it were possible to preserve fish by freezing them when the fisheries are glutted with fish and use the frozen fish later on at a time when fresh fish are scarce or unavailable.

This problem is one of long standing in the pet food industry. It has been realized heretofore that the freezing of fresh fish must be attended with some phenomenon which affects its acceptability to cats but it is not believed that prior to this process there have been any measures for successfully preventing or compensating for the loss of acceptability to cats of thawed frozen fish.

The process described by *J.A. Siregar; U.S. Patent 3,832,471; August 27, 1974; assigned to Liggett & Myers Incorporated* developed from research for the purpose of ascertaining whether there may be any detectable chemical difference resulting from freezing and, if so, whether any such difference is that which is responsible for the aforesaid difference in acceptability to cats. In the course of that portion of the research involving infrared spectrum analyses it was demonstrated that freshly killed fish contains 5'-nucleotides.

It also was ascertained by comparative studies that the 5'-nucleotides found to be present when the fish are freshly killed start breaking down due to enzymatic action as soon as the fish dies. It also was found that the enzymatic degenerative action continues at refrigerated temperature until the 5'-nucleotide content of the fish is lost completely or very nearly completely.

Based on these discoveries a cat food was prepared containing thawed frozen fish and there was included therewith as an additive a quantity of 5'-nucleotide in the form of a blend of disodium guanylate and disodium inosinate in an amount roughly corresponding to that present in the fish when the fish was freshly killed. When thus fortified with the added 5'-nucleotide the fish food prepared from thawed frozen fish was found to be as acceptable to cats as when the corresponding cat food was prepared from freshly killed fish.

A cat food which may be improved by the inclusion of a 5'-nucleotide or a mixture of 5'-nucleotides (including salts thereof) in the practice of this process may be formulated in any conventional or otherwise desirable way. The thawed frozen fish may constitute all or substantially all of the meat content or may be blended with other meats such as chicken or beef or animal parts such as liver or kidney. The cat food in addition to meat may contain various amounts of other ingredients such as meat meal, poultry meal, fish meal, soy grits, cereals, grains, vitamins, minerals, etc.

Usually the thawed frozen fish is the major proportion by weight of a cat food with which this invention is utilized. The total fish content usually is at least about 50% of the cat food. The following is a typical formulation for a prepared cat food in which 5'-nucleotide is present as an additive, the percentages being % by weight:

	Weight Percent
Thawed fish	60
Beef	10
Meat by-products	15
Chicken	14
Vitamins and minerals	0.6
Disodium guanylate and/or guanylic acid	0.01
Disodium inosinate and/or inosinic acid	0.01

Any conventional procedure for preparing the cat food and packaging it in containers suitable for distribution and storage may be employed. In typical preferred practice the thawed frozen fish is ground through a plate having ⅜" apertures. For other meats ⅜" apertures normally are preferable. The other ingredients including the 5'-nucleotide additive are then mixed with the ground meat and the mix is brought to a temperature of 160° to 170°F and cooked at this temperature for about 5 to 15 minutes. The cooked product is then pumped to a filling machine which fills the product into cans. The cans are then sealed and the contents sterilized as by placing the cans in a retort and holding them at 250°F for 50 minutes.

MINK FEED

Rough Fish Formed into Cakes

Species of fish in certain parts of the world have changed gradually over the years. In many regions, fish of high human food-value have almost disappeared and have been replaced by less desirable species known as rough fish, not suitable for human food. However, the animal food market represents a very large outlet for these less desirable fish, but many factors have prevented economic success in this area.

Mink farms provide a potentially large outlet for these fish since raw fish are generally regarded as an excellent food supplement for animals. However, mink ranchers do not routinely accept these raw fish because:

(1) The composition of fish often varies seasonally which necessitates constant changes in ration formulation.

(2) Regular supplies of fresh fish cannot be had owing to the seasonal nature of the fisheries.

(3) Frozen raw fish, stored to compensate for off-seasonal slack, undergoes an undesirable gradual lowering of quality.

(4) Many of these fish contain the enzyme thiaminase, which destroys thiamin (valuable health vitamin B_1). Such fish are avoided by mink ranchers, some of whom cook all their fish as a preventative measure against thiaminase contamination and against other harmful enzymes and bacteria contamination, before serving the fish to their minks. These methods often are not adequate to ensure destruction of all harmful enzymes, and often lower the quality of

even fresh fish. These rough fish also represent a good food source for zoo animals, but the problems, economic and otherwise, that are encountered with the feeding of minks are also encountered with zoo animals, thereby restricting the desirability of these fish in this market, also.

R.H. Gnaedinger; U.S. Patent 3,429,710; February 25, 1969; assigned to the U.S. Secretary of the Interior provides a method and apparatus for economically processing these rough fish in order to produce a form of fish very desirable to the animal food market. More specifically, the object of this process is to provide a quality, uniform, stable, easily storable, non-thiaminase-active product by an inexpensive cooking process and apparatus that can be operated at dockside or aboard ship. The process may be described as follows.

Raw fish is conveyed by a helical screw through a meat-type grinder with, e.g., ¼" diameter holes in a perforated plate. The ground fish is then fed into a cooking chamber by a feed mechanism such as a piston-type sausage stuffer. This stuffer forces the ground fish between two closely spaced perforated metal plates which protrude from the cooking chamber, spaced ½" apart.

These plates extrude the ground fish into a thin layer before it enters the cooking chamber and maintain the fish in the thin layer while it is contacted with jets of live steam which are delivered uniformly through perforations in both sides of the cooking chamber.

Under this arrangement, the ground material can be heated rapidly to a relatively high temperature to destroy harmful bacteria and enzymes including the thiaminase, as opposed to a slow heating process which can result in loss of valuable elements in the fish, and which can increase the formation of undesirable odors and flavors. One pound of raw fish can be heated up to 180°F in less than one minuet during continuous passage through the cooking chamber of this process. Once the raw fish enters the cooking chamber, the steam pressure within the chamber aids in moving the material through the rest of the chamber.

Material emerging from the cooking chamber is ground through a meat grinder similar to the one already described, which further makes it possible to regulate the rate of the fish through the cooking chamber thereby controlling the rate of cooking. The cooked fish emerging from the grinder falls directly into a press cage and is pressed, e.g., for 5 minutes at a pressure of 10 to 15 psig. To minimize loss of heat during filling and pressing, the cage is insulated. Pressed material in the form of a cake is packed in plastic bags and allowed to cool at room temperature for about 1 hour before freezing.

To enable compression of the fish in the press cage the rough fish must be cooked sufficiently. A minimum temperature of 180°F for the cooked product is necessary to achieve this. In order to destroy the undesirable thiaminase enzyme that is present in most rough fish, the cooked fish must be held at 180°F or above for several minutes (e.g., 5 to 10 minutes) during the process.

No matter what particular rough fish is employed as the feed material, the presscake produced by this process is of a relatively uniform composition. Such a non-thiaminase-active, compact uniform product is very desirable to mink ranchers and other animal food outlets.

Only one-half the storage space required by raw fish is required by the press-cake product, and the stored product is more stable than the raw fish since undesirable enzymes, bacteria, etc., have been destroyed by cooking thereby preserving the desirable qualities of the original fish. This capability of the press-cake product makes it highly suitable for storage during glut seasons.

Dry Feed Composition

Being carnivores, mink prefer to feed upon raw meat or fish, and since freshness of the fodder and a well-balanced diet are essential to obtain high quality furs, mink farmers generally prepare their own mink feed from current supplies of fresh offal, comprising surplus fish, and/or offal from slaughter-houses and meat consuming industries, supplemented by various fodder additives, such as vitamins and mineral mixtures.

Unless deep freezing facilities are available, the fish and/or meat components of the feed should not be more than 48 hours old on consumption, since otherwise a substantial deterioration will have set in, greatly detracting the feed value. Thus, the enzyme thiaminase being present in the fresh fodder will eventually decompose vitamin B_1, causing an avitaminosis in the animals, if the feed is not fresh, and offal from slaughter houses may be infected with *Clostridium botulinum,* which produces the dreaded botulinum toxin.

Also, surplus fish or fish offal may, at certain seasons, comprise or consist of fish, such as coalfish, whitings, hakes and turbots, the use of which has proved the cause of certain fur defects, if the fish constituents are not subjected to a heat treatment.

Owing to the above factors, and also because mink stock as well as supplies of the above fodders are subjected to seasonal variations, it has been attempted to supply a feed composition, and more particularly a heat treated dry feed composition, for the feeding of mink, in an attempt to eliminate seasonal variations, at the same time sterilizing the feed composition and destroying the thiaminase and the botulinum toxin, if any.

The results of using such dry feed compositions based upon fish meal or the like, however, have not come up to expectation. Not only are these feeds apparently insufficient, but being in a pulverized form they are of a fineness incompatible with the wishes of the mink farmers since, as is well known, mink fodder is generally placed on a wire netting ceiling of a mink pen. This does not give difficulties if the fodder is minced meat or fish, but pulverized fish-meal or the like will run through the netting of the pen, and if it is mixed with water, it will be unable to form a mass sufficiently viscous or solid to correspond to the consistency of minced meat or fish.

E.N. Riisberg; U.S. Patent 3,529,967; September 22, 1970; assigned to Lovens Kemiske Fabrik Produktionsaktieselskab, Denmark describes a process the main object of which was to provide a dry mink feed composition equalling freshly prepared mink feed as regards quality, but being storable for weeks or months without noticeable deterioration, thus making the mink farmers independent of daily supplies of fresh fish or meat components, and the transportation and refrigeration problems involved in avoiding deterioration.

The mink feed composition comprises from 30 to 60% of fish meal having a protein content of at least 60%, and a content of available lysine of at least 6%, and in addition to the fish meal, the feed composition further contains 1 to 10% of water-solubles, low molecular weight proteins or peptides, or both, whereby the basal demands of the animals are fulfilled as far as proteins are concerned, which is not the case if the fodder contains fish meal only.

Furthermore the composition contains from 10 to 40% of carbohydrates, which substantially or at least partially consists of dried boiled potatoes, or dried boiled grain, such as barley, wheat, oats, or maize, or both, giving the feed a consistency, when mixed with water which permits it to be placed on, and supported by, the wire netting generally used for minced meat for mink feeding. All percentages are by weight.

In a preferred mink feed composition of the process, the fish meal is a spray dried fish meal, which is poor in fats. In producing a fish meal of this kind, a preferred method consists in comminuting fresh fish and gently heating the comminuted fish to a temperature not exceeding 60°C, removing the oil-containing water phase by centrifuging and spray drying the remainder.

Fish meal produced from such fish as cod, plaice, flounder, lemon sole, or from such fatty fish as herring, mackerel, weever, or sand launce may be used. It is important that the fish meal is produced from fresh fish, and that antioxidants are added to prevent the fish meal from becoming rancid, particulary if fatty fish are used.

Example: A mink feed in pellet form was prepared from the following components:

	Weight Percent
Fish meal	47
Dry boiled potatoes	18
Lard	9
Sugar	8
Molasses	5
Casein	4
Fish solubles	2
Barley husk meal	4
Dry yeast	2
Vitamin preparation	0.88
Combined iron preparation	0.12

The vitamin and combined iron preparations were those obtained commercially. This feed in pellet form was compared with a conventional mink feed prepared from fresh fish et cetera, so as to contain substantially the same amounts of proteins, fats, carbohydrates, vitamins, and micro-minerals.

For the experiment, 50 female kits and 50 male kits were used, being equally divided after weight and sex into a control group and an experimental group. The average weight before slaughtering, and the average fur values are given as follows.

	Controls	Pellets
Males, g	2,110	2,190
Females, g	1,140	1,130
Fur value	100	107

FOOD FOR DOLPHINS AND SEALS

Small cetaceans (e.g., porpoises, dolphins) and all pinnipeds (e.g., seals) eat primarily fresh fish. However, due to the perishable nature and limited storage life of fresh fish, it is undesirable as a regular food for such captive animals. Further, raw fish food is a potential health hazard to such animals because of the occasional presence of infectious microorganisms or parasites. Accordingly, there is great interest in a stable pasteurized food ration which will be readily accepted by the animals.

M. Patashnik; U.S. Patent 3,726,689; April 10, 1973; assigned to the U.S. Secretary of the Interior developed such a product and process for producing it. Generally, the product is composed of a protein gel consisting of all the ingredients in a raw, whole fish plus a salt selected from the group consisting of NaCl, KCl and mixture thereof. Waterholding, coagulated myofibrillar protein is the basic gel constituent. Homogeneously dispersed therein are (1) globules of the oil (fat) fraction of the whole fish, and (2) particles of all the other fish constituents and the salt, which particles range in size from ordinary matter ($>10^{-5}$ cm) through colloidal (10^{-7} cm to 10^{-5} cm) to true solutions ($<10^{-7}$ cm).

To produce such a ratio, the aqueous fraction of the fish must first be salinized by the addition of the salt, in which environment myofibrillar protein in the fish is soluble. The fish is finely comminuted so that myofibrillar proteinaceous matter in the fish more readily dissolves in the aqueous fraction. In this dissolved state the protein acts as a surfactant or emulsifying agent whereby globules of the oil fraction of the fish can be dispersed in the aqueous fraction to form an emulsion. Once the emulsion is formed, dissolved myofibrillar protein in the continuous aqueous phase is coagulated to form a gel.

Despite the presence of viscera, skin and other waste portions of the fish, a suitable emulsion and a final gel product are formed from the whole fish. To eliminate potential health hazards and to increase its storage life, the gel product can be pasteurized. By proper control of the process variables, the final product can be prepared to have the textural elasticity of fresh, whole fish, and to float on water.

M. Patashnik and M.C. Keyes; U.S. Patent 3,730,728; May 1, 1973; assigned to the U.S. Secretary of the Interior then found that the inclusion of kelp (seaweed) in this basic product and process yields a gel with textural elasticity more closely approximating that of whole fresh fish. Further, the water-retention capabilities of the gel are improved whereby there is greater gel stability. Finally, the presence of kelp acts as a food supplement if the product is employed as food for captive animals. Generally, the particle size of the kelp in the gel product is in the range of ordinary matter ($>10^{-5}$ cm).

In the practice of the process, kelp and NaCl, KCl or mixtures of KCl and

NaCl are added to the aqueous fraction of the fish preferably by coarsely grinding the raw, whole fish, in the presence of the kelp and salt, to a substantially homogeneous mass having a hamburger-like consistency. The salt is incorporated in amounts of 0.5 to 3.0 wt % of the raw whole fish.

With regard to the kelp, it is preferably added in an amount of 1 to 5 wt % of the whole fish. The optimum kelp content, which may be outside this range, is best determined experimentally for each particular feed material. A food-type screw plate grinder having holes of ⅛ to ⅜ inch and/or a silent cutter-type machine is suitable for this coarse grinding operation. Alternatively, the water can first be extracted from the fish in the prior art manner, and then the salt and kelp added thereto, provided that the emulsifying and coagulating properties of the myofibrillar protein are retained during water extraction.

To dissolve myofibrillar fish protein in the now-saline aqueous fraction of the fish while essentially simultaneously homogeneously suspending and dispersing the oil fraction of the fish in the aqueous fraction along with all the other fish constituents, the "hamburger" is passed through an emulsifying apparatus or colloid mill.

Such a machine comminutes fish solids to colloidal size, and thereby enables myofibrillar protein to rapidly go into solution in the aqueous phase, and to act as a surfactant or emulsifying agent therein. During such comminution, globules of fish oil are formed and become dispersed in the aqueous phase. Further, all the other fish constituents become homogeneously dispersed in the resultant paste-like emulsion.

During the steps of coarse grinding, emulsification, and final mincing, it is important that the operating temperatures be maintained low enough to prevent premature coagulation or gelling of the myofibrillar protein. These temperatures may vary considerably from one operation to another depending on such factors as the fish species, the age of the fish, the time of catch, the bone load, etc. Generally speaking, during coarse grinding the temperature of the feed and product should be maintained below about 40°F, preferably about 20° to 35°F; and during emulsification and final mincing these temperatures generally should be below 60°F.

Once the emulsion is formed, it is gelled or coagulated. This can be accomplished in a number of ways depending upon the particular feed material and the desired use of the end product. The particular gelation technique is best determined experimentally in each case. Some emulsions can be sufficiently gelled by storing for several hours at a temperature of 32° to 70°F. Others require heat (e.g., 80° to 140°F) for gelation.

Some emulsions can be preliminarily gelled at lower temperatures, but then must be heat set at elevated temperatures. In many instances, heating accelerates coagulation, and thus immediately after formation of the emulsion, it can be heat coagulated at elevated temperatures or it can be stored a few hours at 32° to 70°F for preliminary coagulation and then heat coagulated.

Pasteurization or sterilization of the product is important if the long range normal health of the food recipient is of interest. Pasteurization is accomplished generally by heating the gelled material with water or steam so that a temperature of 180°F is obtained at the center of the product for at least 30 minutes. In some instances, the pasteurization step also serves as a heat setting step for a preliminarily gelled emulsion, while in other instances the emulsion, in an ungelled state, can be simultaneously pasteurized and completely coagulated. The temperature of the pasteurization heating medium can vary from 180° to 250°F to effect, as desired, slow or rapid pasteurization (and, perhaps, gelation or heat setting). Sterilization may be achieved by conventional methods.

After pasteurization, the product is cooled at about 32° to 70°F and then stored under refrigeration conditions for future use. To insure against spoilage and to maximize the retention of textural and cohesive properties of the product, it is best frozen and stored at 0°F or lower if it is not to be used within 4 weeks.

In its basic form the gel product after emulsification will float on water. However, by agitating the emulsion, prior to gelation, in a high speed blender or vertical type cutter, it can be substantially deaerated and made nonfloatable.

Example: 15 lb of frozen herring caught at Puget Sound, Washington, was ground to hamburger-like consistency in a food-type cutter blender in the presence of 1.0, 2.0, 0.2 and 1.7 wt % of NaCl, kelp, sodium tripolyphosphate and water, respectively. The temperature of the ground product was maintained below 40°F. Next, the "hamburger" was passed through an orifice type colloid mill, the feed and product being maintained below 55°F. Next, the product was stored overnight at 34°F, and then pasteurized with steam (180° to 190°F). After storing in a refrigerator for 21 days it was offered to a female killer whale and readily accepted.

GEL FOR FISH TANK FEEDING

E. Schneider; U.S. Patent 3,591,389; July 6, 1971; assigned to Wardley Products Co., Inc. has developed a fish food which may be placed in a fish tank in the form of a gel. The fish food contains 80 to 92% by weight of an edible food substance, from 0.5 to 5.0% by weight of an organic binder, 0.5 to 5.0% by weight of guar gum and 0.5 to 5% by weight of a chelating agent and is added to water which is at a temperature of from 120° to 212°F in an amount of one part by volume of powder to 3 to 5 parts by volume of water.

This fish food will be maintained in the form of a gel for at least 4 days. In this manner, the food will be presented in a form that is readily available for the fish in the tank to feed upon. Furthermore, the food will be preserved in the gel in a fresh state. The gel maintains the food in a palatable form such that the fish can feed thereon ad libitum without danger of decomposition even after prolonged periods of standing in the tank. Additionally, there is no danger of the fish food swelling and decomposing after long periods of standing in the tank, which swelling and/or decomposition occurs with existing prepared fish food.

This fish food composition can contain any edible fish food substance in powdered form. Generally, it is preferred to utilize a fish food substance having a high protein content. Among the edible fish food substances that can be utilized in accordance with this process are included shrimp, clam meal, fish meal, oyster meal, crab meal,, etc. Generally the edible food substance is present in the composition of this process in an amount of 80 to 92% by weight of the composition.

Furthermore, the edible food substance is present in the form of a finely ground powder. Any conventional organic binder can be used in the composition of this process. Among the conventional binders are the sugars such as corn sugar, fermented corn sugars, gums such as locust bean gum, sodium alginate, ammonium alginate, alginic acid, starches such as corn starch, pregelatinized wheat starch, carboxymethylcellulose and agar agar.

A preferred binding agent is a mixture of fermented corn sugar and locust bean gum. The organic binder or binding agent is usually present in the composition of this process in the form of a fine powder and in an amount of 0.5 to 5% by weight of the composition.

The composition also contains guar gum in an amount of 0.5 to 5% by weight. The guar gum in combination with the organic binder produces a powder composition which will form a firm gel when added to water. In addition to the edible food substance there is also included in the composition a chelating agent.

Among the preferred chelating agents which can be utilized is a mixture of borax and citric acid. The chelating agent is present in the composition of this process in an amount of 0.5 to 5% by weight. A desiccant is added to prevent the take-up of moisture from binding the ingredients which otherwise could offset the gelling properties of the composition. If desired, the composition can include a preservative.

The gelling fish food composition which is composed of a solid edible food substance, an organic binder, guar gum and a chelating agent is formed by mixing all of the ingredients in their solid form. Any conventional means of mixing can be utilized in preparing the dry fish food composition of this process. When this dry fish food composition is to be utilized, the fish food is added to water which is at a temperature of 120° to 212°F. The fish food composition should be added to water in an amount of about 1 part by volume of powder to about 3 to 5 parts by volume of water.

Upon mixing the 1 part of fish food to 3 to 5 parts of water at 120° to 212°F the amorphous agglomerate mass is formed. This agglomerate mass is in the form of a firm gel. After the firm gel is formed, the gel is introduced into the fish tank.

Example 1: A fish food composition was prepared containing the following ingredients.

Ingredients	Amount, grams
Dried ground shimp	62.0
Kelgum*	3.6
Guar gum	2.7
Citric acid	0.9
Borax	0.9
Methyl paraben	0.7
Sodium propionate	1.4
Iron oxide	0.7
Dicalcium phosphate (or any suitable desiccant)	0.7

*A mixture containing locust bean gum and microbiologically fermented fermented corn sugar.

Each of the above ingredients was dry weighed and mixed by tumbling. The resulting product was a powdery substance. One teaspoon of this substance was mixed with three teaspoons of water at 145°F. A firm gel resulted. After exposure to air this gel was placed in a beaker of water at room temperature.

Example 2: A fish food composition was prepared in the same manner as Example 1 containing the following ingredients.

Ingredients	Amount, grams
Oyster meal	50
Kelgum	3.6
Guar gum	2.7
Citric acid	0.9
Borax	0.9
Methyl paraben	0.7
Sodium propionate	1.4
Dicalcium phosphate	0.6

One heaping teaspoon of the above composition was mixed with three teaspoons of water at 140°F. A firm gel resulted.

FISH PROTEIN CONCENTRATES

Although the FDA-approved number of fish varieties has now been increased to six from the original three, most people feel that a new look at these rules should be given, since under present regulations FPC cannot possibly be produced economically enough to spur commercial research in this potentially valuable field. In fact, not even the Bureau of Marine Fisheries is pursuing research on the subject of FPCs at present, since the pilot plant in Aberdeen, Washington lost its Congressional funding in 1973.

Astra Nutrition AB and the National Biscuit Company have jointly formed a company called Nabisco-Astra Nutrition Development Corporation to do research and to market Astra's FPC in underdeveloped countries. Other commercial research and development in the United States seems to have come to a halt.

It has become obvious to scientists and economists that if seafood is going to make any really important new contribution to the world protein supply, it will have to be manufactured in a form which meets certain requirements. First, currently unused species will have to be utilized, since desirable forms of fish are largely being fished to near capacity at present; second, the fish and/or shellfish will have to be used in other than fresh-caught form, since fresh fish is very subject to rapid deterioration, and prevention of this deterioration is very expensive; third, the material will have to be economically attractive to produce and sell; fourth, the material will have to be capable of being stored and shipped easily, and will have to be acceptable to consumers who do not now eat seafood.

Work has been done in the paste twenty years in many countries on producing a product which would meet these criteria, with the result that substances known as fish protein concentrates (FPCs) have been developed. These materials may be used in several ways. As a supplementary livestock feed, FPC improves the nitrogen intake and growth rate of animals, and as a bland, dry flour it may be incorporated into various human foods such as flours, cereals, baby foods, soft drinks, etc. Very small amounts of FPC, on the order of 10 to 20 grams a day, produce enough protein to adequately supplement an average diet in the protein-poor sections of the world.

Physical methods for producing fish protein concentrates usually produce fish meal, an important and useful product, primarily sold for animal and poultry feed. Fish meal made from anchovies was Peru's largest export until two years ago, when the anchovy harvest almost ceased. In 1972, about 752,000 metric tons of fish meal and solubles were used in U.S. feeds, but this had shrunk to 348,000 metric tons by 1974.

The largest producer of FPC in the world, Astra Nutrition AB of Sweden, makes an eviscerated fish protein from fish meal which is produced on a huge factory ship and then processed by a four-stage solvent extraction in their Swedish plant.

Chemical methods for preparing FPC by solvent extractions have also been used commercially by Alpine Marine Protein Industries and by Cardinal Proteins Incorporated of Nova Scotia. The method developed by the U.S. Bureau of Commercial Fisheries uses isopropyl alcohol to extract the solids from red hake in a three-stage extraction system. Biological methods for the production of high quality protein hydrolysates by yeast and enzymes have also been developed.

The Food and Drug Administration set up very strict rules for the manufacture of FPC for human consumption, specifying the varieties of fish which could be used, limiting the amount of fluorides and mercury allowable, and forbidding the use of bone or viscera unless complicated marking procedures for packaging were followed.

PREPARATION BY PHYSICAL MEANS

Apparatus for Making Fish Meal

Fish meal and oils are presently produced from inedible portions of fish, though some industrial or scrap fish, such as menhaden, are caught only for fish meal production purposes. Fish meal is currently produced by a dry-rendering and by a wet-rendering process. Solvent extraction methods are also used.

In the dry-rendering operation, the raw fish are ground and dried under vacuum in a steam-jacketed dryer. The dried meal is pressed hydraulically to remove excess oil which is then purified by use of a settling tank or centrifuge. The raw fish for wet-rendering are ground, cooked and pressed in a continuous screw type device to remove excess oil and water. The fish meal is dried in direct flame, steam-jacketed or steam-tube dryers.

The press liquor is centrifuged to separate the oil from the other constituents of the liquor and purified for various industrial uses. The disadvantages of these processes for producing fish meal include relatively high energy consumption during the preparation of the fish meal, inefficient drying methods and cumbersome processing equipment. Further, excessive liquor material is retained within the fish after the pressing operation, which must be removed for the attainment of the desired fish meal. In addition, excessive time and equipment is presently required to produce fish meal and extracted oils therefrom.

G. Overton; U.S. Patent 3,497,364; February 24, 1970 provides an apparatus for continuously producing protein-meal and liquor extracts from various protein-containing materials by subjecting the protein materials to a chemical action such as cooking or treatment with chemicals to disrupt the material, and change their natural order, applying a relatively high pressure to the disrupted protein materials and separating the protein materials from any skeletal matter; thereafter abruptly releasing the pressure to cause cellular rupture of the protein-containing materials to form a slurry thereof and then dispersing the protein slurry into a pond on a heated rotating drying surface where the slurry is kneaded or worked with the application of regulated pressure to separate liquor extracts from the protein-meal; removing the liquor extracts and spreading a relatively thin layer of the protein-meal onto the drying surface and removing the protein-meal as it becomes dried.

In Figure 7.1a, a continuous process for the production of protein-meal is shown, where protein-containing material, such as fish, etc., is stored in a storage container **11**. The protein-containing material, which may include not only whole fish but waste scraps thereof, including heads, tails and other inedible portions, may be conveyed directly to the storage container from a suitable source of supply or, if desired, it may be first processed through a hammermill, chopper or other similar device to reduce it to uniformly sized particles which are then suitably conveyed to the storage container.

The storage container has a suitable discharge mechanism (not shown) which permits the material to pass continuously into a cooker unit **12** where it is substantially uniformly rendered pliable, to be readily freed from skeletal matter or to render the skeletal matter edible. The protein-containing material is maintained within the cooker unit for a period of time ranging from 20 to 150 seconds to partially cook it at a temperature of 100° to 160°C. Chemicals such as nontoxic acids or bases can be used to soften it.

The cooked material is then continuously passed into a pressure applicator **13** where relatively high pressure, generally in the range of 200 to 1,000 psi, is applied. The pressurized protein-containing material is then continuously passed through a screener **14** to remove any skeletal or bone matter. The skeletal or bone matter is collected in the appropriate station **15** for utilization as fertilizer or the like.

FIGURE 7.1: APPARATUS FOR MAKING PROTEIN-MEAL

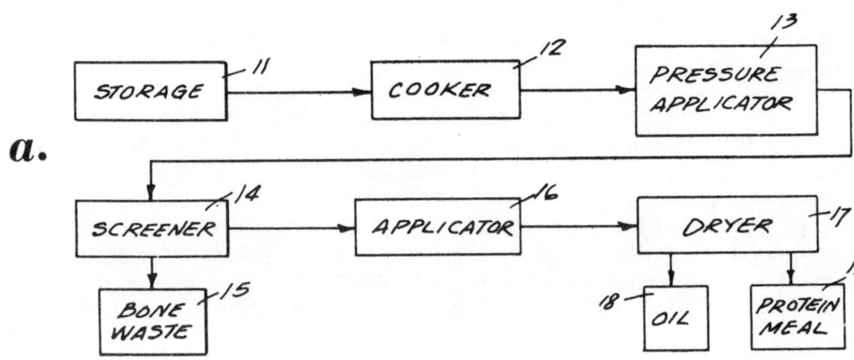

(continued)

FIGURE 7.1: (continued)

b.

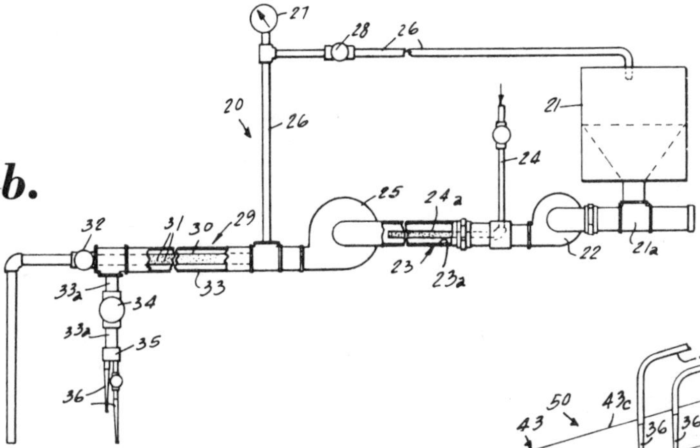

c.

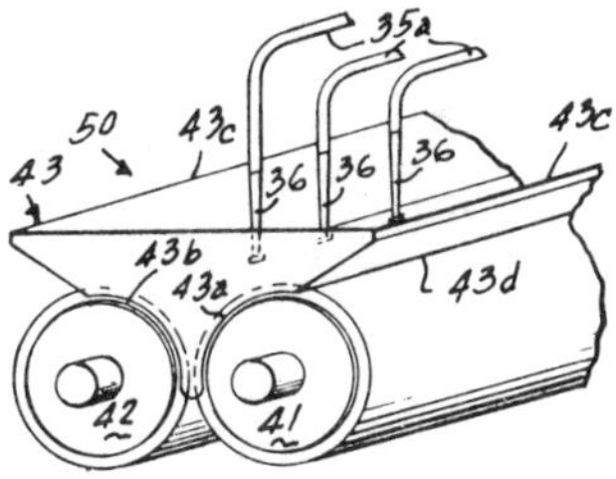

Source: G. Overton; U.S. Patent 3,497,364; February 24, 1970

The protein containing slurry is then continuously passed to an applicator **16** whereat the pressure is abruptly and suddenly released causing substantial cellular rupture of the protein material, releasing liquors, oils, etc. normally present within these protein-containing materials, so as to form a protein slurry.

The protein slurry is next passed to a dryer **17** in which substantially simultaneously regulated pressure is applied to the protein slurry causing a separation of the protein-meal from its liquors. The liquors are removed at an appropriate station **18** for oil extraction and the like, and the protein solids are efficiently and controllably dried for removal at a station **19** to become relatively highly concentrated protein-meal.

Figure 7.1b (elevational view), a first portion **20** of the apparatus for the production of protein-meal is shown. A storage container **21** receives whole fish, fish scraps, fish waste and similar protein-containing material from an appropriate source (not shown), such as a conveyor belt or an appropriate particle sized reduction device. The storage container may include or be preceded by a hammermill (not shown) or a similar device to render the fish into a suitable uniform mixture containing particulated protein-containing materials, natural juices, skeletal matter, etc.

The storage container is connected by an appropriate conduit means **21a** to a pump **22** which removes the protein-containing mixture from the storage unit and forces it into a cooker unit **23**. The pump **22** is of a standard involute-type pump capable of moving solid containing mixtures and preferably it develops a pressure of 10 to 100 psi. The cooker unit has a relatively large diameter heat-insulated imperforated hollow cylindrical body **23a**, the interior of which is preferably stainless steel or similar nontoxic material.

The cooker unit has means **24**, such as a steam inlet, to heat the protein-containing mixture substantially uniformly to about 100° to 160°C in a time range of 20 to 150 seconds. The cooked mixture is then continuously forced into a relatively high pressure pump **25** which preferably develops a pressure of at least 200 to 1,000 psi.

The pressurized mixture is then force fed into a screener unit **29**, with an excess portion of the mixture passing through conduit **26** to activate pressure gauge **27** and continue through valve **28** back to the storage container **21** to again pass through the cycle. A substantial portion of the pressurized mixture is pressure fed from pump **25** into one end of a hollow perforated inner cylindrical body **30**, which has substantially uniform spaced perforations **31** extending substantially uniformly over its peripheral surface in a size range of 0.50 to 0.0156 inch. The opposed end of the hollow perforated cylindrical body is suitably sealed by a pressure valve **32**.

Surrounding the inner perforated body is an outer imperforated body **33** which is substantially coextensive with the inner perforated body at least from the delivery end of the pump up to the pressure valve. As the mixture is forced into the inner peripheral body, the pressure forces the deteriorated cooked fish through the perforations **31** and into the outer imperforated body, thereby separating skeletal or bone matter from the cooked material. The screened protein-containing material is then pressure fed through the outer cylindrical imperforated body through conduit means **33a** and pressure valve **34** into the distributor unit or dispenser **35**.

The skeletal matter remaining within perforated body **30** is packed against the pressure valve and at intervals, passed through for collection and use as fertilizer, bone-black, etc. The pressurized mixture is distributed by dispenser **35** into a plurality of applicator nozzles **36**, whereat the pressure is abruptly released, causing substantial cellular rupture of the protein-containing material, releasing various oils, water, and other liquors within the protein-containing material from the solid fibers thereof. The ruptured protein-containing material is applied by the nozzles to appropriate drying means.

The pressure from the nozzles will be about 500 to 2,000 psi, depending upon the diameter of the discharge port in the nozzles. This relatively high pressure is abruptly and suddenly released as the screened protein-containing material is shot out of the nozzles onto the drying means causing the liquors, etc. within the protein-containing material to stretch and burst the cellular walls thereof, so as to rupture the cells and expose the oils, moisture, liquors, etc. thereof for subsequent removal. The ruptured material thus forms a protein slurry for further processing.

In Figure 7.1c, a second portion (end view) of the apparatus for the production of protein-meal and liquor extracts is shown. A pair of dryer drums **41** and **42** are mounted for rotation in a conventional manner to define therebetween a working nip. The dryer drums are provided with adjustment means to vary and control the spacing at the nip. Preferably, the dryer drums are internally steam heated to obtain a peripheral surface temperature of 100° to 150°C to dehydrate the protein solids so that they contain less than 10% moisture and at least up to 70% proteins.

Mounted above the dryer drums is a housing shield or trough **43** which forms end-dams and side-dams to prevent the protein slurry from escaping beyond the axial peripheries of the dryer drums.

Immediately above the nip and in close-running relationship with the peripheral surfaces of the dryer drums, a straining means is appropriately mounted. This may be in the form of a substantially rigid elongated cylindrical member, which is hollow, and provided with a multiplicity of substantially uniformly spaced fine perforations. The straining means may be of any other appropriate form having a fine-screened peripheral surface which is capable of passing the oils, moisture and other liquor extracts from the protein slurry so as to substantially separate the protein liquors from the solids thereof. The perforations are generally in the size range of 0.01 to 0.005 inch.

Centrally mounted along the axis of straining means is a conduit for directing and removing the protein liquors within the strainer away from the working nip. The conduit means may be subjected to reduced pressure, such as partial vacuum, to aid in the liquor removal. The removed protein liquors may then be appropriately purified, separated and classified to obtain the desired protein oils, fish meal, and so forth for numerous industrial uses and commercial products.

The strainer means is provided with a plurality of reciprocably movable agitators to displace protein solids that might otherwise tend to cake or adhere to the peripheries of the strainer. Doctor blades are provided to remove the dried protein-meal from the surfaces of the dryer drums. In working relationship with the doctor means are suitable apron means for directing the particulated protein-meal into suitable troughs for conveyance to packaging stations or the like.

As the dryer drums are rotated toward each other, they will tend to circulate and knead or work the protein slurry to cause a separation of the various protein oils and other liquors from the protein solids thereby enabling the strainer means to gather the separated liquors and remove the same for further processing as desired.

The protein slurry is thus subjected to regulated pressure, squeezing, kneading, and otherwise working the protein slurry mass to substantially separate the liquids therefrom. The liquids tend to gravitate toward the center and top of the protein slurry puddle where they are removed by strainers as described.

Stabilizing Fish Meal by Antioxidants

S. Arakawa, S. Tominaga and T. Terase; U.S. Patent 3,437,489; April 8, 1969; assigned to Nichiro Gyogyo KK, Japan describe a process to produce fish meals of stable and nutritious quality at low cost by preventing the oxidation of fat and oil in them. The process consists of the following steps.

(1) Fish are ground by a grinder with about 4 to 8 mm plate mesh, about 10 to 30% water is added so that the mixture may be of a proper viscosity, and it is fed to a heating apparatus by a constant volume pump.

(2) The fish meat gruel fed in to the apparatus is mixed with injected steam so that heat exchange may occur and the temperature of the ground fish may reach the desired temperature.

(3) The fish meat gruel thermocoagulated by steam injection is heat exchanged in a very short time, therefore little protein is digested by enzymes.

(4) Undiluted hydrogen peroxide water of 30% concentration may be poured into the receptacle tank with stirring, but usually diluted hydrogen peroxide water of about 0.3% is added with mixing and stirring. Although the amount of hydrogen peroxide added to fish meat is somewhat variable according to the kind of fish, the content of fat and oil and the freshness of fish meat, the optimum amount of 100% equivalency of hydrogen peroxide per kilogram of fish meat ranges preferably from 0.5 to 0.65 gram.

The proper temperature during the reaction of hydrogen peroxide with fish meat is 50° to 65°C. The fish meat is heated by the heating apparatus, and the required amount of hydrogen peroxide solution is immediately added to fish meat with mixing and stirring. If hydrogen peroxide of above concentration is added to small migrating fish such as sardine, anchovy, herring, pike, mackerel and horse-mackerel, their scales and skins turn light green. When they are treated at reaction temperatures of 50° to 65°C, no pronounced fishy odor will be noticed, but above 65°C a strong fishy odor develops. Below 50°C no color reaction will be seen.

(5) Solids and liquids are separated.

(6) The raw presscake is fed to a pneumatic conveying dryer and kept at a temperature below 70°C for no longer than 2 hours. After this operation, no trace of hydrogen peroxide can be found in the fish meal since it is decomposed completely during the drying.

Example: 100 kg of sardines are ground with a meat grinder having a plate mesh of 6 mm and mixed together with 30 kg of water. The mixture is stirred in a hopper to reduce its viscosity, and is fed to a steam injector by a pump. The fish meat is instantaneously and uniformly heated to 55°C and falls into a receptacle having a stirrer. Prepared hydrogen peroxide water produced by diluting 120 grams of 30% hydrogen peroxide to 10 times as much is gradually added to the fish meat at a consistant rate with stirring, and the mixture falls into a continuous solid bowl centrifuge by pump to separate liquids from wet solids.

The wet solids are mixed with a previously dried fish meal at a ratio of 1:1 to reduce the water content and are dried in a pneumatic conveying dryer to obtain 18 kg of a fish meal having a mild smell.

Grinding, Centrifuging and Spray Drying

E.E. Anderson; U.S. Patent 3,586,515; June 22, 1971; assigned to Marine Technology, Inc. describes an apparatus and process for the conversion of whole fish into fish meal and oil by grinding the whole fish to form a slurry; flash heating the ground fish to a temperature under its boiling point; holding the slurry for a time; centrifugally separating the slurry into a solid phase and a liquid phase; centrifugally separating the liquid phase into oil and a liquid residue; removing the oil; combining the liquid residue with the solid phase; and spray drying the combined material with a pulse jet engine to form fish meal particles.

Referring to Figure 7.2, there is shown a boat **10** having removed from the hold thereof the whole fish by means of a fish pump **12** through a suitable pipe **14** to a dewatering screen **16**. In order to efficiently utilize the fish pump, seawater is introduced into the system by means of a seawater pump **18**, which assists in the transporting of the whole fish through the pipe. This water, along with the seawater taken in with the whole fish, is ultimately removed at the dewatering screen.

The whole fish is then transported through a magnetic separator **20** for removing any metal objects, such as fish hooks, onto a weighing conveyer **22**, and into a suitable storage facility such as a fish tank **24**. A gate **26** is provided at the lower end of the fish tank for continuously feeding the whole fish onto a suitable transporting means **28** such as a screw conveyer.

The fish are transported through the screw conveyer to a grinder **34** where the bodies are ground into a slurry. The ground bodies are transported through suitable conduits **35** and **36** by a positive displacement pump **38** from the grinder to a receiving hopper **42**. From the hopper the slurry is pumped by means of another positive displacement pump **46** through a suitable conduit **48** into a flash cooker, generally designated **50**.

The fish slurry is transported in coil **52** through the cooker during the heating operation. The term slurry is used to describe a watery mixture comprising particles of ground fish solid, water, blood and oil. The flash cooker comprises a series of stainless steel coils through which the slurry is transported.

The coils are heated by suitable heating means such as by two four-million Btu per hour pulse jet engines **54** or other suitable flash heating means. The term flash cooker is used to indicate an apparatus that heats the fish slurry very quickly, on the order of a few seconds, to a temperature between 190° and 210°F, just below the boiling point of the slurry, which is primarily water based, so as to cook the fish slurry. The heated slurry is then discharged through a conduit **56** into an insulated holding tank **58**.

Temperature control of the cooker is achieved by placement of a thermocouple **59** in the discharge conduit. The thermocouple controls a throttle valve **60** which in turn regulates the quantity of fuel fed to the pulse jet engines. The slurry remains in the holding tank a sufficient time, generally on the order of several minutes, for removal of oil from the fish by allowing the breakdown of the fish fat cells so as to release the oil. From the holding tank the slurry is pumped at a fixed rate of flow by means of a pump **61** through a conduit **62** to a decanter **64**.

The decanter centrifugally separates the slurry into a solid phase and a liquid phase, the liquid phase comprising moisture, oil and suspended fine particles of solids.

FIGURE 7.2: APPARATUS FOR PREPARING FISH MEAL

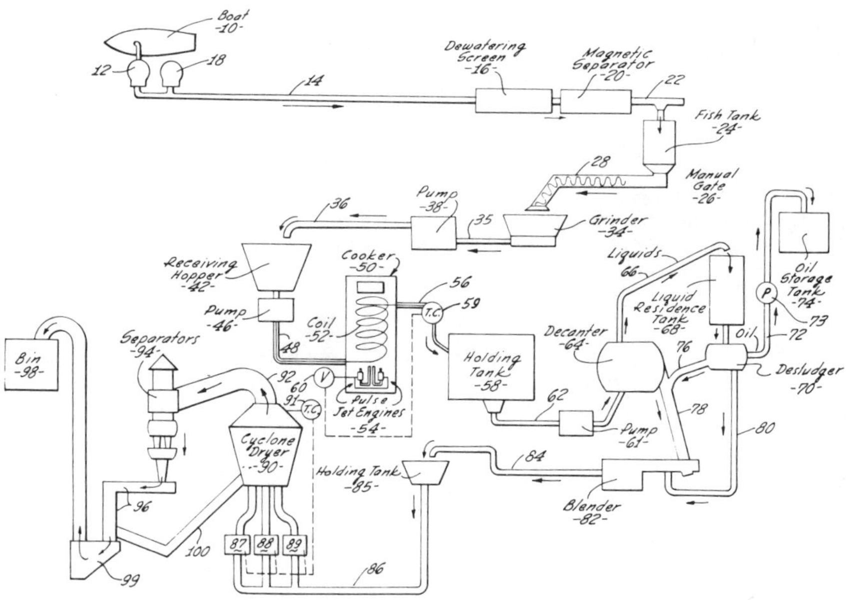

Source: E.E. Anderson; U.S. Patent 3,586,515; June 22, 1971

The liquid phase is then transported through a conduit **66** into a liquid residence tank **68** from where it is then fed into a desludger **70**. The desludger centrifugally separates the oil from the liquid phase with the oil being transported by pump **73** through conduit **72** to a suitable storage tank **74**. Suspended fine particles in the liquid phase, called sludge, are removed from the desludger through conduit **76** to be combined with the solid phase from the decanter **64** in a conduit **78**.

The remaining part of the liquid phase, known as stickwater, is conveyed through a conduit **80** and is combined with the solids from conduit **78** in a blender-sizer **82** to make a uniform slurry containing the proteins and nutrients from the stickwater. The slurry is then ejected through a conduit **84** into a holding tank **85**.

From the holding tank the slurry is transported through a conduit **86** by three positive displacement pumps **87, 88** and **89** and fed to three corresponding pulse jet engines (not shown) located at the base of a cyclonic or spray dryer generally designated **90**, which has one pulse jet engine at its base.

The slurry is injected into the jet's exhaust pipes where the oscillating flow of hot gases breaks up the slurry into particles which are then effectively dried by the heat of the gases. Four additional pulse jet engines are displaced around the periphery of the dryer **90** at various angles to create a swirling or cyclone effect to assist in the rapid drying of the particles. The pumping rate of the pumps is controlled by a temperature signal from a thermocouple **91** located at the upper end of the dryer so that the dryer exhaust temperature is kept at a relatively constant level assuring proper drying of the particles.

Finer particles are removed through the exhaust conduit **92** and are transported into a centrifugal separator **94** so as to separate the particles and the moist air. Coarser particles too heavy to be airborne out of the dryer fall to the bottom of the dryer where they are removed by a suitable conveying means **100**, combined with fine particles removed by a suitable conveying means **96** such as a screw conveyor and deposited into a feeder bin **99** which feeds a pneumatic conveyor means **101** in communication with a storage bin **98** where the particles (fish meal) are deposited until packaging.

High Protein Material with Reduced Fishy Taste and Odor

The process described by *K.R.M. Hogstedt, T.L. Knutsen, E.T.L. Svensson and S.-O. Osterman; U.S. Patent 3,804,964; April 16, 1974; assigned to AB Astra Nutrition, Sweden* is particularly interesting in that it produces high protein concentrate with substantially reduced fishy taste and odor, and a considerably lower fat content by the use of physical treatment only.

The fish is first cut into pieces by a slicing means including a plurality of rotating blades which cut the individual fish along lines perpendicular to their longitudinal axes. The fish pieces are then advanced to a tank where they are washed in a countercurrent flow of water. Next, the pieces are moved to a rotating drum on which they are held while they are sprayed with water emitted by nozzles disposed along the rotational axis of the drum.

The pieces then pass through a direct-steam boiler in which coagulation takes place. Bones and other nonfibrous hard pieces are removed in a separator, and the fish is then homogenized, after which it is transported to a centrifuge. Fat, soluble substances, and skin fragments are removed by centrifuging, and water is then removed from the fish by pressing and drying. The homogenizing and centrifuging steps may be repeated in a second homogenizer and second centrifuge before the final product is moved to the presser and the dryer.

Figure 7.3 is a pictorial diagram of an apparatus for processing the fish by this method. Fresh fish to be processed is brought from a fishing boat to a raw material silo at a processing plant. Preferably, this is recently caught fresh fish, and it is stored in a suitable aqueous environment such as a mixture of 80% fish and 20% water.

Next the fish are moved from the raw material silo to an elevator tank where the transporting water is removed, and the fish is then transported by a stepless, variable-speed elevator to a slicing means **1** shown in Figure 7.3. At the slicing means, the fish are cut into pieces. The slicing means includes a conveyor belt **2** across which are mounted a series of parallel carriers **3** that function as guide means for positioning and holding the fish perpendicular to the direction of motion of the belt. The guide means also includes a group of guide bars (not shown) obliquely positioned above the belt which push the fish into their proper position against the carriers **3**.

FIGURE 7.3: FPC WITH REDUCED FISHY ODOR AND TASTE

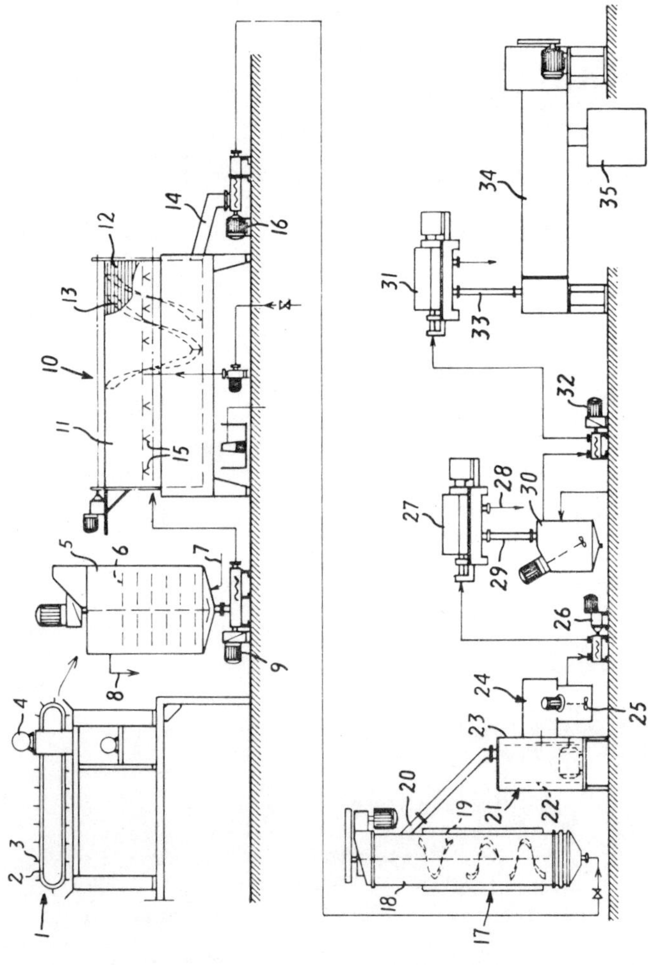

Source: K.R.M. Hogstedt, T.L. Knutsen, E.T.L. Svensson and S.-O. Osterman; U.S. Patent 3,804,964; April 16, 1974

The fish are moved by the belt **2** and associated carriers **3** past a plurality of rotating circular knives **4** which pass through slots in the carriers and cut the fish into pieces along lines perpendicular to their longitudinal axes (the longitudinal axis of a fish being a line connecting the head and tail). As the fish is cut, the blood rich regions are laid bare, and blood is carried away by a water spray.

From the conveyor belt the fish pieces move to a diffusion tank **5** which functions as a first means for washing the fish pieces. Here the pieces are contacted with a countercurrent of water, the ratio of water to fish in the tank being at least one-to-one. A plurality of stirrers **6** are disposed within the tank to constantly agitate the fish pieces and mix them with the water. The tank is also provided with an air inlet **7** that emits a constant stream of air to its lower portion. The motion of the stirrers plus the constant air stream moving through the tank increases the contact between the fish in the water and provides desired diffusion at the bottom of the tank.

While the fish pieces are in the tank, blood, gut fragments, and fat deposits are removed via a spillway **8**. The fish pieces should be retained in the tank long enough to soak out most of the blood but not long enough to damage the protein. The optimum time of retention varies from approximately 15 to 120 minutes. A typical optimum time is about 60 minutes.

After countercurrent washing in the tank, the fish pieces are pumped, with a quantity of transporting water equal to about 50% by weight of the mixture, by a first pump **9** to a second washing means **10**. The primary purpose of the second washing step is to separate the viscera from the fish pieces. The second washing means comprises a rotatable drum **11** which is provided with a plurality of closely spaced open slots **12** extending longitudinally along its surface.

A plurality of guide bars **13** are disposed about the surface of the drum, each guide bar **13** forming an open helix. The guide bars feed the fish pieces toward an outlet **14** as the drum rotates. As the pieces move toward the outlet, they are sprayed by water emitted from a plurality of nozzles **15** positioned along the rotational axis of the drum. By the time the pieces reach the outlet approximately 95% of the original viscera and blood content has been removed by the water spray. The water carrying this waste product passes out through the slots and is discharged.

The water used in the first and second wash means may be either fresh water (tap water) or saline water. If saline water is used, fresh water should be supplied to the nozzles which are closest to the outlet to reduce the salt content of the fish pieces, and thus the ash residue of the final product.

The pieces emerging at the outlet are mixed with fresh water, and moved by a second pump **16** to the bottom of a direct-steam boiler **17** where coagulation takes place. Fresh transporting water should be present in an amount equal to between approximately 0.5 and 1.0 times the weight of the fish pieces. When the pieces enter the boiler, this transporting water facilitates efficient heat transfer from the steam to the fish.

The direct-steam boiler includes a vertically disposed steam-jacketed cylinder **18** and a plurality of stirrers **19** that form an open, broken helix. The stirrers are disposed within the cylinder and are arranged to rotate about its vertical axis.

In the boiler **17**, concurrent flow is utilized, i.e., the fish pieces are introduced at the bottom of the cylinder **18** along with the steam and both flow upwardly together. The quantity of steam is equal to about 20 to 30% by weight of the fish and it is injected along the bottom of the cylinder. This causes immediate denaturation of the surface of the fish pieces, and, in this way, a fixation of the melanin pigment of the fish skin is obtained.

The boiler includes thermostatic control means that varies the amount of direct steam introduced so that the final temperature is about 92° to 95°C. The temperature of the fish is raised gradually as the pieces move upwardly through the cylinder and completely coagulated fish material is withdrawn from the boiler through a spillway **20**.

The fish material flows from the spillway to a deboning separator **21** in which most of the remaining bones, fins, scales, gills, eyes, and gut fragments are separated. The separator includes a rotating element **22** which carries slightly inclined rotor blades that force the relatively soft fish material through perforations in the surface of the drum **23** that surrounds the rotating element. The harder undesirable parts of the fish are retained within the drum and disposed of. 70% of the bone content of the fish material is removed by the separator which results in a commensurate reduction in the ash content of the final product. The fluorine content is also greatly reduced in the separator.

The fish material flows directly from the separator to a first homogenizing means **24** in which the coagulated fish material is disintegrated by a rotating stirrer **25** of the Ultratrurrax type. This disintegration causes the melanin pigment to be released. The fish material remains in the homogenizer for a period of about 1 to 2 minutes during which the stirrer rotates at a speed of about 2,700 rpm.

The ratio of water to dry substance in the homogenizer is about 5 to 1. This water derives mostly from that added immediately before the fish pieces are transported to the boiler and from the condensation of the steam within the boiler. Part of the water in the homogenizer is attributable to the natural water content of the fresh fish.

The fish material is moved from the first homogenizer to a first centrifuge **27** by a third pump **26**. The released melanin pigment is then removed from the fish material along with an aqueous phase containing stickwater and fat. After centrifuging, the fish material has a dry substance content equal to about 33% by weight and it has a pronounced fibrous structure. Skin fragments which are disintegrated by the homogenizing step are not retained with the more fibrous fish material. Fat and water-soluble compounds in the form of proteins and salt are also removed by the centrifuging process after disintegration in the homogenizer and are discharged through an outlet **28** as part of an aqueous phase. The removed portion also includes most of the odor producing compounds such as trimethyl amine.

The fish material from the first centrifuge is discharged through an outlet **29**, preferably to a second homogenizer **30** in which water, at a temperature of approximately 90°C, is added in an amount equal to about 82% by weight of the total water-fish material mixture. The second homogenizer is similar in construction to the first homogenizer.

The fish material is transported from the second homogenizer **30** to a second centrifuge **31** by a fourth pump **32**. In the second centrifuge additional melanin pigment, fat, water-soluble compounds, and water are removed. The fish material is then transported to a press **34** in which additional water is removed. From the press the material moves to a dryer **35** from which it emerges as a finished product suitable for use as a substantially fat free, tasteless, and odorless protein material of light color which can be used as an ingredient in foodstuffs for human or animal consumption.

Although it is preferable to include the second homogenizing and centrifuging steps as set forth above, these steps and the corresponding apparatus may be omitted. A final product taken from the outlet **29** of the first centrifuge **27** and fed to the press is not of the same quality as the final product obtained by the preferred process. It does, however, compare favorably with products obtained by previously known processes.

The fish material that has been treated by a process including the use of the second centrifuge contains only a very small quantity of water-soluble compounds such as trimethylamine and other odor producing compounds and has a low fat content. When fish having an original fat content of 39% by weight of the dry substance content is treated, the output of the second centrifuge typically has a fat content equal to only 5% based on the weight of the dry substance. If the second homogenizer and the second centrifuge are omitted the fat content is approximately 8%.

A fat content of 5 or 8% compares favorably to a fat content of 12% which is present when the same fresh fish is disintegrated, boiled, freed of bones, and pressed. The fat contents given above were determined by a chloroform-methyl extraction.

The method and apparatus of this process yields a high protein fish product which is particularly suitable for use as a starting material in the preparation of high protein, nutritious food products. The final product can be used immediately as a feed stuff or food ingredient or it can be dried and stored. In the latter case, it must, of course, be protected against oxidation. The fat content and the trimethylamine content are greatly reduced. The color is lightened by the removal of the melanin pigment.

Small traces of elements such as iron and copper are substantially washed out, thereby inhibiting oxidation which is catalyzed by the presence of these elements. Iron is normally present in fish in about 100 ppm, mainly in the form of myoglobine. The quantity of pure, raw, digestible protein present is considerably higher in a product prepared according to this method as compared to a product yielded by previously known methods. It is of considerable importance that the fat content is reduced by 50 to 75%. Ash residue is reduced about 50%.

PREPARATION BY CHEMICAL MEANS

VioBin Processes

Dr. Ezra Levin holds a number of patents for the defatting and dehydrating of fatty substances. The basic process for the production of FPC by azeotropic distillation is U.S. Patent 2,619,425. The defatting and dehydration steps it describes are set out in detail in U.S. Patent 3,468,674, which will be reviewed below.

Dr. Levin's U.S. Patent 2,972,542 describes the process of deodorizing FPC after azeotropic distillation by washing the defatted proteinaceous material with an alcohol having less than 5 carbons, or by oxidation with standard oxidation agents such as sodium hypochlorite, hydrogen peroxide, etc., or by volatilization by steam, or with a combination of these methods.

E. Levin; U.S. Patent 3,468,674; September 23, 1969 has found that whole or ground fish may be preserved at ambient temperatures when maintained in the presence of an atmosphere of sulfur dioxide gas for periods exceeding a month. Sulfur dioxide gas and sulfites have long been known for their preservative properties. Sulfites have been used to preserve sausages and canned meats, but such preservation methods have been discontinued because the concentration of residues left in the product are deemed sufficient to be harmful for human consumption. His method of processing wet, fat biological substances which have been preserved or processed in an atmosphere of sulfur dioxide or contain a concentration of sulfurous acid, results in a product containing only harmless residues in small concentration.

Figure 7.4a illustrates a typical conventional fishing trawler **1** which has been equipped for carrying out the process. The trawler has a hold **2** which contains a number of vessels **3** which are adapted to house and contain freshly caught fish. Each of the vessels is provided with an air tight cover **4** which protrudes through the deck of the ship and is accessible from the deck. In addition, each of the vessels has a concave bottom **5** with an aperture at the lowest point thereof connected through a manually actuatable valve **6** to a pipeline **7**. The pipeline is connected to a pump **8** located below the vessels and the pipeline, and communicating with a port **9** located above the water line.

In accordance with the process, fish are harvested from the sea and thrown directly into one of the vessels. The fish are not preprocessed in any manner whatever, and the fish need not be dead when placed in the vessel. The fish when placed in the vessel, like all fish found in nature, are not sterile, and contain large quantities of seawater bacteria, particularly adjacent to the skin of the fish.

As each vessel is filled, the vessel is sealed and filled with sulfur dioxide gas. A particularly suitable way in which to accomplish this operation is to add bisulfite above the mass of the fish within the vessel. Any bisulfite which releases sulfur dioxide when contacted with water is suitable, such as sodium bisulfite, metabisulfite, pyrosulfite and other bisulfites. Sodium bisulfite is preferred, and may merely be poured in granular form on the mass of fish within the vessel. Sodium bisulfite forms as aqueous solution with the water in and on the harvested fish with the release of sulfur dioxide, as follows:

$$2NaHSO_3 \longrightarrow SO_2 + Na_2SO_3 + H_2O$$

After supplying the granular sulfite, the cover is sealed on the vessel and the valve at the bottom of the vessel is opened. Sulfur dioxide gas also may be directly used from tanks containing the gas under pressure, and it is to be understood that this process is fully equivalent to that of applying granular sulfite.

The pump **8** is utilized to remove any water or other liquid which may be present in the vessel **3**. When sulfur dioxide gas begins to be present at the port **9**, liquid having ceased to be ejected therefrom, the valve **6** located at the bottom of the vessel is closed, and the sulfur dioxide atmosphere within the vessel is secured. The sulfur dioxide penetrates the fish and combines with the water in and about the fish to form sulfurous acid, as follows:

$$SO_2 + H_2O \rightleftharpoons H_2SO_3$$

The reaction is reversible, hence requiring an adequate sulfur dioxide atmosphere to maintain the preserving sulfurous acid. Sulfur dioxide gas in adequate quantity may be formed by sodium bisulfite in the manner described above when applied in a quantity no greater than 4% by weight of the quantity of fish within the vessel. Under normal conditions, sodium bisulfite in an amount equal to 2% of the weight of the fish has been found adequate and in cold weather, the quantity of sodium bisulfite may be reduced to 1% by weight of the fish. In warm weather, or hot weather, an additional quantity of sodium bisulfite is usually necessary, generally 3% by weight of the fish present.

When the fish is removed from the vessel, it is not fit for consumption by man or animal and it must be processed to remove the sulfurous acid and sulfur dioxide. A particularly suitable manner of processing the fish or meat containing sulfurous acid and sulfur dioxide is to grind the tissue into comminuted particles and subject the particles to azeotropic distillation in an apparatus such as set forth in Figures 7.4b, 7.4c and 7.4d.

FIGURE 7.4: VIOBIN PROCESS

a.

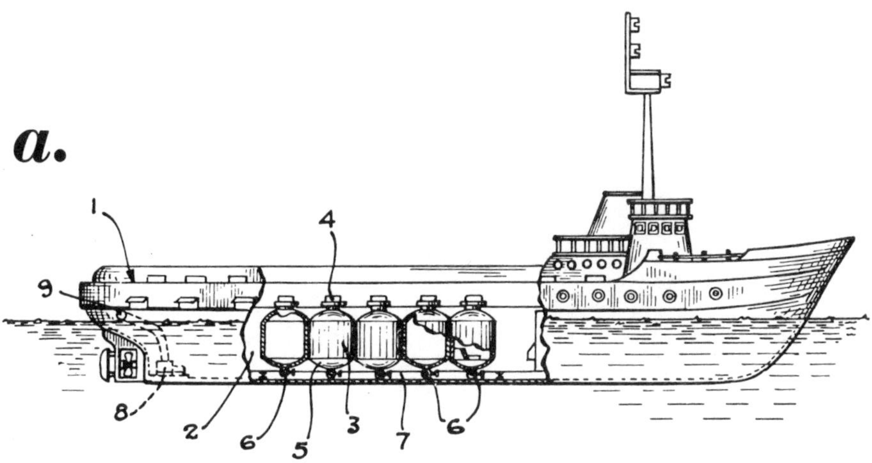

Fishing Vessel for Use in the Process (continued)

FIGURE 7.4: (continued)

b.

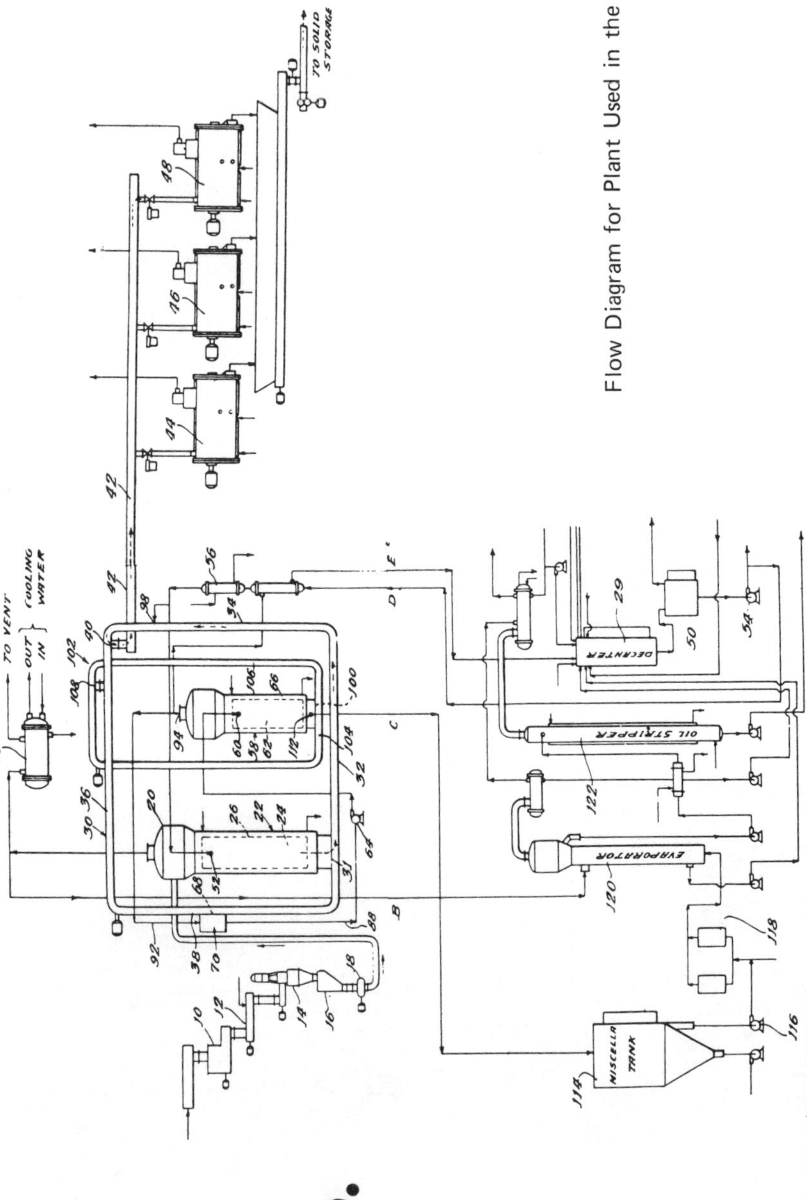

Flow Diagram for Plant Used in the Process

(continued)

FIGURE 7.4: (continued)

c.

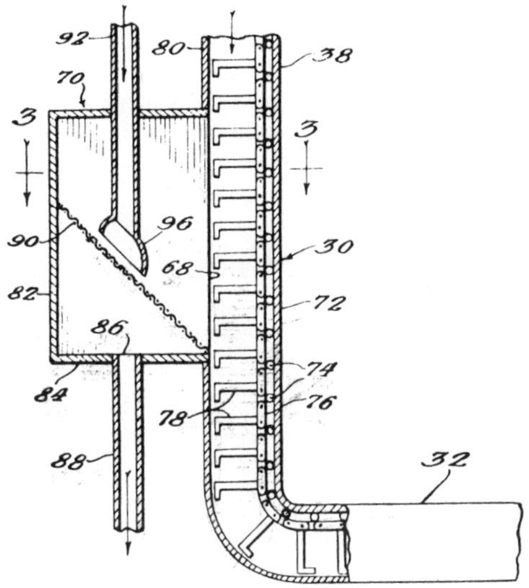

Sectional View of Part of Main Conveyor

d.

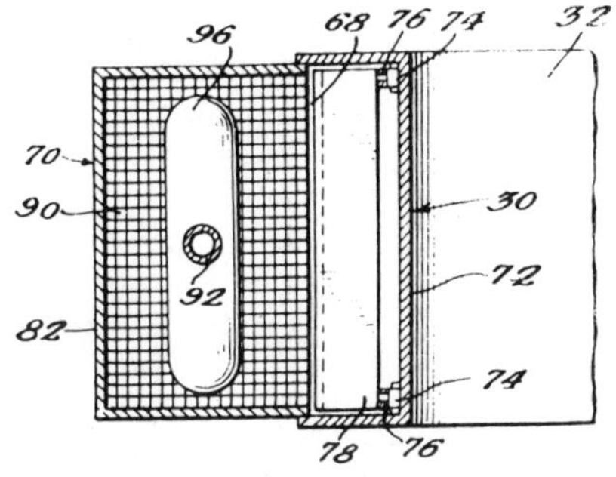

Sectional View Along Line **3–3** of Figure 7.4c

Source: E. Levin; U.S. Patent 3,468,674; September 23, 1969

As indicated in Figure 7.4b, fish, whether whole or chopped, is transformed into a pumpable fluid by means of a prebreaker **10**, feed screw **12**, disintegrator **14**, and hopper **16**. The small particles of meat or fish are subjected to low pressure steam which is introduced prior to the hopper, specifically into the feed screw. A feed pump **18** is then utilized to pump the particles into the upper portion or head **20** of a primary cooker **22** or first distillation vessel. The primary cooker **22** is a vertically disposed elongated vessel with the head **20** positioned at the top thereof. A body of substantially water-immiscible fat solvent is disposed within the primary cooker below the head. The body of solvent, designated **24**, must have a boilint point under the conditions of operation of at least 65°C and must form an azeotrope with water preferably boiling below 100°C under operating pressure. The solvent should be selected to form an azeotrope which will remove substantial portions of water in relation to the amount of solvent distilled at the operating temperature selected. Among solvents of this class, ethylene dichloride is preferred.

Ethylene dichloride has a boiling point at atmospheric pressure of 83°C, and a water-ethylene dichloride azeotrope boils at 71.5°C. Another example of a particularly suitable solvent is heptane which boils at 98.4°C at atmospheric pressure. A water-heptane azeotrope boils at 79.5°C under atmospheric conditions. Other suitable solvents include propylene dichloride, trichloroethylene, perchloroethylene, and other low boiling chlorinated solvents. Suitable chlorinated solvents may include the bromine, iodine or fluorine derivatives of aliphatic hydrocarbons. In general, a suitable solvent must boil below 120°C under standard conditions. The hydrocarbon fat solvents including benzene, hexane, toluene, cyclohexane, heptane, and others are suitable.

The solvent must not be reactive with the tissue constituents or the preservative under operating conditions and must be capable of being removed by evaporation from the fat without leaving harmful or toxic residues. The moist particles of fish are continuously introduced into the boiling solvent within the primary cooker by spraying the particles into the solvent adjacent to the upper level thereof, or some other suitable means which will avoid formation of an agglomeration of the particles.

The particles are dehydrated and defatted in the primary cooker at least to the extent that they cannot coalesce into lumps or stick to the vessel, but the particles all remain wet due to the fact that raw tissue particles are continuously introduced, thus maintaining a water-solvent azeotrope within the vessel. The particles of biological substances do become heavier than the solvent due to partial drying, however, and tend to settle toward the bottom of the primary cooker, in spite of the violent boiling of the solvent.

As indicated in Figure 7.4b, the primary cooker is provided with an internal heater **26** which utilizes low pressure steam as a heat source in order to maintain the body **24** of solvent under rapid boiling conditions. Vapor from the body of solvent rises through the head **20** of the primary cooker and is conducted to a condenser **28** which is provided with a flow of cool water. Both the solvent and the water vapor are condensed to liquid form, and the solvent and water vapor are separated by a decanter **29**, as is well-known in the art. The hydrogen gas and water vapor are discarded, and the recovered solvent is returned to the system.

The bottom of the primary cooker is in communication with a runaround main conveyor **30**. Granules formed from the fish settle through the boiling body of solvent, pass through an opening **31** at the bottom of the primary cooker and enter the bottom leg **32** of the conveyor. The conveyor has four legs **32, 34, 36** and **38** forming a continuous rectangular path in a vertical plane, and a belt-type conveyor is continuously translated within the four legs to elevate the granules passing through the opening **31** at the bottom of the primary cooker **22**. The top leg **36** has an opening **40** which communicates with a horizontal conveyor **42** for transporting particles to one of a plurality of desolventizers **44, 46** or **48**. The desolventizers are used in sequence and operate on the batch system.

A relatively large quantity of solvent is maintained in storage in a work tank **50**, and this solvent is continuously introduced into a port **52** located near the upper level of the boiling body **24** of solvent in the primary cooker. A pump **54**, and solvent heater **56** are connected in the path between the solvent work tank and the port to provide an adequate supply of heated solvent to the primary cooker.

Operation of the primary cooker results in a slurry of solvent, granules of wet-fat partially dried tissue, and fat which has been extracted from the tissue. Since raw tissue is continuously being introduced into the primary cooker, and the solvent is violently boiling, the tissue is not permitted to dry. The relatively wet tissue is present throughout the body of slurry because the azeotropic temperature is maintained while wet tissue is being added continuously. Under these conditions, the miscella is wet, and filtering the miscella would clog the filter. Thus, it is not possible to filter the miscella directly from the primary cooker.

A portion of the slurry which is formed in the primary cooker is continuously introduced into a secondary cooker **58** through a port **60** located in the secondary cooker near the top of a vigorously boiling body **62** of slurry from the primary cooker disposed in a secondary cooker. A pump **64** located between the primary and secondary cooker maintains the solvent at a relatively fixed level above the port and a heater **66** located within the secondary cooker maintains boiling conditions.

The pump is coupled into an opening **68** in the upper portion of the leg **38** of the conveyor. The conveyor travels in a counterclockwise direction and is hence traveling down the leg. The figures illustrate a filter **70** which is utilized to retain as many of the solid particles as possible in the primary cooker and pass only a minimum number of solid particles to the secondary cooker.

Figure 7.4c is a sectional view of the filter **70** showing a portion of the leg **38** of the conveyor. The inner wall **72** of the tube carries a plurality of rollers **74** which support a number of links **76** of a continuous chain, each of which carries an L-shaped shoe **78** which catches solid particles passing through the opening from the primary cooker and carries the solid particles through the legs **32** and **34** of the conveyor **30** to deposit them in the outlet **40** in the leg **36**.

The conveyor also has an outer wall **80**, and the opening **68** which permits the slurry from the primary cooker to flow to the pump is disposed in this outer wall well below the level of the slurry in the primary cooker. A rectangular fluid-tight box **82** is sealed about the perimeter of the opening, and the box has a bottom **84** with an aperture **86** sealed to a tube **88** which communicates with the inlet of the pump.

A screen **90** is sealed within the box on a plane at an angle to the horizontal in order to filter large particles from the flow of miscella to the pump, and hence to the secondary cooker. In practice, approximately 90% of the particles introduced into the primary cooker are removed by means of the primary conveyor, and the screen contributes substantially to this result. If the screen were not present, experience proved that 30 to 40% of the particles would be reworked by pumping the particle-laden miscella to the secondary cooker **58** for drying.

The wet miscella causes a glaze to develop and build on the screen **90**, and unless some means is provided, the glaze will clog the screen, even if the perforations of the screen are very large. A 20 mesh per inch screen will permit flow to the miscella by utilizing dry solvent vapors such as are evaporated from the secondary cooker to keep the screen clean. A tube **92** communicates with an outlet **94** in the head of the secondary cooker and conducts the hot solvent vapors from the secondary cooker to a nozzle **96** confronting the side of the screen. The vapor pressure from the secondary cooker is maintained at about 5 lb/in^2. The flow of pressurized solvent vapor sweeps the screen clean and open and prevents clogging of any kind to permit the miscella, including the sticky finer particles, to flow through the screen.

The solid particles passing from the primary cooker **22** are only partially dried and partially defatted. In addition to the fat within the particles, the particles carry with them a quantity of occluded fat. The occluded fat is washed from the particles by a flow of clean solvent introduced into leg **34** of the conveyor. This flow of clean solvent also extracts fat from the particles, since the quantity of moisture in the particles has been reduced to a level permitting extraction by conventional processes.

All of the particles in the conveyor **42** are of low fat content, but those particles from the primary cooker contain substantial moisture. This moisture is removed with the solvent in the desolventizers **44, 46** and **48** to produce a solid product which is granular and contains very little moisture or fat, and hence has great stability. Further, the desolventizers subject the particles to steam for a period of from one-half to four hours in order to remove the solvent, and all remaining sulfurous acid is converted to sulfites and sulfates.

A residue of sulfite principally in the forms of calcium sulfite and sodium sulfite is present in the granules after desolventizing, and in addition small quantities of calcium sulfate and sodium sulfate are present in the residue. The granules are then placed in a vessel containing methyl alcohol in an amount equal to approximately five times the volume of the granules. The granules are then stirred in the methyl alcohol and the alcohol removed. The process is repeated once again, and the alcohol is removed. Thereafter, the product is dried by the application of heat and vacuum.

The use of alcohol is effective to deodorize the product, and at the same time reduces the concentration of the sulfite and sulfate residues. Fish meal produced in this manner has been found to contain approximately 2,000 ppm of sulfite residues.

The secondary cooker is provided with an outlet port **112** near the bottom and a miscella is withdrawn through it. It is to be noted that the miscella is highly concentrated in fat, since no fresh solvent is introduced into the secondary cooker. This miscella is collected in the miscella tank **114**, pumped by a pump **116** through 1 of 2 filters **118** to a vacuum evaporator **120**. The evaporator evaporates the solvent from the miscella, and the fat is thereafter conducted through an oil stripper **122** to a fat storage tank **124**.

Example: One ton of hake is placed in a vessel having a diameter of 3 ft and a height of approximately 9 ft, the vessel containing approximately 63 ft^3. The fish will fill approximately 50% of the vessel. The fish may be inserted into the vessel without washing or other preliminary treatment. 40 lb of commercial grade sodium bisulfite is poured on top of the fish when the vessel has been loaded with fish, and the cover of the vessel is then sealed. As an alternative, the atmosphere of the vessel is replaced by an atmosphere of SO_2 gas drawn from a tank containing SO_2 gas under pressure.

All water and other liquid is removed from the interior of the vessel through the port located in th bottom of the vessel, and the port is left open until sulfur dioxide gas begins to flow freely from the port. It takes less than 1 hour for the sulfur dioxide gas to penetrate the mass of fish within the vessel and begin to flow from the port when sodium bisulfite is used and an even shorter time when applying sulfur dioxide gas from a tank, and at this time the port is closed. The vessel is maintained closed for 4 weeks. Thereafter, the vessel is opened, and the fish removed.

The sulfurous acid content of the fish on opening of the vessel is approximately 0.5% by weight. The nutritive value of the fish remains substantially unchanged, that is, the protein biologic quality remains substantially unchanged and the total nitrogen quantity of the fish also remains substantially unchanged throughout the period of storage. There is a slight increase in free fatty acids of the fish during the storage period approximately the same as occurs in deep-freezing over the same period. The fish must be thereafter processed, for example, in a continuous azeotropic dehydrating and defatting process.

The fish is then subjected to pretreatment with steam for approximately 5 seconds to convert part of the collagen to gelatin, and thereafter the whole fish is ground in a prebreaker to particles of less than 0.5 inch diameter. The 2,000 lb of hake are introduced into the primary cooker in a continuous stream over a period of approximately 14 minutes. The primary cooker is approximately 65% full of boiling ethylene dichloride, and the pressure within the vessel of the primary cooker is maintained at substantially atmospheric pressure.

The vessel of the primary cooker contains about 800 gal of ethylene dichloride, and the heating coils of the vessel supplied sufficient heat to maintain boiling at the boiling point of azeotrope, namely approximately 71.5°C. The body of solvent within the primary cooker is kept vigorously boiling. The secondary cooker is identical to the primary cooker in construction and contains a slurry of approximately 800 gal which is transferred from the primary cooker to the secondary cooker.

The defatted particles are then desolventized in the presence of steam for 4 hours and thereafter deodorized in a methyl alcohol bath with agitation for about 10 minutes. The methyl alcohol bath is repeated, and thereafter the particles are dried by heat and vacuum.

The particles are then ground to uniform size. The end product is an odor-free particle of about 1% fat, 1% water, 70% protein, and contains about 2,000 parts per million of sulfite in the form of sodium sulfite and calcium sulfite. The particles are white in color.

FPC from Oily Fish by Use of Antioxidants

Although in the case of cod or haddock, which contain only 1 to 2% oil, it is possible to obtain a fish flour having a satisfactory flavor and a shelf life of a few months; this has not heretofore been possible in the case of fish having an oil content. High oil-containing fish are typified by menhaden which has an oil content of 5 to 20%, usually 12%.

The advantages of using high oil-content fish, such as menhaden, for use in preparing a high protein fish flour are manifold. Menhaden is the most abundant fish caught in the United States. In 1945, for example, the fisheries of the Middle Atlantic States landed 368.1 million pounds of menhaden fish out of a total catch of all fish of 433.1 million pounds. Thus menhaden represented 85% of all fish caught in that area, and similar percentages are found in other areas. In 1958 approximately two billion pounds of menhaden fish were caught in the United States.

Because of the defects of flour prepared from high-oil type fish, such as menhaden this catch has found its way almost exclusively into animal feed and fertilizer, despite its high protein content and low price. It would, therefore, be highly desirable to obtain these proteinaceous flours in suitable condition for human food use.

W.M. Whaley and R.J. Moshy; U.S. Patent 3,252,962; May 24, 1966; assigned to General Foods Corporation provide a technique for producing high-protein fish flour of good quality from oily fish. They describe a process which comprises comminuting the fish, forming the water slurry, adding antioxidant to the slurry, acidifying the slurry to precipitate the protein, and then heating the slurry to about 155° to 212°F to thereby denature the protein.

The protein is then separated from the slurry and extracted with a bland solvent. The solvent extract is removed from the protein, the protein is dried at 90° to 132°F, moistened to 10 to 30% moisture, and then redried at 90° to 132°F. The use of antioxidant may be minimized by conducting the extraction process in an inert gas atmosphere. However, even in this latter case an antioxidant is preferably incorporated at some stage of the process to improve the shelf life of the protein.

The fish which may be treated in accordance with this process are preferably the oily fish including generally those fish having an oil content of at least about 5% and usually upwards of 10% to as high as 20%. Included within this group are, e.g., herring (e.g., 11% oil content), mackerel (e.g., 13% oil content), king salmon (e.g., 16% oil content), sardine (e.g., 13% oil content), pilchard, and tuna. However, it is found that the advantages of this process are obtained to an outstanding degree when the fish to be treated is menhaden, typically at an oil content of about 12%. It is a particular feature of this process that the entire fish may be used, although scrap, offal, or trimmings can also be used if desired.

Preferably the charge of fish is either fresh fish (fish which has been stored at room temperature for not more than one day); fish refrigerated at 35°F for less than 5 to 9 days; or fish refrigerated at 0°F for less than 4 months and preferably less than 6 weeks. It is a particular feature of this process that if the fish charged be stored for at least about one day and preferably up to about three days, under refrigeration at about 35°F, subsequent treatment including filtering, denaturing, etc., may be effected more easily. For example, the time of filtration may be decreased by as much as 40%. Under these conditions the product possesses the superior features herein set forth.

The so-treated fish, preferably whole, may be comminuted in e.g., a Hobart mill, a Fitzpatrick mill (typically having a #5 screen), or a Waring blender type mill. Comminution is effected to provide a product most of which will pass through about a 16 to 20 mesh screen on which the bulk of the bones and scale may be retained.

The comminuted fish is slurried with 2 to 10, preferably 2, volumes of water which contain sufficient acid, typically as sulfuric acid, to adjust the pH of the resulting slurry to about that of the mean isoelectric point of the proteins (of which actomyosin muscle protein may be the main one), i.e., to a pH of about 3.5 to 6.0, preferably 4.0 to 5.5. Generally, a lower pH will yield a lighter color product, and when this is the prime desideratum, the pH may be as low as 3.

In carrying out this process antioxidizing conditions should be observed. This is readily accomplished by incorporating an antioxidant such as ascorbic acid, butylated hydroxy anisole or butylated hydroxy toluene to the slurry in an amount ranging from 0.001 to 0.2% by weight of the slurry. The particular antioxidant used is not critical and various antioxidants may be used.

Suitable antioxidants for use in this process include the following: tocopherols, gum guaiac, nordihydroguaiaretic acid, gallic acid and its esters, ascorbic acid and isoascorbic acid and their esters, as for example, ascorbyl or isoascorbyl palmitate, stearate, etc., thiodipropionic acid and its esters, as for example, the dioctyl and the didodecyl esters, phenol derivatives, as for example, butylated hydroxy anisole, quinoline derivatives, such as 2,2,4-trimethyl-1,2-dihydroquinoline, and so forth.

Typically for example, the antioxidant may be added in an amount of 1 ml of 9% alcoholic solution of butylated hydroxy toluene per pound of fish. Use of antioxidant may be eliminated or minimized when subsequent operations are effected very rapidly under conditions which preclude or minimize oxidation of the fish oils. This may be accomplished by conducting the operations in an oxygen-free environment, i.e., under nitrogen, carbon dioxide, or other inert gases. However, even when the process is conducted in an oxygen-free atmosphere, it is preferred to incorporate an antioxidant in the fish slurry.

The slurry or paste is preferably screened through a 16 to 20 mesh screen to separate out bones and scale thus lowering the resulting ash content of the final product from a starting ash of 15 to 20% down to 4 to 5% or less on a dry basis. The filtered bones and scales are water washed, the washings typically being combined with the slurry-filtrate and the bone and scale disposed of as by conversion to meal.

The slurry-filtrate, containing the deboned and descaled protein content of the fish, will be heated to 155° to 212°F, preferably 175°F. Preferably heating is accomplished by adding the paste of slurry-filtrate to preheated water at the desired temperature. The mixture may be maintained hot for a period of 2 minutes up to 30 to 90 minutes, but preferably about 20 minutes.

The slurry may be filtered, preferably by centrifugal filtration, to give (a) a semisolid wet cake of fish flour, typically in amount of about 10 to 30%; (b) an oil, typically in amount of 2 to 8% (this comprising 95 to 99% of the oil originally in the fish); and (c) an aqueous solution of fish solubles containing amino acids, vitamins, proteins, etc. (which may be concentrated and sold as such), typically in amount of 50 to 90%. Preferably filtration is effected quickly to produce a semisolid wet cake which may contain (a) 20 to 60%, typically 40 to 50% solids; (b) 50 to 75% water; and (c) 3 to 8% oil. This cake may then be mixed with a bland solvent to form a slurry or paste.

The bland solvents which may be employed are those which can be used to extract from the filter cake a substantial portion of the lipoid (i.e., fatty and oily) material present therein and which permit attainment by the process of a protein-flour characterized by its blandness, freedom from odor, freedom from taste, good color, and extended storage life. The preferred solvents are those which do not react chemically with any of the components of the fish under the conditions prevailing during the practice of the process and which are substantially nontoxic and not liable to produce toxicity in the product.

The bland solvents which may be employed include ethers, including water-miscible ethers, such as 1,4-dioxane, and ether alcohols, including water-miscible ether alcohols such as methyl propylene glycol, or the methyl ethers of dipropylene and tripropylene glycols. The preferred specific materials include tertiary butyl alcohol and isopropyl alcohol.

In the extraction operation, the semisolid wet cake is slurried with solvent in ratio of one part of solvent to 0.2 to 1, say 0.5 part based on the weight of the original charge raw fish. The mixture may be heated, preferably under reflux for e.g., 2 to 20 minutes or longer. The reflux temperature for the water-isopropyl alcohol system is about 175°F and that for water-tertiary butyl alcohol is also about 175°F. It will be apparent that various extraction techniques may be employed, such as a Soxhlet-type extraction.

The extraction operation is followed by a filtration to separate the solid from the solvent, the latter containing many of the undesirable flavors and odors and their precursors together with much of the oil and water. The slurrying, extraction, and filtering operations may be repeated a plurality of times, e.g., six times, either continuously or batchwise, to produce a solvent-wet cake (a flour) containing typically (e.g., when the solvent is azeotropic isopropyl alcohol) about 30% solids, 6% water, and 44% solvent.

It may be desirable under certain conditions to use an antioxidant during the extraction. Typically butylated hydroxy anisole in amount of 0.001 to 0.2% by weight of the solvent may be added to the last solvent, the antioxidant thus being plated or deposited on the flour during the final solvent extraction operation.

In accordance with this process, the wet flour may be desolventized at temperatures of 90° to 132°F, e.g., 110°F, and preferably under pressure of 5 mm Hg to 400 mm Hg e.g., 25 mm Hg. Operation at temperature less than about 132°F minimizes deterioration of flavor components remaining in the flour. The vacuum-dried flour possesses physical and chemical properties which are generally comparable to the best grades of commercial protein concentrates. However, it is characterized by a residual fishy odor and flavor, particularly when moistened to form, e.g., a dough, which proscribes its use in bland products.

Elimination of the fishy odor and aroma is also effected by the process described by *W.M. Whaley and R.J. Moshy; U.S. Patent 3,164,471; January 5, 1965; assigned to General Foods Corporation.* Treatment of the proteinaceous flour to eliminate the residual fishy odor, taste, and aroma, as well as any residual solvent to obtain a light, bland product may be effected in accordance with this method by humidification to a moisture level of below 30% and then redrying at a temperature of below 132°F.

The partially desolventized fish flour may be moistened from a moisture level of 4 to 12%, say 6 to 8%. Humidification or remoistening may be effected by adding water in amount sufficient to raise the moisture content to 10 to 30%, and preferably to 20 to 25%. When humidification is effected to less than 10%, say to 5%, some improvement is noted as hereinafter described, but full deodorization and dearomatization are not obtained; however, a plurality of humidifications to less than 10%, e.g., each to 5% can be employed to obtain the desired product.

For example, a fish containing 4% moisture may be humidified to about 15%; a fish containing 12% moisture may be humidified to about 30%; and in each case the desired bland product may be obtained. Typically the increase in moisture content during humidification will be at least about 5 to 6% and preferably about 10 to 20%.

Humidification to moisture contents above about 30% tends to discolor the otherwise light product and if the humidification be to about 60 to 80% the product will be very dark in color and unacceptable from a color standpoint. It will be otherwise totally satisfactory with respect to odor, aroma, and taste, and if the lightness or whiteness of color in the fish flour is not essential in the final product, humidification levels of about 30% will be effective in removing solvent and fishy traces while leaving a fish flour of acceptable quality from a flavor standpoint, but of dark color.

However, the fish flour should not be humidified to a level of about 100% moisture since high levels of moisture, i.e., water wash and water slurries are not effective in completely removing solvent and odor constituents at below 132°F. The moistened flour or protein, at the moisture content of below 30%, will be dried under mild conditions, preferably vacuum dried at pressure of 5 mm Hg to 400 mm Hg, say 25 mm Hg and at temperature less than 132°F, preferably 90° to 110°F, say 100°F. Drying is effected to a moisture content of less than 10% and preferably less than 8%, typically to about 5 to 8%.

Humidification (including adjusting the moisture content and redrying) may be done once or as many times as desired. Each humidification decreases the amount of odor and aroma. Although commonly one humidification will remove virtually all of the residual flavor, taste, aroma and odor of the product, two or three humidifications may be employed if desired. Naturally the number of humidifications will depend upon the type of raw material and the degree of purification desired.

The fish protein or fish flour obtained by the process is characterized primarily by a lightness of color and an odor, taste, and aroma which ranges from very faint to nonexistent. Even under the least favorable conditions, the color is substantially white, the odor is nonfishy, and it reflects neither the raw material from which it was prepared nor the solvents employed in its preparation. The residual solvent concentration in the product is substantially zero.

Typically the taste of the flour is bland, it is free of undesirable sharp or pungent flavor or aroma notes, and it is nutritionally equivalent to or superior to high quality proteins such as milk or beef protein. The storage life is indefinite and it may be stored for indeterminately long periods. For example in one test at 90°F and 85% relative humidity (very severe storage conditions) it was unchanged after 2 months. In comparison, the best product heretofore available when stored 30 days at room temperature and humidity, gave off an odor which was both fish-like and solvent-like in character.

Typically the improved fish protein product may have the following composition on a dry basis: protein 92 to 98%, typically 95.3%; ash 4 to 6%, typically 4.57%; fat 0.02 to 0.4%, typically 0.17%. On addition to water to form a paste, the product developed no odor and its taste was bland. The best product heretofore available, when similarly tested, had a distinctly fish-like taste and aroma. Nutritionally, the product is fully the equivalent of the best protein available on the market and its low cost, high availability, and long storage life make it a uniquely superior product.

Among the uses to which this high-quality protein product may be put the following may be noted: (a) as a protein-enriched supplement for (wheat) flour in any and all of its uses including manufacture of bread and cake of the highest quality; (b) in the preparation of a proteinaceous whipping agent; (c) as a protein enrichant in connection with, e.g., cereals, etc.; (d) as an adhesive; (e) as a binder, etc.; (f) as a base for protein fibers and films; and (g) as a key protein ingredient in fabricated foods. It may be further modified to provide a wide variety of uses including, e.g., specific adhesives, food uses, etc.

Example: 100 lb of menhaden fish flour were treated. The charge material was characterized by an off-white color, by a nonbland (a blandness rating of 5.5) aroma, taste, and odor, its fishy nature when tested according to the various tests hereinbefore noted. 100 lb of this protein which had been prepared by solvent extraction of ground whole fresh menhaden, and which had a moisture content of 6% was mixed with 20.5 lb of water which raised the moisture content to 22%.

The mixture was passed through a Fitzpatrick mill to ensure homogeneity and it was then dried at reduced pressure in a rotary double cone vacuum dryer at 25 mm Hg. Temperature of the material being dried was maintained at 90° to 110°F until moisture content of 6% was obtained. The protein product had a blandness rating of 3.2 which is comparable to spray-dried egg albumen which has a blandness rating of 3.2.

The product was substantially white in color, had very little taste, odor, and aroma and it was substantially completely free of any fishy or solvent-like character when subjected to aroma and taste tests. It had a very long shelf life. When tested under accelerated storage conditions after 2 months at 90°F and 85% relative humidity, the protein still failed to evidence any fishy or solvent-like odor or aroma and it was still substantially as good as it was at the beginning of the shelf life-storage test.

Protein Hydrolysate Low in Fish Oil

Recovery of the protein content of fish or fish products such as fish meal in a stable and palatable form normally involves a digestion of the fish in subdivided form in an acid or neutral medium at elevated temperature and pressure followed by a separation of the resulting layers from the digestion step and subsequent concentration, including in some processes a further treatment to remove acid and/or salt contaminants.

However, many processes of this kind produce a product having a substantial percent of residual fish oil, for instance, from 5 to 10% based on the dried solid protein content, oil, and ash.

O.A. Roels; U.S. Patent 3,836,686; September 17, 1974 has found that conducting the digestion at a pH of about 1, followed by separation and purification of the resulting aqueous phase will yield a protein hydrolysate having a concentration of oil of less that 0.15%; yet obtaining a high recovery of the original protein content. At pH of 0.5 to 1.5 a recovery in excess of 50% of the protein value present in the originally treated product was obtained. At pH 1.5 the oil content was 0.12% and at pH 0.5 the oil content was below 0.1%.

In general, the process comprises digesting finely subdivided fish or fish products at a pH of about 1. Preferably hydrochloric acid or sulfuric acid is used. When the mass is at the proper pH, it is then subjected to elevated temperature. A superatmospheric pressure may be used. A temperature of 120°C and pressure of 15 psig for a period of 15 minutes have been found suitable though a range of temperature, pressure and period of treatment may be employed.

When whole fish or a residue thereof is used, subdivision may be attained by grinding and macerating the fish in an aqueous medium, usually but not necessarily containing an appropriate amount of antioxidant and preferably in an inert atmosphere.

If desired, when whole fish is employed, the grinding and maceration step may be followed by a separation, such as, centrifugation, to remove bones and scales, the supernatant liquid being comprised mainly of water, protein and fish oil. Alternatively, the precipitate from this separation (bones and scale) may be subjected to an aqueous extraction to remove further amounts of protein and oil. The supernatants from each of the separations may be combined and treated as described above.

Following acid digestion the mass may be cooled, if desired, and the resulting insoluble material, usually a flocculent solid, is separated from the subnatant liquid, for example, by skim-centrifugation or filtration. If the bones have not been removed, they may settle as a bottom layer. The acidity of the liquid phase, containing dissolved protein hydrolysate, may then be neutralized by the addition of base or preferably by passing through a suitable ion exchange medium to the desired pH.

If it is desired to recover the protein hydrolysate in solid form, the neutralized liquid may be subjected to a drying step, e.g., spray drying. The product resulting from the drying step is usually pale yellow to brown. The product may be further decolored by passing the neutralized aqueous protein solution, preferably before drying, through a charcoal adsorption or ion exchange. An oil and mineral product may be obtained as by-products of the aforedescribed process. The oil may be refined by procedures such as solvent extraction, heating and pressing or other methods known to those skilled in the art.

The mineral product is obtained by subjecting the precipitate from the first centrifugation step, which precipitate consists mainly of bones and scale, to an acid treatment under elevated temperature and pressure. Examples of suitable acids are hydrochloric and sulfuric.

A further feature of this process is the direct preparation of a protein hydrolysate having a dietetically useful mineral content, particularly with respect to the elements calcium and phosphorus. In experiments on Peruvian fish meal, the mineral content of the protein hydrolysate prepared by heating at 121°C for 15 minutes at pH 1, was respectively, 16.3 and 16.4% when resin Amberlite IRA 93 was used; and 18.0 and 19.2% when resin Amberlite IRA 68 was used. This may be favorably compared with the mineral content of the original Peruvian fish meal, which was 14.1% on a dry weight basis.

To prepare a mineral-rich protein hydrolysate from whole fish the solids resulting from the initial grinding and maceration are not removed but are subjected along with the liquid to the acid treatment at about pH 1. During this step a substantial quantity of the mineral content will dissolve in the liquid and be retained therein until the drying step. Undissolved solids remaining after the acid treatment may be removed during the oil separation step.

Example 1: 240 grams of fresh herring having the following composition:

	Grams
Water	188.2
Oil	8.8
Crude protein (total nitrogen x 6.25)	35.8
Ash (prepared at 650° to 700°C)	5.54
Unidentified components	1.66

were mixed with 480 ml of deionized water and 0.5 gram of butylated hydroxy anisole in a high speed blender for 8 minutes under a nitrogen atmosphere.

This mass was centrifuged in an angle rotor at a relative centrifugal force varying from 2,000 to 5,000 times gravity for 5 minutes. At the end of centrifugation, the supernatant liquid was decanted. After the addition of 240 ml of deionized water, the precipitate was again subjected to blending for a period of 2 minutes and then subjected to centrifugation and the supernatant separated. The supernatant layers from each centrifugation were combined and subjected to an acid treatment as more fully described hereafter. The precipitate analyzed as follows:

	Grams
Ash	2.4
Crude protein	3.9
Oil	0.2
Unidentified material	0.2

To the combined supernatant liquid was added 206 ml concentrated hydrochloric acid (38%) to achieve a final concentration of 3 N hydrochloric acid. The pH of the mass was less than 1. The mass was heated at 120°C, at a steam pressure of 15 psi for 60 minutes. Following this, the mass was cooled to room temperature. At room temperature, two phases appeared, one a clear liquid and the other a flocculent solid floating in the liquid. The mass was filtered through a Whatman No. 12 filter paper to separate the liquid and solid. The solid (filter cake) was repeatedly washed with deionized water to remove the excess acid and filtered, the filtrates being combined with the initial filtrate and subjected to an ion exchange treatment as described below. The filter cake analyzed as follows:

	Grams
Oil	8.6
Crude protein	0.9
Ash	0.04
Unidentified material or loss	0.36

The filtrates from the acid treatment step were passed through a bed of a weak base anion exchange resin, namely Amberlite IRA-93, for the purpose of removing any residual hydrochloric acid and to further purify the product. The effluent from the ion exchange bed was freeze dried and yielded 35.2 grams dry material analyzing as follows:

	Grams
Crude protein	31
Ash	3.1
Unidentified material	1.1
Oil, not more than 0.002%	

The freeze dried product was a pale salmon colored powder, completely soluble in water and with a pleasant shrimp-like odor.

Example 2: A Peruvian fish meal was prepared from anchoveta. 250 grams of the fish meal, 625 ml of deionized water, 1.5 grams of butylated hydroxy anisole and 200 ml of concentrated hydrochloric acid (38%) were mixed and heated at 120°C at a steam pressure of 15 psi for 60 minutes. The pH of the mass was less than 1.

The mixture was then cooled to room temperature and filtered through a No. 12 Whatman filter paper. The filter cake was washed with 6 portions of deionized water to a total of 1,500 ml. All filtrates were combined including the filtrate from the initial filtration. The filter cake was dried and analyzed:

	Grams
Oil	26.2
Crude protein	13.4

The combined filtrates were passed through a bed of Amberlite IRA-93 and the effluent freeze dried; the freeze dried product amounted to 152.7 grams and analyzed as follows.

Crude protein	128.7 g
Ash	21.0 g
Unidentified material	4.0 g
Oil	0.02 %

The product was a light brown hygroscopic powder, completely water-soluble and had a pleasant bouillon-like aroma.

Preservation of Wet Fish Press Cake

In the conventional wet reduction process for producing fish meal, raw fish are cooked, pressed to separate a major portion of the water and oil, and the press cake thereby obtained is then further defatted and/or dried to form fish meal. Cooking and pressing is one of the most efficient means for partial dehydration available. The primary product of this operation, wet press cake however is a perishable intermediate in the production of fish meal, and the final drying step requires heavy capital investment for equipment of sufficient capacity to process large quantities of press cake. In addition, raw fish, like the press cake itself, is highly perishable and, therefore, must be rapidly converted into fish meal. Furthermore, the supply of raw fish fluctuates greatly.

Accordingly, a serious economical problem in the art of fish processing involving the conventional wet reduction process of cooking, pressing, and drying lies in the fact that plants utilizing such a process must be built with overcapacity in the final drying stages compared to the average load of raw fish material in order to accommodate the large quantity landings which occur from time to time. Thus, there is a real need for a convenient method of preserving fish press cake prior to drying. Such a method of preservation would provide significant reduction in overall processing costs since the cake could then be stored in relation to supply fluctuations. This would allow for appropriate reduction in drying processing capacity with attendant economic benefits.

In accordance with that need, *T.L. Meade; U.S. Patent 3,712,822; January 23, 1973* developed a process whereby wet fish press cake can be preserved prior to drying for long periods, for example, periods in excess of 60 days, and can then be conveniently dehydrated. More specifically, the process involves mixing an alcohol with wet (water-containing) fish press cake and then storing the press cake-alcohol-water mixture in a silo or other suitable storage vessel for a period of time.

Subsequently, when it is desired to convert the preserved cake into fish meal, a hydrocarbon solvent is added to the mixture which is then subjected to an azeotropic extraction where the alcohol, water and hydrocarbon are removed from the press cake in the form of tertiary and/or binary azeotropic vapors. This extraction substantially completely removes the alcohol from the press cake as well as effecting partial or complete dehydration. After the extraction, the dehydrated cake can be passed through a desolventizer to remove the hydrocarbon residue and to provide the fish meal product. Any oil present in the press cake can also be removed during the extraction in the form of a hydrocarbon-oil miscella.

This process in combination with a conventional wet reduction process for obtaining fish meal allows the cooking and pressing operations to be carried out in direct relation to the supply of raw fish (landings) and yet, because of the ability to preserve the fish press cake, the final defatting and/or dehydration step can be suspended from time to time and performed only when the accumulation of press cake matches installed processing capacity.

Figure 7.5 shows the steps of the preferred process and how it can be used to produce dry fish meal from raw fish. In the process, raw fish **11** are cooked **12** and then pressed **13** where a major portion of the oil **14** and water **15** are removed. An amount of water and some oil remains associated with the press cake. The wet oil-containing press cake **1** remaining after the pressing step is preferably ground **16**, or is transferred directly **17** to a mixing stage **2** where alcohol **3**, **18** is mixed therewith. The press cake-water-alcohol mixture **4** is then stored **5** for a period of time in a preserved state.

Subsequently, the preserved press cake-water-alcohol mixture **19** is placed in an azeotropic extraction zone **6**, a hydrocarbon **7** is added, and the alcohol, water and hydrocarbon **8** are removed in the form of azeotropic vapors. These vapors are condensed **20** and ultimately separated **21**, e.g., by decantation, into an alcohol-water phase **22** and a hydrocarbon phase **23**. The alcohol-water phase is distilled **24** whereby alcohol vapor **25** is driven off, condensed **26** and reused in combination with the original alcohol feed stream and the water residue **27** is discarded. The decanted hydrocarbon is combined with a hydrocarbon-oil miscella **28** which is withdrawn from the extraction zone **6** and the combined mixture **29** is evaporated **30**.

The hot hydrocarbon vapors emitting from the evaporation preferably serve as the hydrocarbon feed **7** to the extraction zone and provide heat for the extraction. The oil residue **31** remaining after the evaporation step can be stripped of any remaining hydrocarbon to provide pure oils **32** which in turn can be combined with the oil **14**. The hydrocarbon **33** so stripped can be recycled by adding it to the alcohol-water-hydrocarbon condensate **20**.

The azeotropic extraction can substantially completely remove all traces of water and alcohol from the press cake but leaves some hydrocarbon residue associated therewith. This residue **34**, comprising dry press cake plus hydrocarbon, is removed from the extraction zone and desolventized **35**, e.g., in a rotary vacuum evaporator, where the hydrocarbon **9** is removed and recycled leaving dry press cake **10** which is the desired fish meal product. The following example will describe the process in greater detail.

Example: Preparation of Fish Press Cake — 10 liters of water was added to 20 kilograms of raw sea herring *(Clupea harengus)* (Linnaeus) 1758, taken in the North Atlantic in the vicinity of Block Island, and the mixture cooked by holding in a 125 liter steam jacketed kettle for 10 minutes. Free water was decanted from the above mixture and 1,500 gram portions of cooked herring were wrapped in muslin cloth and pressed in a Carver laboratory hydraulic press at a maximum pressure of 10,000 psi.

The resultant press cake portions were removed from the cloth and totalled 7,850 grams. Of this total, the water content was 55%, and the oil content was 4.25%. The press cake was passed through a Hobart meat grinder equipped with a discharge plate die having one-fourth inch diameter openings to obtain the cake in particulate form such that it would pass through a #3 inch screen.

FIGURE 7.5: AZEOTROPIC REMOVAL OF WATER-HYDROCARBON ALCOHOL MIXTURES FROM PRESS FISH CAKE

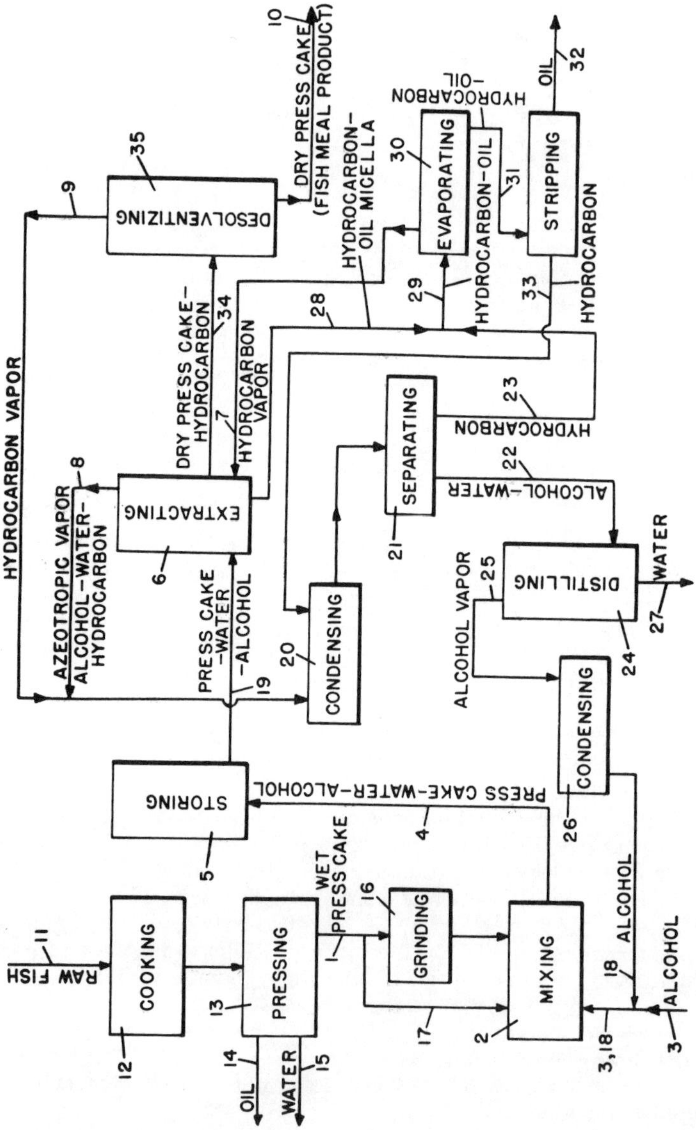

Source: T.L. Meade; U.S. Patent 3,712,822; January 23, 1973

Preservation – 950 grams of ethanol (95%) was added to the ground press cake-water mixture to obtain a press-cake-water-alcohol mixture containing 12% alcohol. This mixture was mixed in a twin shell liquid-solids blender to provide a uniform dispersion and intimate contact between the press cake particles and the alcohol. 2,500 gram portions of the press-cake-water-alcohol mixture were placed in each of three 4-liter tin can containers which were then sealed at approximately atmospheric pressure and stored at ambient temperature (68° to 86°F) for 5, 30 and 60 days respectively. At the time indicated, the containers were opened and the alcohol preserved press cake was processed into fish meal which was then analyzed for fish protein characteristics. These data are shown in the following table.

Fish Protein Characteristics

	- - - - -Time Preserved with Ethanol - - - -		
	5 Days	30 Days	60 Days
Protein	84.8	85.3	85.7
Fat	0.2	0.5	0.2
Ash	9.0	9.9	8.9
Moisture	8.1	8.9	8.0

These data indicate essentially no difference in samples and also show total preservation.

Azeotropic Extraction – 2,500 grams of the preserved press-cake water-alcohol mixture obtained as described above was placed in a 12 liter extractor along with 4 liters of n-heptane. The apparatus comprised a 12 liter, stainless steel, stirred reactor flash, heated with a Glas-Col electric mantle. A three-neck glass reactor flask cover permitted the continuous addition of hydrocarbon (n-heptane) from a 2-liter separatory funnel equipped with an equalizer arm, and the continuous removal and condensation of the azeotropic vapors.

The temperature of the material in the extractor was brought up to 155.4°F, which is the boiling point for an ethanol-heptane-water ternary azeotrope comprising 33.0% ethanol, 60.9% heptane and 6.10% water. This temperature gradually increased as the amount of ethanol remaining diminished. When all the ethanol had been depleted from the press cake material, the boiling temperature again became constant at 175°F, which is the boiling point for a heptane-water-binary azeotrope comprising 87.1% heptane and 12.9% water. Water was removed from the press cake material through the formation of the binary azeotrope with heptane, until the amount of water in the press cake material was reduced to less than 10%.

Simultaneously, the amount of oil in the press cake material was reduced to less than 0.5% via the formation of an oil-heptane miscella. At this point the temperature of the materials in the reactor reached 185°F. During the extraction, n-heptane at ambient temperature (77°F) was continuously added to the extractor at a rate equivalent to the removal of n-heptane in the azeotropic vapors.

During the initial formation of the ethanol-heptane-water ternary azeotrope, the vapor was condensed and cooled to between 68° and 70°F. The condensate separates into two layers with the upper amount 65% by volume and the lower 35%.

The upper layer contained ethanol 5%, heptane 94.8%, and water 0.2%. The lower layer contained ethanol 76%, heptane 9%, and water 15%. The condensate recovered during the rapid transition state between the ternary azeotropic and the higher boiling binary azeotrope was characterized by a continuous change in ratio and composition of the upper and lower layers.

With the attainment of a stable heptane-water-binary azeotrope vapor, condensation and cooling to 68° to 70°F resulted in phase separation with the upper layer consisting of 99.95% heptane and amounting to 90% by volume, the lower layer consisted of water essentially free of heptane and amounting to 10% by volume.

After the extraction, the oil-heptane miscella was decanted and the press cake was washed twice with 2,500 ml volumes of n-heptane. The resultant press cake contained less than 10% water, less than 0.5% oil, less than 0.005% alcohol, and about 50% heptane. This heptane-wet cake was then placed in an evaporator flask which was connected to a condenser flask. Heating to 125°F at 200 mm Hg for 1 hour while rotating the flask at 25 rpm provided a heptane-free, substantially dry, fish meal product.

In this example, heptane and alcohol recovery and recycle are accomplished as described hereinbefore. Also in this example, substantially equivalent results are obtained in that wet fish press cake is preserved by adding an alcohol, and the alcohol and the water are subsequently removed by azeotropic extraction with a hydrocarbon, when methanol, n-propanol, isopropanol, decyl alcohol, or a mixture of these alcohols and/or n-hexane, isooctane, or a 50:50 mixture of n-hexane and n-heptane are respectively substituted for the ethanol and n-heptane of the example.

Preservation of Fish Meat with Solvent to Be Used in Extraction

The objects of the process developed by *T.L. Knutsen and S.-O. Osterman; U.S. Patent 3,852,260; December 3, 1974; assigned to Astra Nutrition AB, Sweden* are to obtain a method for preparing protein from fish during periods when the supply of fish is limited or decreased, and to store the fish material for a long time without bacterial growth, autooxidation of the fat present in the fish material, and protein destruction.

The Bureau of Commercial Fisheries has described (1967) the preservation of whole, ground red hake in 91% isopropanol. However, this process has the drawback that already after 11 days of preservation, 5% of the total protein has been destroyed by means of enzymatic activity. It has also been proposed that one should prepare a valuable fish meal from fresh fish by boiling and pressing it. Such a fish meal may be stored for long times, but has the drawback that the fat of the material is autooxidized and that the phospholipids present in the fat give the meal a very unpleasant odor and taste. A marked decrease in nutritional value is also obtained.

It has now, however, surprisingly been found that very good results are obtained by a process in which the fish in a fresh state after disintegration, boiling, deboning, separation, pressing and partial dewatering to a remaining proportion of water of 12 to 20% by weight whereby bones, stickwater and certain fats are removed, is mixed with the same kind of solvent which is used in a following extraction step, and in an amount of 10 to 30%, preferably 15 to 25% by weight calculated on the amount of the water remaining in the material. The process provides, in addition to an inhibition of bacterial growth and fat oxidation, that all enzymatic activity has ceased before storage, so that no hydrolysis of the protein takes place and the protein will retain all its nutritional value, and that the protein yield will increase.

Suitable solvents for the extraction, and thereby the preservation, are lower alkanols, having at most 4 carbons, such as ethanol, isopropanol, n-butanol, sec-butanol, and isobutanol. Mixtures of these alkanols may be used as well. The essential feature is not that the protein material per se is mixed with a solvent, but that the water present in the protein material is mixed with such an amount of a solvent that it cannot serve as a medium for bacterial growth using protein and fat as a nutritional source.

Fresh, newly caught fish is boiled in a vertical boiler by means of indirect as well as direct steam while stirring. After the boiling, which takes 10 minutes at 90° to 95°C, the fish mass is pumped to a deboning station where bones, big gut fragments, eyes, gills, etc. are partially removed. By this removal of the bone fraction the ash residue of the final protein material may be decreased by 50%. The deboning operation takes place at a suitable temperature of 90°C.

The fish material thus boiled and freed from bones etc. is then centrifuged and pressed, so that stickwater and some fat is removed. Such a material is called press cake. This is then transported to a drying station where a partial drying is carried out, whereby the amount of water is reduced from 50 to 12 to 20% by weight of the total material. By breaking off the drying at this amount, the beginning of the autooxidation may be retarded remarkedly.

The stickwater containing fat may be separated in order to obtain water-soluble protein and fat. The partially dried fish material is then transported to a mixing vessel. In this vessel the protein material is mixed with 15 to 25% by weight, calculated on the weight of water present in the material, with the same kind of solvent which will be used in the extraction process. Thereafter the material is transported to a sacking station where the material is put into air and gas impervious sacks, for the storage before it enters the final extraction process, where all taste and odor-producing substances and substantially all the fat is removed from the material.

By the addition of the solvent a preservation is obtained so that the material may be stored in a buffer storage for a very long time. The material thus preserved is brought into an extraction process, e.g., when the supply of fresh material is low, where the material is extracted with a solvent, preferably in countercurrent fashion in a number of steps. After each step of the extraction process the solvent is separated from the material by centrifugation, whereupon the material is transported to the following step and the solvent is transported to the previous step. After the last step the material is transported to a solvent removing station where the residual amount of solvent is removed, while the solvent from the first extraction step is transported to a recovery plant.

The removal of the solvent is carried out by means of direct steam countercurrent to the material flow, all solvent being removed. At the same time a deodorizing procedure takes place by means of the direct steam on the volatile taste and odor-producing substances, preferably amines.

Protein material thus freed from solvent is then brought to a drying station where the amount of water is reduced to about 5% by weight, whereupon the material is ground and sieved. The material prepared contains 90 to 93% by weight of raw protein calculated on the dry contents, of which protein 95 to 96% is digestable raw protein. The total fat amounts to 0.1 to 0.2% by weight only, and the remaining amount of solvent is only 10 to 40 ppm.

The solvent used in the extraction was isopropanol but other solvents such as n-, sec-, and isobutanol may be used as well, as mentioned above. The preservation of the material partially dried is carried out suitably with the same kind of solvent as will be used in the following extraction process. The solvent used in the preservation is either pure or is taken from the second step of the extraction process, where it is removed by centrifugation. The preservation ought to take place immediately, but shall take place before any bacterial growth occurs. Besides the solvent, an antioxidant may be added to the material in order to improve the antioxidation effect.

In order to determine the effect obtained by preservation by means of isopropanol and sec-butanol, an analysis of the contents of bacteria was carried out after 40 days of storage at room temperature and two different contents of each alcohol. In the table below, the amount of alcohol is the total amount in the liquid around the material during the preservation process.

Isopropanol, percent	sec-Butanol, percent	Total Number of Bacteria per Gram Material After 40 Days
15		100
30		200
	15	300
	30	400

As is evident from the table, there are very few bacteria present per gram of material after storage for 40 days, using either isopropanol or sec-butanol. In order to show the difference in autooxidation between a conventional fish meal and a fish material prepared according to this process, before extraction and without addition of an antioxidant, comparative tests were carried out in which the two materials were stored in dark at 25°C totally for 40 days. Samples were taken continuously, the amounts of peroxides and the benzidine values being determined.

It was found that there was a very high proportion of peroxides in fish meal after 9 days, while the proportion of peroxides in the preserved material was consistently very low. The same was true for the benzidine value, which in fish meal rises very markedly after 7 days and reached its maximum after 10 days. In the preserved material there is a minor increase after 10 days, but the total value was 80% below the value for the fish meal.

As there are very large distances to the fishing grounds, difficulties may arise in getting fresh fish ashore. Under such circumstances it is therefore suitable to produce a semifinished product according to the described process on board a ship. As preservation takes place at the fishing grounds, the fish is in a warranted fresh condition when the process starts, whereby an extremely qualitatively pure protein product is obtained in the extraction process.

PREPARATION BY BIOLOGICAL MEANS

It is desirable that fish protein concentrates in the form of flours have a digestibility above 92%, a consistent protein content, and high values of available lysine and methionine. Predigestion or hydrolysis of proteins by chemical means, whether acid or alkaline, is imperfect because of the racemization of the amino acids, which then lose their biological value. Something similar happens with the processing of fish meal by conventional means, since the coagulation, precipitation and cooking with high heat causes the final product to be poorer in its protein values.

Protein Hydrolysis by Yeast

V.H. Bertullo and F.P. Hettich; U.S. Patent 3,000,789; September 19, 1961 describe a proteolytic yeast, isolated from the liver of a common Brazilian fish, the croaker corvina *(Micropogon opercularis),* which was given the name *Saccharomyces platensis proteolytica.* This yeast was used for hydrolyzing proteins by mixing proteinaceous material, such as whole fish, fish remains, etc. with it, adding molasses, and fermenting the mixture. The product was used as silage for animal feeding, or for fertilizer.

V.H. Bertullo and C.R. Pereira; U.S. Patent 3,516,349; June 23, 1970 have provided a method for the production of protein hydrolysates by means of a proteolytic yeast, the hydrolysates serving as a food for human consumption. The form of the product may vary considerably. It may be a dehydrated and degreased powder, a paste, resulting from concentration of the hydrolyzed, unattacked protein, up to a total of solids not under 50%; or a liquid obtained by separation of the hydrolyzed from the nonhydrolyzed protein by mechanical means.

The particular proteolytic yeast used was named *Hansenula montevideo.* According to the process, protein of fish, mollusks and crustaceans is hydrolyzed to between 60 and 75% depending upon the species worked upon. Hydrolysis is effected by acting on protein material of the type mentioned with the specific yeast at temperatures between 32° and 34°C for a period ranging from 18 to 24 hours with constant stirring.

This hydrolysis is effective with oily species such as, for example, anchovy *(Engraulis ringens)* of the Pacific Ocean; lacha *(Brevoortia tyrannus)* of Southwestern Atlantic waters, and lean species such as the hake *(Urophycis)* of the North American Atlantic waters, or the whiting *(Merluccius merluccius hubbsi)* of the Southwest Atlantic, or the Pacific hake *(Merluccius gayi).*

According to the process, hydrolysis is also effected in a similar percentage if the meat of cetaceans *(Balaenopterae)* of the South Atlantic is processed. The process provides a method for the production of polypeptide and amino acids, to which incidentally should be added members of the vitamin B complex, which are processed per se by the yeast, apart from the trace elements so important in modern nutrition, and the hormones that the animal forms possess, and the presence of the fish factor which is maintained wholly in the final product.

The process, upon effecting the proteolysis by means of *Hansenula montevideo,* does not cause crystallization of the hydrolyzed products as they are formed during hydrolysis, as occurs with other methods. When hydrolyzing by proteolysis, for example, fish meat with a solids content varying between 25 and 32%, putrefaction that might be occasioned by means of marine bacteria of the fish or of such bacteria contaminating the fish during capture, storage, handling, and marketing, is controlled from the very start.

Bacteria in the mucus of fish skin, gills, and intestines start disappearing after 8 to 10 hours once the yeast has developed vigorously in a medium favorable thereto. Action of the yeast is carried out both on gram-negative bacteria of marine origin, and on gram-negative or gram-positive bacteria which may contaminate fish with the presence of a 7 to 10% of sucrose with a content of 97 to 98% as energy source. Because the process is developed at temperatures not above 35°C, it does not racemize any free amino acid, allowing for example that lysine be recovered in not less than 80%. The proteolysis processing of fish for animal use is illustrated by the following steps:

(1) Grinding of the product, namely fish or residues thereof.
(2) Adding of a fermentable carbohydrate in 7 to 10% by weight of the total amount.
(3) Mixing of the ingredients and placing them in a fermenter.
(4) Adjusting the temperature to between 28° and 35°C.
(5) Adding of the *Hansenula montevideo* yeast culture after reaching the right temperature.
(6) Stirring of the mass at 40 to 55 rpm.
(7) Carrying on hydrolysis for 18 to 24 hours or more.
(8) Filtering and separating of solids.
(9) Degreasing by mechanical means.
(10) Concentrating the solids to 50%.
(11) Drying.
(12) Packing in sacks and storing.

It is mandatory that the fish or residues of fish be absolutely fresh, so as to obtain a product suitable for animal consumption. The pH should not be in excess of 7. Grinding [Step (1)] for a satisfactory hydrolysis must be as fine as possible, so as to expedite the proteolytic action of the yeast. The addition of fermentable carbohydrate [Step (2)] may be carried out on the basis of commercial sucrose of 98% purity, or with any other fermentable sugar, except lactose which is not used nor fermented by the yeast. The fermentable carbohydrate may be incorporated in a percentage varying between 7 and 10% according to the type and nature of the fish to be hydrolyzed.

Speaking in general terms, the process requires that for each metric ton of fish or any other protein it is intended to hydrolyze, apart from the addition of the fermentable source provided by a carbohydrate, for example commercial sucrose, at least one liter of the culture of *Hansenula montevideo* with a yield of not less than two million viable cells per milliliter should be added, a greater amount being admissible.

The culture may be added either with the nutrient medium in which it was produced, or else prepared in the usual commercial manner, previously diluted so as to facilitate dispersion. Cultures can be maintained viable for a considerable period of time, the only requirement being that they should be kept at a suitable temperature, for example, not above 4°C.

The mixing of fish and carbohydrate [Step (3)] must be intimate, and it should reach the desired temperature before seeding of the yeast culture [Step (5)] is effected. Continuous stirring [Step (6)] is essential to expedite hydrolysis which will be completed in about 30 to 48 hours [Step (7)] according to the type of fish employed. Filtering [Step (8)] is effected to separate scales and bones not digested in the above period of time; these by-products may be used for manufacture of organic fertilizers.

If desired, the product may be degreased by centrifugation at a temperature of 40°C [Step (9)]. The hydrolysate, consisting of protein, amino acids and polypeptides, is then concentrated [Step (10)] to about 50% of solids by vacuum concentration at a temperature not above 40°C, and then dried by any of several known methods [Step (11)]. The resulting final product is a fine, pleasant smelling powder having a crude protein yield varying between 54 and 60%.

For proteolysis processing for human consumption, the hydrolysis process is fundamentally the same, its technology varying according to whether it is desired to prepare the product as a liquid, a paste or a powder. The process is the same as already explained through Step (8). In the processing of liquid proteolysis products, after Step (8), to obtain the liquid phase of the product consisting of free amino acids and polypeptides with a nitrogen content varying between 15 and 22 g/l, gravity sedimentation may be performed at an environmental temperature not exceeding 25°C but this procedure takes a lot of time.

Another method is to use centrifuges which separate the solids in continuous production at a speed not below 2,800 rpm, a liquid being thereby produced, having a specific gravity varying between 1.055 and 1.115, with a color between light yellow and dark brown, depending on the species of fish used. The product has a specific odor, is stable at ambient temperature, similar in taste, physical appearance and nitrogen yield to the nuoc-man or fermented fish sauce widely used in the Far East.

The final product may be salted to conform to the taste and dietary customs of the people of the Far East although this is not necessary for preservation purposes. The proteolysis method permits manufacture of the product similar to the nuoc-man in no more than 3 days, while following the classical method production takes between 3 and 18 months. It is obvious that the method of the process is considerably more practical, economical, and requires a lower outlay of initial capital.

The product thus obtained is rich in free amino acids and polypeptides, can be preserved indefinitely without additive, or else may have added sodium chloride in the proportion of 10% weight-volume so as to provide it with the taste to which the consumer of nuoc-man is accustomed. The color will range between light yellow and dark brown, depending on the type of fish used.

For production of fermented fish sauces of the baggong type, widely used in the Phillippine Islands, processing is continued after Step (8), that is to say, after filtering and separating of solids, concentrating the resulting product down to a solids content of 55 to 60%, or also concentration of the residue remaining after obtaining the liquid product is carried out up to a similar percentage. The fish paste can be preserved by itself, although addition of sodium chloride in the proportion of 10% weight-volume is recommended on account of the fact that consumers are accustomed to eat it with salt. In both products, liquid phase and paste, salting of the product is carried out after the process is completed, this being an additional advantage over classical methods where the fish, to prevent starting of putrefaction, is salted from the beginning, which requires a larger consumption of salt, on the one hand, and leaves a highly salted residue on the other hand, while by means of the process, on salting only the portion to be used for human consumption, the saltless residue may be used in different applications without the limitations imposed by the salt.

For the processing of fish protein concentrates, the process uses the same operations listed before until a concentration of solids of 50% is reached, whereafter dehydration and full degreasing are carried out by means of polar solvents of the type of 99% isopropanol which extracts fatty matter at low temperatures, dehydrating the product at the same time. The solvent does not affect essential amino acids, or choline, or the phospholipids, or the phosphoproteins; a final product is thus attained having a protein content between 70 and 80% depending on the fish species used, with a grayish to creamy white color, also depending on the type of fish.

The product is odorless, has a lightly pungent taste due to the free amino acids, with a fatty content not exceeding 0.05%. Impairment of taste due to elimination of volatile amines is not observed. The moisture content of the product is between 4 and 5% and may be preserved indefinitely under normal storage conditions in cellophane or bags of similar material.

The product leaves a residue of crude fatty matter which may be used for various industrial applications, as well as a residue of scales and bones which, adequately dried, may be used as organic fertilizer. The product thus obtained, in a percentage of 7%, has the following composition: N:P:Ca:K 7:20:20:2, plus a series of microelements all of which remain in part of the digested protein, adhering to the scales and bones.

The proteolysis product in powder form made according to the process has been tested from a toxicologic and microbiologic point of view in laboratory animals for extended periods of time. No toxicity has been found and the microbial count remained beneath the limits set by the FAO for these products. Test animals used were white mice, albino rats, and chicks. Microbiological techniques were effected in the conventional slide counting.

The proteolysis product in powder form for human consumption has been tested by pediatricians in premature and distrophic children; in a proportion of 1.5 to 2 g/kg weight of the premature children, with promising results as to tolerance and fixation of organic nitrogen; and in larger proportions in the case of distrophic children, with similar results.

Primary tests show ingestion and retention tolerance with no secondary disorders of any kind. The product has also been used on old people, with good results, and it is thus considered that it is a useful, active aid both in pediatrics and in geriatrics.

By *Aspergillus Oryzae-Flavus*

In the process described by *G.A. Jeffreys and A.J. Krell; U.S. Patent 3,170,794;*

February 23, 1965; assigned to Reliance Chemicals Corp., the enzymes used for treating fish to prepare high quality FPC are those produced by the oryzae-flavus group of molds of the genus Aspergillus.

The amount of enzyme culture or extract added is based on the fish solids, dry basis, and on the length of time of digestion which is to be employed. The preferred type of enzyme used is that produced by a 30-hour culture of *Aspergillus oryzae-flavus* on wheat bran dried to 6% moisture or its water extract equivalent. For 8 hours digestion time the amount preferred to obtain an odorless product is 1% based on the fresh fish, or 5% based on the dried fish solids. More or less enzyme material may be used, if desired. For example, doubling the amount of enzymes will reduce reaction time by approximately one-half. Also, if it is desired to produce a fish meal with only a slight taste and odor of fish, then only enough enzyme is used to liquefy the fish material, after which it can be immediately drum dried.

The stepwise or batch process takes place with the following steps. The fish, if for human consumption, is first eviscerated and cleaned. The cleaned fish is then conveyed through a comminuter where it is cut up into small pieces. From the comminuter the fish is discharged into a cooking and blending kettle or vessel. This vessel is jacketed for heating and cooling and is equipped with a variable speed agitator. The fish is heated directly with steam to the desired temperature and agitated at relatively slow speed while cooking. Sufficient water is added to make the concentration about 20% fish solids. The amount of water for filets or cleaned fish will average about 20% or for 500 lb of fish will require 100 lb water.

A small amount of additional water will be added through the steam cooking and the subsequently added enzyme and buffer solutions. The cooking may be conducted either at the higher temperature of 100°C or at the preferred lower temperature of 60° to 65°C. In either case, a cooking time of 10 minutes will usually be sufficient. The cooked fish is next cooled to the desired temperature for either 55°C temperature activation or for digestion at a low temperature of either 34° or at 42° to 45°C. Upon cooling to the selected temperature, the enzyme slurry is added in an amount equal to ½ to 2%, by weight, based on the wet fish, the exact amount depending on the processing time and degree of conversion desired.

As soon as the enzyme slurry or extract is added, a buffer salt of monocalcium phosphate is added in the amount of 0.25% by weight of the fish solids and citric or other suitable acid is added to the mixture to acidify the mixture to a pH of about 5.5. The fish material is agitated or blended at high speed until it becomes liquefied and assumes a creamy consistency. This will usually take from 20 minutes to 1 hour depending on the amount of enzyme used.

As soon as the fish is liquefied, it is pumped over into an activating or digesting kettle. This is a jacketed vessel equipped with a slow speed agitator which keeps the fish material in suspension. The temperature of the vessel is maintained at the selected temperature previously mentioned, namely, either at a low temperature of 34°C when yeast is used or for 54° to 56°C without yeast. Where processing is without yeast, intermediate temperatures in the range of 42° to 45°C can be used for a short time.

With low temperature activation, dried bakers' or brewers' yeast is added in the amount of 1% based on the liquid fish and 1 to 2% dextrose or other fermentable sugar. Processing is continued in the activating vessel for 1 to 8 hours or more depending on the degree of conversion desired. Usually a completely odorless fish protein can be obtained in 6 to 8 hours.

The reaction in the vessel can be stopped at any point by heating to 70°C or higher. The liquid fish is thereupon delivered to the drum dryers and dried thereby. The dried fish is scraped from the drums and pulverized in a grinder such as a conventional hammermill, from which the meal is delivered to a conventional collector and bagger apparatus, where the product is suitably packaged. The process is further demonstrated by the following example.

Example: The method described herein was based on the following formula and processed in a stepwise manner:

Ingredients	Parts by Weight
Fish comminuted	500.0
Water	100.0
Enzyme or equivalent water extract	5.0
Monocalcium phosphate	1.25
Citric acid or equivalent	3.0
Dextrose	10.0
Compressed bakers' yeast	5.0

500 parts of eviscerated and cleaned fish was comminuted to small pieces and then discharged into a jacketed cooking kettle. About 20% or 100 parts of water was added and the fish was cooked directly with steam to a temperature of 60° to 65°C. Slow agitation was used while cooking. The cooking temperature was maintained for about 10 minutes. The fish was next cooled to a temperature of 34°C; then the enzyme extract equal to 1% of the raw fish or 5 parts was added.

Also, 1.25 parts of monocalcium phosphate and 3 parts citric acid were added. This made a pH of 5.5. Bakers' yeast was then added in the amount of 1% or 5 parts and 2% dextrose or 10 parts. The mixture was agitated at high speed for 1 hour. In 30 minutes the mixture became liquid and in 1 hour it assumed a creamy consistency. The liquid fish was next transferred by a pump to a holding or activating vessel where mild agitation was maintained for a period of 7 hours at a temperature of 34°C.

At the end of this time the pH was adjusted with sodium hydroxide to pH 6.2 and the mixture was heated to about 82°C which rendered the material inactive and prepared it for drying. The liquid fish was then pumped to drum dryers where it was dried at atmospheric pressure, using 40 lb steam pressure on the drums, then pulverized, collected and bagged. The final product assayed as follows:

	Percent
Protein	80.0
Fat	0.8
Moisture	5.6
Ash	13.6

Speeding Up the Fermentation

G.A. Jeffreys; U.S. Patent 3,547,652; December 15, 1970 has accomplished a speeding up of the yeast fermentation of proteolytically liquefied cooked fish meat by adding to the fish meat a yeast which has started fermenting in the presence of the reaction product of previously liquefied fish and yeast.

Figure 7.6 shows a flow diagram of a plant suitable for the practice of the process. As shown, raw fish may be brought from a storage vessel **10**, which preferably is refrigerated if the fish is to be held for any extended period of time, through a comminuter **12** in which the fish is cut up into small pieces. The comminuter may comprise a hammermill or some other shredding device suitable for reducing the fish to small pieces which are readily cooked and digested. From the comminuter the fish is discharged into a cooking vessel **14**, which preferably is elongated and has conveying means therein, such as a screw conveyor, so that the fish pieces may be moved through the cooking vessel while they are being cooked.

The cooking may be performed at temperatures of 100°C or even higher, and a preferred method is by direct heating with live steam. Preferably the fish is agitated slowly while cooking. A cooking time of about 10 minutes is usually sufficient at 90° to 100°C. In addition to making the fish easier to digest by the enzymes, the cooking destroys many undesirable bacteria.

If fish with a low fat content is being processed, it may be passed from the cooker through line **16** directly into a cooling blender **18**. However, if high fat content fish is used, it may be desirable at this point to extract the major proportion of the oil. This is accomplished, by transferring the fish through conduit **20** into a filter press **22**, which may include a screw conveyor to remove the filter cake.

The wet filter cake may be taken off by means of a pump **24** or a transfer conveyor through a line **26** leading to line **16** and the filtrate may be transferred through the line **28** to a conventional centrifuge **30**. The centrifuge separates most of the oil, which is taken off through a line **32**, and the stickwater which is transferred to line **16** by means of a pump **34** and a pipe **36**. From this point the wet cake and the stickwater, now substantially free of oil, are passed together to the cooling blender.

The cooling blender has means therein for conveying the cooked fish longitudinally of the blender, as for example, a screw conveyor, and also has refrigeration means for reducing the temperature of the cooked fish to a temperature suitable for digestion by the enzymes. Upon being removed from the cooling blender by means of pump **38**, the cooked fish is transferred to a digester **40**. At this point the desired enzymes are transferred from a container, **42** by means of a pump **44** into the digester.

In the digester the cooked fish and the enzymes are mixed together and move from the left end to the right end of the digester, as by means of a screw conveyor or the like, for a period sufficient to substantially complete the digestion of the fish solids to liquid form. Alternatively, several digester tanks may be used alternately. For example, one tank may be filling while a second is on the digestion cycle and a third is being emptied.

FIGURE 7.6: FPC PROCESS

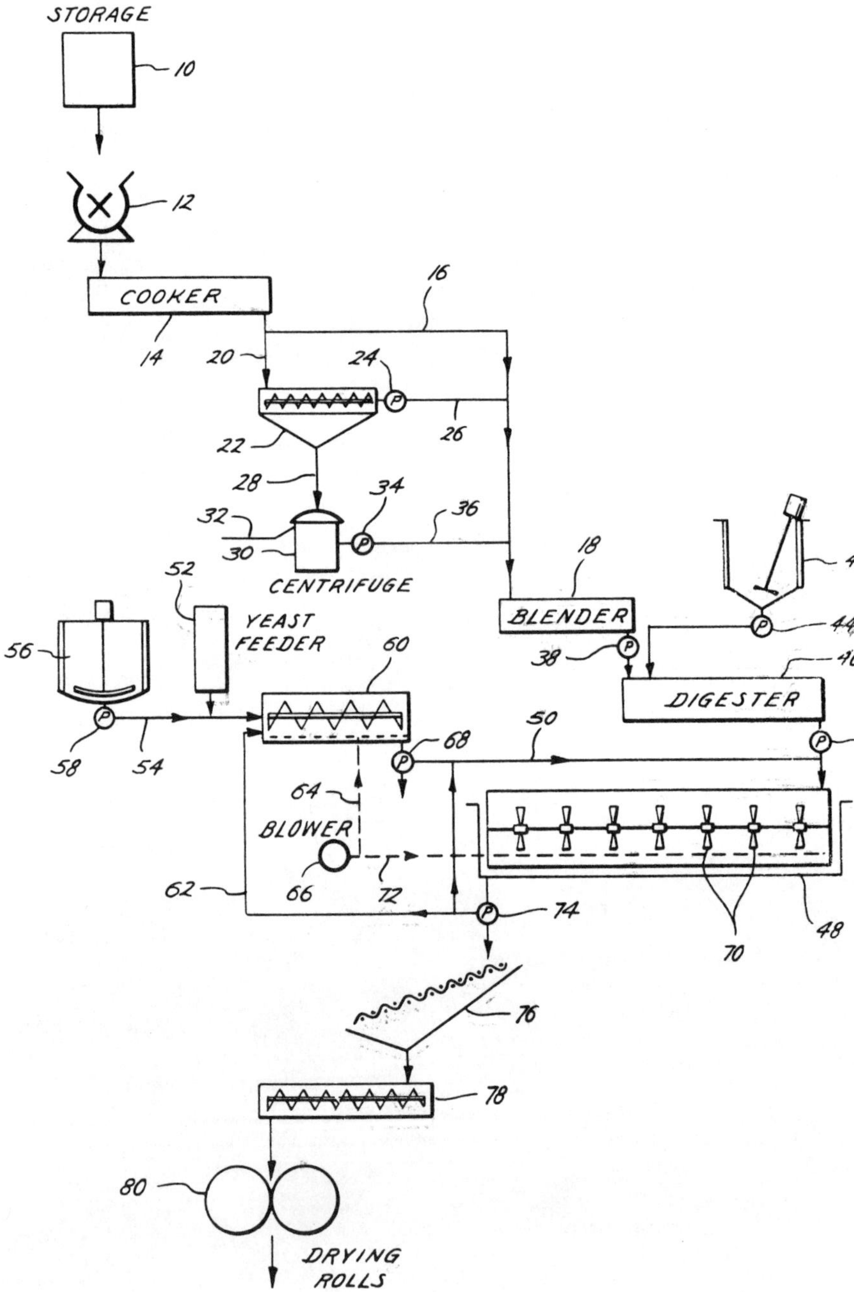

Source: G.A. Jeffreys; U.S. Patent 3,547,652; December 15, 1970

The liquefied fish is then transferred by means of a pump **46** into a reactor **48.** At this point the liquefied fish is combined with activated yeast which is fed in through a line **50.**

Yeast may be provided from a dry yeast feeder **52** which feeds the yeast into a line **54** leading from a concentrated wort storage container **56.** A pump **58** may be used to transfer the concentrated wort together with the dry yeast into a yeast reactivator **60.** In this device a conveyor such as a screw conveyor is used to slowly move the growing yeast from the left end to the right end of the reactivator as shown in the drawing.

At the same time extract from the outlet of the reactor is fed into the left end of the yeast reactivator by means of a line **62**, and the composition in the reactivator is continuously aerated by means of air supplied through a pipe **64** from an air source **66.** The activated yeast is removed from the right end of the reactivator by pump **68**, which transfers the yeast at a desired rate through the conduit **50** into the right end of the reactor where it is mixed with the liquefied fish.

In the reactor the liquefied fish and yeast mixture is stirred by blades **70** and is aerated by air provided through a line **72** while slowly being moved from the right end of the reactor to the left end, where it is extracted by means of a pump **74**, with a small portion being fed back through the line **62** to the yeast reactivator. However, the major portion of the reacted material is passed through a screen **76** which serves to remove any undigested bones and other solid materials. The liquid is then fed through a homogenizer **78**, and from this to a dryer **80**, where the material is dried until most of the liquid has been removed.

From this point the product may be used as is or it may be further dried or solvent extracted to remove any remaining oils. It may then be ground to a suitable powder or granular form as desired. It will be appreciated that in order to economically utilize a continuous process, it will be highly desirable to speed up the various reactions as much as possible, without sacrificing the desirable effects sought. It is, therefore, highly desirable that an enzyme be used which is capable of a high liquefication rate.

It is appreciated that the enzyme used for digestion must be proteolytic, that is, one which is capable of hydrolyzing proteins. Such enzymes include those produced by the oryzae-flavus group of molds of the genus Aspergillus. Other enzymes are proteolytic enzymes from bacterial sources, such as *Bacillus subtilis,* proteolytic enzymes from vegetable sources such as papain, bromelain, ficin, etc., other fungal proteolytic enzymes such as *Aspergillus niger,* and enzymes from animal sources such as pancreatin and trypsin.

The temperature used in the digester will vary depending upon the particular enzyme used and the concentration of that particular enzyme. Also, higher temperatures usually speed up the digestion process. Preferably temperatures no higher than about 60°C should be used with a bacterial enzyme although they can be used at temperatures up to about 90°C. Pancreatin cannot be used at temperatures above 60°C and preferably is used at temperatures below about 52° to 53°C. Other enzymes are reasonably active at temperatures as low as about 30°C. The lower temperature limit is governed by the desired for a high reaction rate and by the fact that it must not be so low as to allow excessive growth of undesirable bacteria. Upon the addition of the enzyme to the fish slurry in the digester, the pH in the digester is adjusted to improve the liquefaction activity. Each enzyme has its optimum pH for best results, and pH of the fish must therefore be adjusted accordingly. Bacterial and pancreatin enzymes work best at 6.5 to 7 pH while fungal enzymes are most efficient at 5.5 and vegetable enzymes may be used at pH's from 5 to 7. The enzymes are used in such concentration as will modify the fish to a creamy liquid state in a short time without breaking it down completely to amino acids. For a continuous process, economics dictate a liquefaction period preferably no greater than about 30 minutes.

When any of the above enzymes are used, liquefaction may be obtained in less than 30 minutes. Thus, a digester may be designed to retain the mixture of fish and enzyme for 30 minutes before it moves out the end of the digester into the reactor **48**. Alternatively, the digester may discharge into the reactor in batches, or two or more digesters may be used with one of them discharging into the reactor while the digestion step is taking place in the other.

In the reactor, the liquefied fish is fed into one end and the product is fed out the other end. In the typical continuous process, a load from the digester would be fed into the reactor every 20 to 30 minutes and the material will stay in the reactor from 1 to 3 hours. The reaction may therefore have several digester loads in it at the same time. Suitable baffling may be used to cause the material to move laterally along the reactor as additional material is fed in at one end and the completely reacted material is taken out at the other end. Alternatively, three or more reactors may be used in sequence.

In the preparation of the yeast for the yeast reactivator, it is combined with concentrated wort or nutrient from the container **56**. The wort is a liquid nutrient for the yeast whose principal ingredients are sugar or dextrose plus nutrient salts and nitrogen. Malt phosphates and other well-known nutrients for yeast may also be included in the wort. Preferably, it is in concentrated form so as to provide a relatively large amount of nutrient in a condition that does not spoil readily, which can be diluted upon feeding into the yeast reactivator **60**.

Thus, if the wort contains 30% average sugar, it can be diluted by the addition of 10 parts of water to make a diluted nutrient in the yeast reactivator containing about 3% sugar. Preferably the yeast is activated in a wort solution containing no more than about 5% sugar.

Either a yeast emulsion or a dry yeast may be combined with the wort upon feeding into the yeast reactivator. The preferred yeasts to use are the usual bakers', brewers' and distillers' yeasts.

Advantageous results are obtained by the addition to either the yeast reactivator or the reactor of whey, which is a standard commodity, being a by-product in the manufacture of cheese. It is composed principally of lactose or milk sugar and a small amount of protein, usually 12 to 13%. It also contains several unidentified growth factors, and further, it has the ability to mask fish odors and to improve the flavor of the fish protein concentrate.

Alternatively or in addition to whey, corn steep liquor, which is a by-product from the manufacture of starch and dextrose, may be used for the same purposes.

This material also would be of some value in the wort to accelerate yeast fermentation. The amount of whey and corn steep liquor used may be varied to suit particular requirements. A typical operation may use 5 to 15% of these materials in the concentrated wort.

As has previously been stated, a small amount of the product from the reactor is recycled or fed back into the yeast reactivator. This speeds up the activation of the yeast and reduces the amount of yeast required. In addition, some of the organic material from the fish is thereby fed into the yeast reactivator, and the yeast is therefore pretrained to work on the fish. Thus, fermentation of the yeast is begun before it is fed into the reactor, and it is obtained under conditions highly conducive to fast reproduction of the yeast cells.

Such conditions include an abundant supply of air or oxygen and a rather low concentration of nutrients, preferably not in excess of about 5%. By this process, the production of metabolites which inhibit spoilage, that is, which decrease the activity of putrefaction bacteria, is greatly increased. At the same time, the activity of beneficial bacteria, such as lactics, is increased.

When it is necessary to shut down the plant for any reason, the feedback material can be held in the line running to the yeast reactivator or in a separate container, preferably at a temperature of 50°F or lower, and can be used when the plant starts up again to regenerate the yeast feed. Thus, the feedback method permits a startup of the plant without any delay while waiting for the yeast to be activated. This feedback method reduces the total time required for the yeast reaction to about one-third to one-fourth the time which would be required if the dry yeast were fed directly into the reactor.

Preferably, the temperature in the yeast reactivator and in the reactor should be held between 25° and 37°C in order to insure rapid proliferation of the yeast cells. It will be appreciated that the yeast fermentation continues in the reactor, and that in doing so it consumes some of the flavor and odor-producing factors of the fish. In addition, this multiplication of yeast cells improves the nutritional value of the fish protein concentrate.

It will usually be found desirable to continuously add yeast, in the amount of from ¼ to 1% of the weight of fish solids processed. Cooked fish usually contains about 25% of fish solids. The time of reaction in the reactor is determined chiefly by the flavor desired and the degree of destruction of undesirable bacteria. Usually a time of 1 to 3 hours is sufficient to obtain a substantially fish flavor free product. The speed of reaction depends on a number of factors, including the concentration of yeast used, the amount of the feedback, and the temperature and pH of the emulsion in the reactor.

The addition of yeast in the process adds to the nutrient value of the product, helps to neutralize fishy flavors, and keeps down spoilage by acting against putrefaction bacteria. It is desirable to maintain a low pH in the reactor, that is, a pH of no greater than about 5.2, in order to stimulate the activity of the yeast in reducing spoilage. Yeast grows better at the low pH and is therefore particularly desirable for reduction of spoilage under conditions which are not substantially sterile.

Following the reaction in the reactor, the liquid product may be screened and then delivered to the dryers and dried to a desired moisture content. It may in some instances be desirable to dry it to about a 6% moisture content. This would be particularly desirable if, because of the use of an oily fish, it is necessary to extract the oil with isopropyl alcohol or the like. Following the drying, and extraction if used, the product is ground to a flour or meal, as desired. The following specific example illustrates the practice of the process.

Example: In a pilot plant operation for processing low fat containing fish, whole fish are continuously conveyed from a cold storage tank at a rate of 2 lb/min and washed by passing under clean water sprays. The washed fish is discharged into a hammermill where it is cut into small pieces. The shredded fish is next discharged into a standard continuous direct steam cooker. The fish is continuously cooked with live steam for 10 minutes while being conveyed by flights from one end of the cooker to the discharge end. The average cooking temperature is 100°C. After cooking, the fish is continuously discharged into a cooling blender which has cold brine circulating in a jacket around it. In this blender the fish is cooled in less than 10 minutes to 55°C.

From the cooling blender, the fish is continuously discharged into a digestion tank along with a water mixture of enzymes. The enzymes are continuously pumped into the digester at a rate of 45 g/min and at a temperature of 10°C, which at this temperature helps to cool the fish down to 45°C. The enzyme mixture consists of pancreatin 3X and a fungal culture of *Aspergillus oryzae* on bran.

The concentration of the enzymes is such that a 5% addition of the enzyme slurry to the cooked fish, or approximately 45 g/min, adds 100 mg of pancreatin and 1 gram of fungal culture per pound of fish. The pH of the enzyme mixture is adjusted to 6.5 by the addition of citric acid, and the final fish mixture in the digester has a pH of 6.4. Other food acids, such as acetic acid or phosphoric acid, might be used to adjust the pH.

The mixture in the digester is moderately agitated while being conveyed through it. The mixture stays in the digester for 20 minutes, at the end of which time it has been liquefied to a consistency of heavy cream. The liquefied fish is then discharged by means of a proportioning pump into a continuous reactor. The reactor is of such size that it will hold 240 lb of fish for 2 hours. The capacity of the pump at the discharge end is so regulated that it pumps out 2 lb of fish per minute.

As the liquid fish is pumped into the reactor, it is corrected to a pH of 5.4 by the addition of citric acid for the purpose of making it more suitable to yeast fermentation and to a fungal enzyme reaction. A concentrated wort is prepared composed of 30% dextrose, 5% molasses, 5% corn steep liquor, 5% dried whey, and 1% monocalcium phosphate. The wort is pumped continuously at the rate of 9 ml/min and to this is added water at the rate of 81 ml/min.

1 gram of active dry yeast per minute is fed into the yeast reactivator by means of a percentage dry yeast feeder. This provides approximately ½% of dry yeast based on the solids of fish processed, or 0.110% based on the raw fish. The yeast is maintained in the yeast reactivator approximately 1 hour at a temperature of 33°C.

The liquefied fish in the reactor is inoculated with the active fermenting yeast from the yeast reactivator by means of a proportioning pump. The fermenting yeast solution is added at the rate of 90 g/min. After 2 hours in the reactor, the product is discharged, and about 2% of the discharged fermented liquid fish is fed back to the yeast reactivator.

Each of the yeast reactivator and the reactor are continuously aerated by means of an air pump. The main portion of the product from the reactor is discharged into a homogenizer which pulverizes the fish bones and other particles into a fine state for conditioning it to be dried on a drum dryer. The liquid is then continuously passed to a drum dryer for drying, and is subsequently ground to a powder and is ready for use. The product produced is of light tan color with substantially no flavor or odor of fish.

This process results in the production of extremely high quality fish protein concentrate in a continuous operation which is far more economical than conventional systems. Because of the preactivation of the yeast, the yeast fermentation in contact with the fish is greatly speeded up so that a much smaller reactor can be used. The separation of the reaction of the yeast from the liquefaction operation also allows each of these operations to be speeded up and also improves the flavor masking, the destruction of putrefaction bacteria, and the rate of liquefaction.

By Lactic Acid Bacteria

H.M. Ehlert; U.S. Patent 3,041,174; June 26, 1962; assigned to A/S Lumino Feed Company, Denmark has developed a process for treating oil containing animal materials, such as fish or fish offal, for recovering a large yield of animal oil and for producing foodstuff having a high nutritive value. This is achieved by adding to the pulp formed by mechanically disintegrating the materials a substrate of fermented dead culture of lactic acid bacteria.

Such a substrate contains proteolytic enzymes which convert the animal material in such a manner that the oil can easily be separated out by centrifuging. Thereby a very large yield of the oil is obtained and by the centrifuging a pulp is simultaneously obtained containing the original proteins in hydrolyzed state and of a very large nutritive value.

It has been found that this result is a consequence of the use of the special culture of bacteria, while a fresh culture containing living lactic acid bacteria cannot be used. This is probably due to the fact that the active enzymes effecting a quick hydrolysis of the proteins and facilitating the separation of the oil from the pulp are not formed or released until after the final fermentation of the culture of bacteria and the death of the cells. When a culture of bacteria after the final fermentation stands for a long time without being supplied with nutrient for the bacteria, the culture will die, but, of course, the bacteria may also be killed by the addition of any bactericidal means.

It has been found preferable to carry out the treatment of the pulp with enzymes for a period of at least 3 hours at a temperature of 20° to 50°C. Under these conditions the proteolytic enzymes in the culture of bacteria will have maximum effect. The minimum of time for the treatment will to some extent depend upon the temperature. Thus, the process may be carried out in 5 to 6 hours at a temperature of 30° to 40°C, but a longer time for the treatment will often be of advantage.

The mechanical disintegration may be executed by means of a vacuum gear pump or other appropriate disintegrating devices, for instance choppers, impact mills, centrifugal mills and rolling mills. It has been found appropriate to mix the pulp with acid to obtain a pH of 3.6 to 4.2, the decomposition being thereby accelerated, and the final product will have an excellent storage resistance and maintain a high nutritive value, the presence of the fermented and dead culture of lactic acid bacteria having the effect that the pH value will remain practically constant.

The pulp has a very high nutritive value owing to a large content of free amino acids. The product is therefore very well adapted as foodstuff either alone or mixed with other foodstuff. The produced foodstuff may in many cases be used without separating out the oil, because the extraordinary durability and great nutritive value of the material still bring about considerable advantages. The produced pulp is well adapted for treatment by atomization drying, the powder formed thereby being easily soluble in water.

In addition to the enzymes supplied with the fermented cultures of dead lactic acid bacteria, further enzymes may be added to the pulp, such as papain, katepsin, lipases or corresponding hydrolyzing enzymes. In this manner the decomposition of the proteins to free amino acids is accelerated. The process is further described by the following example.

Example: By means of a meat chopper herrings are ground to a mass which is further pulverized by means of a flail impact mill. The pulverized mass is transferred to a container in which it is mixed with a fermented culture of dead lactic acid bacteria. The mixture is heated in the container at a temperature of 35° to 40°C and is circulated by means of a vacuum gear pump being thereby further pulverized and homogenized.

This treatment is continued for 5 to 6 hours, and after this period the cells are so disintegrated that the oil may be separated out by centrifuging. Then the pulp is heated in a stirring device with a steam jacket at a temperature of 50°C and passed over a vibratory screen to one or more sludge centrifuges where it is divided into three phases, viz oil, an aqueous solution, and a sludge phase.

The oil is finished in a purificator, and the aqueous solution and the sludge phase are brought together in a container and mixed by means of a pump. This pulp, the oil content of which is reduced to 0.3 to 0.5%, is adjusted at a pH of 3.9 and is now durable for months.

By treating raw herring containing 65.5% water, 14.0% oil, 15.5% proteins and 0% free amino acids with the culture of bacteria as stated, a pulp is obtained containing 67% water, 13.8% oil and 15.0% proteins, 10% thereof being free amino acids. By centrifuging 13% oil is recovered, i.e., a yield of 94 to 95%.

Adding Fat and Surfactant and Digesting with Bromelain

M. Rutman; U.S. Patent 3,561,973; February 9, 1971; assigned to Instituto de Fomento Pisquero, Chile describes a method for the preparation of a water-soluble, stable, odorless fish protein concentrate containing added fat which is useful as a food supplement. The process comprises cleaning, washing, deboning and pulping fresh fish, mixing the pulp with an aqueous emulsion of an edible fat prepared with an edible surfactant, digesting the fat-fish mixture with a proteolytic enzyme for a short period at a controlled, low temperature, the enzyme bromelain being preferred, heating the digested mixture to terminate the enzyme activity and pasteurize the mixture, and thereafter homogenizing the mixture and drying it to recover the product.

One object of this process is to drastically reduce the reaction time for digestion by employing isolated proteolytic enzymes. In this way, rapid multiplication of microorganisms cannot take place and the production cost is lower. A further object is to provide a method for preparing a product without bitter flavor. Enzymatic hydrolysis of vegetable or animal protein develops, in many cases, bitter products or polypeptides. Proper selection of a proteolytic enzyme, and of a correct short time of reaction, provides an odorless, and only slightly bitter product.

Careful research has shown that an extended enzymatic digestion of fish proteins causes a reduction of their nutritional value and a partial destruction of essential amino acids. Therefore, by digesting the fish meat during a very short period of time, a method is provided by which the nutritious value of the fish meat remains almost intact.

The process seeks to provide an instant product, namely, one which easily dissolves in water, forming an emulsion, i.e., a colloidal suspension. The product will be referred to as water-soluble, with this understanding. The fine division of the protein fibers through the enzymatic hydrolysis of the process produces a final product that can be easily dispersed in water and which will remain stable therein.

For dietetic and other reasons it is most desirable to add lipids to the proteins. The method of this process introduces th lipids in an early stage in order to facilitate the dispersion of the protein concentrate, to form a lipid-protein complex that can be easily dried by conventional methods (i.e., spray-drying, drum-drying, etc.) and to avoid the production of an excessively hygroscopic powder which would be unstable during storage. Because of the lipids, a gross agglomeration of the protein particles is avoided and stability toward decantation is accomplished.

The addition of lipids has a further advantage. It is possible to add antioxidants to a stable fat and when this mixture is homogenized with the meat proteins that carry a small fraction of the original unstable fish fat, total protection against oxidation is obtained. The requirement of an easy dispersibility of the final product determines an additional object.

This is the provision of a method in which the reaction temperature of the enzymatic hydrolysis is programmed, and which assures a uniform division of the fibers, and hence, a more stable dispersion. Different changes of the physical structure of the fish meat occur when it is heated. It is known, for instance, that the proteins of cooked fish become insoluble in the tissue fluid or in dilute salt solution. In order to avoid the curdling and coagulation of the minced fish meat, the temeprature of the enzymatic reaction is carefully controlled and, as noted above, a more uniform fractionation is accomplished.

The method is applicable to any type of lean fish. The fresh fish should be eviscerated and cleaned with heads and entrails removed. The fillets so obtained are passed through a meat separator machine to separate the meat from the scales, skin, and a large percentage of the undesirable lipids. The comminuted fish meat is then washed with water. Washing accomplishes several important features. It removes the blood and most bacteria, and it reduces the concentration of other compounds which can lead to an off-color and off-flavor product (desired color is white and desired flavor is neutral and, specifically, not fishy). Further, washing removes a significant fraction of the total lipids contained in the fish. Such lipids are reactive (i.e., lead to spoilage). After washing, the comminuted fish meat is white, odorless and tasteless.

This meat slurry is conveyed through a fish bone separator machine and then it is centrifuged to eliminate part of the water. The preparation of a fat-water emulsion can be carried out in parallel with the washing step. There are several reasons for adding suitable fat to the mixture. As noted above, it is desirable from a dietetic point of view, as it provides a high-energy component to the final product. However, the fat also has several other functions. The added fat and natural lipids remaining in the fish are mutually soluble, but are insoluble in water.

By adding an antioxidant to the added fat the entire fat content is rendered stable. Further, without the fat, it would not be possible to produce a nonhygroscopic material. Lastly, a stable, homogeneous mixture with the fish slurry is produced, for the reasons noted below.

Hydrogenated fish oils, vegetable oils and the like are all suitable as the added lipid. In order to incorporate the lipids in the reaction mixture it is necessary to emulsify them in water by conventional methods such as colloid mills, with the addition of edible surfactants. The amount of oil that is desired to emulsify depends on the type of surfactant and on the total lipids percentage one desires in the final powder. In general, a 6 to 10% stable oil-in-water emulsion can be prepared by adding 1 to 5% (based on total lipid weight) of surfactants.

The next step of the process consists in combining the meat slurry with the oil-in-water emulsion. This operation can be done in a colloid mill or in any other machine of equal efficiency which could also mince the fibers. The proportion of volume of emulsion to weight of meat depends, of course, on the percentage of proteins that are contained in the meat and on the percentage of lipids in the emulsion. But one also has to take into consideration the desired percentage of proteins in the final dried powder. In general, if the meat slurry is prepared in the way described above, it will contain 9 to 15% protein.

In essence, the method is flexible and in order to calculate the proportion of emulsion volume to meat weight it is enough to specify the relation in weight of lipids:proteins of the final powder. The method has been applied to ratios in weight of lipids to proteins of 1:5 to 1:1 in the final dried product. A preferred ratio of lipids:proteins is 1:2. When dissolved in water to form an emulsion, the product looks and tastes like milk, but is much more nutritious.

The combination of the water-fat emulsion with the fish meat slurry has a synergistic beneficial effect. The emulsion per se is not stable over long periods, and will eventually separate. Likewise, the meat-water slurry tends to separate. When the two are mixed, however, the protein of the meat coats the individual fat globules. This prevents both kinds of separation from taking place and produces a stable, homogeneous mixture. The choice of an operable enzyme for use in the process depends on many factors, not the least of which is cost. Intended end use of the product is also important. Thus, the cost of an enzyme rises with its purity, but the tendency to produce a bitter or off-flavor product decreases with increasing purity of the enzyme. So if flavor is not an important consideration, a less-pure, cheaper enzyme can be employed.

Also, some enzymes produce a protein-lipid particle which is more stable (i.e., will not decant) than others. The enzyme bromelain, derived from pineapple juice, is preferred in the process. However, many enzymes can be used, subject to the foregoing considerations (ficin and papain, for example, produce a more bitter product) such as: ficin, papain, trypsin, pancreatin, Rhozyme, HT Proteolytic, Prolase, etc.

The homogenized lipids-protein mixture is then subjected to hydrolysis. The enzyme, dissolved in a buffer of pH 6, is poured in the homogenized mixture at room temperature. The amount of enzyme added depends on its proteolytic activity and on the percentage of proteins of the reaction mixture.

On the other hand, the reaction time is fixed by organoleptic considerations and by the stability sought for the dispersion. The temperature of the reaction mixture is slowly raised and the system is gently agitated. Typical values for temperatures and reaction time are given in the example below. The buffer can be chosen from a list that complies with the requirements of good additives.

Generally, the digestion takes place in the range of 25° to 70°C, and reaction time is less than about 15 minutes. The liquid hydrolyzed product is filtered or screened to remove any gross solid residue and then it is pasteurized at approximately 80°C for 15 minutes and, simultaneously, the enzyme is denatured and inactivated. The hot liquid product is homogenized (i.e., in colloid mill) and then dried by any of the conventional methods (spray-drying, drum-drying, etc.). Yield of the process is better than that obtained in known prior art processes. The following example is given by way of illustration.

Example: Fresh samples of Pacific hake were eviscerated, beheaded and cleaned with water. The resulting fillets were conveyed through a meat separator machine. The comminuted meat without the skin, scales and bones was washed with water and in this way blood and other materials were eliminated. The clean, white, odorless meat pulp was then passed through a fish bones separator machine and centrifuged to eliminate a fraction of the water. An approximate analysis of the meat pulp gave 88% water, 11% protein, 0.2% ash and 0.4% fat.

To 10 parts of hydrogenated fish fat, melted at approximately 65°C, a mixture of antioxidants (BHA, BHT and dodecyl gallate) was added, 0.15 part of sorbitan monostearate was dissolved in the melted fat, the whole mixture was stirred for 15 minutes and then poured into 90 parts of hot water that contained 0.35 part of polyoxyethylene (20) sorbitan monostearate. The resulting hot mixture (60°C) was then homogenized in a blender and left overnight. Next day the stable oil-in-water emulsion was separated from the cream of excess fat.

10 parts of bromelain of low activity was mixed with 100 parts of buffer pH 6.0 (citric acid + sodium citrate, 0.1 M in water), gently stirred and then centrifuged at approximately 700 grams for 10 minutes in order to settle the undis-

solved particles. The bromelain solution was kept at 5°C. 100 parts of meat pulp, 80 parts of emulsion and 20 parts of water were homogenized in a blender at room temperature for 5 minutes. To the resulting paste, 6 parts of the bromelain solution was added with slow agitation. The temperature of the reaction mixture was slowly raised from 28° to 54°C using slow agitation (the reaction kettle was kept at 57°C).

After 14 minutes, the already liquefied mixture was rapidly screened and heated to about 80°C using a heat exchanger and high speed agitation. The cooking temperature was maintained for 12 minutes which rendered the material substantially sterile and the bromelain inactive. The hot liquid was homogenized in a high speed blender (5 minutes) and then rapidly cooled to about 5°C (in about 7 minutes).

Drying of the material was accomplished in a spray drier. The product was white in color, stable, nonhygroscopic, odorless and it imparted a very light bitter taste that could easily be masked. The powder was readily soluble in water and gave an emulsion that did not decant in several days. The final (dry) product had a PER (protein efficiency ratio) of 3.5, (casein was 2.6), and the following approximate composition:

	Percent
Protein	65
Fat	27
Moisture	4
Ash	2

Fish Proteins Recovered as Protein-Phosphate Complex

Prior processes for preparing fish protein concentrates (FPCs) include solvent extraction, extraction and precipitation, and chemical or biological hydrolysis of the fish. Depending on the process and the fish species, the prior FPCs are powders or pastes which range from light to dark in color and from bland to intense in flavor. They have high nutritional value and may be used to protein-enrich baked goods such as bread. However, they generally lack functional properties such as the ability to emulsify lipids and consequently are not usually incorporated into meat products, beverages or pastries.

J. Spinelli and B.J. Koury; U.S. Patent 3,826,848; July 30, 1974; assigned to the U.S. Secretary of Commerce have found that organoleptically stable fish protein concentrates and isolates possessing desirable functional properties may be prepared by enzymically modifying whole fish or the muscle protein fraction and then recovering the modified proteins as protein-phosphate complexes.

The complexes preferably are extracted with a polar solvent to remove lipids. The dried products are white to light-colored, are practically tasteless and odorless, moderately soluble in water, and are capable of emulsifying over 200 times their weight of oil.

In this process whole or eviscerated fish are comminuted in a grinder or blender and slurried with 1 to 5 parts, preferably about 1 part, of water. The pH of the fish slurry should be 6.0 to 7.0; if it is less it may be adjusted upward to a maximum of about 8.0 to 10.0. A proteolytic enzyme of animal or microbiological origin, such as papain, pancreatin, bromelin, ficin, or Rhozyme P-11 is added to the reaction mixture at a ratio, depending on the enzyme activity, of 1 part enzyme to about 25 to 500 parts protein substrate (dry weight basis). The preferred enzyme:substrate ratio is about 1:50 to 1:200, as it has been found that higher ratios produce excessive amounts of nonprotein nitrogen (NPN) and correspondingly lower protein yields. The reaction mixture is held at 20° to 60°C, preferably 25° to 30°C, and slowly stirred for 15 to 90 minutes, preferably 30 to 60 minutes.

The partial hydrolysis modifies the fish proteins by splitting them into smaller protein subunits and into a small amount of potent flavor-producing NPN (various amino acids and peptides). In contrast, a complete hydrolysis would split the protein almost entirely into NPN. It has been found that a high quality organoleptically stable functional FPC is obtained when the NPN content of the reaction mixture is 4 to 20%, preferably 6 to 10%, of the protein substrate (dry weight basis).

Upon completion of the partial hydrolysis, 2 to 10%, preferably 4 to 6%, based on the dry weight of the protein, of a linear condensed phosphate in the form of a 5 to 15%, preferably 10%, solution is added to the reaction mixture. Examples of suitable condensed phosphates are sodium or potassium metaphosphate, trimetaphosphate, hexametaphosphate, or other linear condensed phosphates containing 1 to 20 phosphate groups in the molecule. The preferred phosphate is sodium hexametaphosphate.

The resulting protein-phosphate complexes are precipitated by adding about 0.1 N sulfuric, hydrochloric, phosphoric, lactic, or acetic acids, preferably sulfuric acid, to the reaction mixture in an amount sufficient to lower the pH to about 5.5 and then adding about 1 N similar acid to reduce the pH to 2.0 to 5.0, at which values the protein-phosphate complex is completely insoluble (all of the protein is recovered). The pH preferably, however, is reduced to about 2.5 to 3.5, because it has been found that a pH in this range sufficiently inactivates the proteolytic enzyme to prevent further hydrolysis of the complex during the subsequent extraction and drying steps.

Recovery of the partially hydrolyzed proteins as a protein-phosphate complex at a pH of 2.5 to 3.5 thus (1) obtains 100% of the protein and (2) avoids further hydrolysis during the additional processing. In contrast, recovery of the proteins by precipitation at the isoelectric point of pH 4.0 to 5.0 for example, recovers a maximum of 90% of the protein, and the unrecoverable protein in solution (10% at pH 4.0 to 5.0) rapidly increases if the pH is lowered below 4.0 to inactivate the enzyme.

The precipitated acidic complex is separated from the aqueous solution by any suitable separation technique, preferably by centrifugation, and is extracted about 1 to 5 times, preferably about 2 times, with water to remove excess acid and phosphates. The unneutralized complex is then extracted about 1 to 5 times, preferably about 3 times, with a polar solvent at 20° to 70°C, preferably 50° to 60°C, to remove residual lipids. The extracting solvent may be methanol, ethanol, isopropanol, isomers of butanol or combinations thereof, and preferably is azeotropic isopropanol.

The ratio of solvent to wet complex may range from 1:2 to 2:1, preferably 1:1, and the residence time of the complex in the solvent may be 2 to 30 minutes, preferably 4 to 6 minutes. The residual solvent may be removed by drying but preferably is removed (and recovered) by means of 1 to 6, preferably 2 to 4, aqueous extractions at 0° to 60°C, preferably 15° to 30°C, using a water:complex ratio of 1:2 to 2:1, preferably 1:1.

The complex should be extracted while it is in the acidic or unneutralized state, as it has been found that the resulting dried products retain more functionality than products obtained by neutralizing the complex before extraction. For example, a protein-phosphate complex that is precipitated at pH 3.0, extracted with isopropanol at 50°C, washed with water, neutralized to 7.0 and spray dried will emulsify about 268 grams of vegetable oil per gram of protein. The same complex, if neutralized to 7.0 before the same alcohol and water treatments and drying, will emulsify only about 172 grams of oil per gram of protein.

The alcohol- and water-washed complex is then neutralized by adding an alkali such as sodium hydroxide, sodium carbonate or bicarbonate, calcium hydroxide, calcium carbonate, or calcium sucrate to the wet complex in an amount sufficient to raise the pH to about 5.0 to 8.0, preferably about 7.0. To retard the loss during storage of functional properties such as the ability to emulsify lipids, about 5 to 25%, preferably 10 to 20%, of a carbohydrate such as a mono- or disaccharide sugar or a degraded starch such as corn syrup solids, Morex, or Maltrin may be added to the neutralized complex.

It has been found that complexes containing 15% carbohydrate suffer only a 7% loss of emulsifying capacity after 2 months as compared to a 20% loss with preparations containing no carbohydrate.

The neutralized complex, with or without the carbohydrate additive, is drum, spray, or freeze dried. Freeze dried samples are about 20% soluble in both water and 5% salt and will emulsify about 220 grams of oil per gram of protein. Spray dried samples have about 5% less solubility and emulsifying capacity than freeze dried samples. However, the residual solvent in the spray dried samples is only about 0.005% (50 parts per million or ppm) as compared to about 0.5% (5,000 parts per million) for freeze dried samples.

The Food and Drug Administration currently allows a maximum of 250 ppm alcohol residue in FPCs for human consumption. Desolventizing the protein-phosphate complexes by water extraction and spray drying thus reduces the residual solvent well below the permissible maximum. In addition, this desolventization avoids the denaturing effects of the steam stripping or prolonged vacuum drying usually employed to reduce the solvent in FPCs to an acceptable level.

A myofibrillar protein isolate may be prepared by preliminarily extracting the comminuted fish, preferably eviscerated and freed of skin and bone, for about 10 minutes at 0° to 35°C in 0.1 M saline (sodium chloride). The extracting solution preferably contains 0.01 to 0.05%, based on the weight of comminuted fish, of an antioxidant such as propyl gallate, butylated hydroxytoluene, or butylated hydroxyanisole.

The solution:fish ratio may range from 1:1 to 10:1, preferably about 4:1. The solids, consisting mostly of myofibrillar protein, are separated from the sarcoplasmic and lipid fractions by any conventional means, preferably by centrifugation. The solids preferably are resuspended in additional saline solution and recentrifuged. The myofibrillar fraction is then resuspended in tap water at a water:fish ratio of 1:1 to 4:1, preferably about 2:1, and recentrifuged to further reduce the sarcoplasmic and lipid fractions. The thus prepared myofibrillar protein fraction is then partially hydrolyzed and recovered as a protein-phosphate complex in the manner described previously.

Example 1: Whole Pacific herring are ground through a food grinder equipped with a ⅛" plate and are mixed with an equal part of water. The pH of the slurry if between 6.0 and 7.0 is unadjusted and Rhozyme P-11 is added to the reaction mixture at a ratio of 1 part enzyme to 150 parts of protein substrate (dry weight basis). The reaction mixture is slowly stirred for 60 minutes at 30°C. 5% (based on the weight of the dry protein) of sodium hexametaphosphate as a 10% solution is added to the partially hydrolyzed proteins and the pH of the mixture is lowered to 5.5 with 0.1 N H_2SO_4 and then to 3.0 with 1 N H_2SO_4.

The insoluble protein phosphate is collected by centrifugation and washed twice by resuspending in water. Residual lipids are then removed from the complexed proteins by extracting 3 times with azeotropic isopropanol at 50°C. A ratio of 1 part isopropanol to 1 part wet complex is used with a residence time of 5 minutes during each extraction. Residual isopropanol is removed by 4 aqueous extractions at 20°C, using a 1:1 ratio of water to complex. The complex is neutralized to pH 7.0 with sodium hydroxide and spray dried. The composition of the final product is: protein, 82%; ash, 14%; moisture, 4%.

Example 2: Eviscerated Pacific Ocean rockfish are freed from skin and bone fractions by the use of a mechanical deboner such as a Bibun or Yanigiya flesh separator or Beehive deboner. The comminuted tissue is added to 0.1 M NaCl containing 0.01% propyl gallate (based on the dry weight of the fish protein). A 4:1 ratio of solution to fish is used and the mixture is held at 30°C and slowly stirred for 10 minutes.

The mixture is centrifuged in a bowl-type centrifuge for 10 minutes to separate the myofibrillar solids from the sarcoplasmic and lipid fractions. The solids are resuspended once in the 0.1 M salt solution and once in tap water both at a 2:1 ratio of liquid to fish.

The resulting myofibrillar protein is adjusted to 5 to 10% solids with water and papain is added to the mixture at a ratio of 1 part enzyme to 250 parts protein substrate. Then, as described in Example 1, the mixture is allowed to hydrolyze and the modified proteins are recovered as a protein-phosphate complex and extracted with azeotropic isopropanol and water. The product when neutralized and spray dried has the following composition: protein, 90%; ash, 6%; moisture, 4%.

The myofibrillar protein isolate prepared as described in Example 2 is about 40% soluble in water and 5% salt solution and will emulsify about 270 grams of vegetable oil per gram of protein. As a comparison, soya isolate and sodium caseinate will emulsify respectively about 230 and 140 grams of oil per gram of protein. Furthermore an emulsion of myofibrillar protein isolate, vegetable oil, and water will break in about 120 minutes, whereas corresponding emulsions of soya isolate and sodium caseinate will break respectively in about 90 and 30 minutes.

PREPARATION FROM STICKWATER

Fish stickwater is a by-product of fish canning which has been difficult to handle. Viewed as a waste product, its disposal is troublesome because of its tendency to putrefy, to give rise to noxious odors and to pollute bodies of water into which it is dumped. The odoriferous quality of fish stickwater is due to the decomposition of protein material by enzymatic and bacterial action.

Also, fish stickwater contains potentially valuable protein material. A major problem has been the efficient utilization of this protein material. The fish with which this process is concerned may be of any kind but are primarily commercial, marine types of fish which are processed and canned on a large scale; e.g., tuna, sardines, mackerel, anchovies, salmon and the like.

Fish stickwater is normally produced as follows. Fish scrap (including heads, tails, viscera and even whole fish which are purposely rejected or which find their way into scrap by accident is cooked, e.g., at steam gauge pressures up to 5 psi for periods of 7 to 15 minutes. The cooked scrap is then pressed, e.g., in a continuous screw press, to remove most of the water and oil. The press cake or solids separated in this manner usually contains about 50% moisture and it is dried to form the fish meal of commerce. The filtrate separated from the press cake is what is known as fish press water.

It usually contains about 5 to 10% solids, including some oil as part of the solids. The nonoleaginous solids consist chiefly of insoluble and soluble proteins; of other water-soluble nitrogenous compounds such as proteoses, peptones and amino acids; and of vitamins and other growth factors.

In conventional processes fish press water is heated, e.g., to 190°F and centrifuged to remove the residual oil. This leaves an aqueous portion containing, in a typical case, about 1% oil, 0.75 to 1.25% insoluble protein particles and 3 to 5% soluble proteins and other soluble nitrogenous materials, vitamins and various growth factors. This aqueous product is called fish stickwater. It is usually acidified to a pH of 4 to 5 and held at 180°F to preserve it until it is evaporated to a solids content of about 50%. This concentrate is the condensed fish solubles of commerce.

Peptizing Insoluble Proteins to Prevent Gelling

Condensed fish solubles are a valuable supplement to animal feeds because of the broad spectrum of proteins, protein components, vitamins and other growth factors. However, condensed fish solubles have been difficult to utilize because of several properties. For example, condensed fish solubles contain finely divided, insoluble protein particles which settle out and give rise to a nonuniform product. They also have a marked tendency to gel at temperatures at 50°F and less. Gels so formed are difficult to remelt because of a hysteresis effect.

Also fish solubles contain residual oil in the form of very fine particles and which is highly unsaturated and sensitive to oxidation. It therefore oxidizes very rapidly and, in doing so, it not only gives rise to offensive odors but the oxidation products, such as hydroperoxides and peroxides, cause oxidative deterioration of nutritive values such as vitamin A, carotene and certain of the B complex vitamins contained in feedstuffs with which the fish solubles are mixed.

Previously it has been proposed to improve fish press water, fish stickwater or condensed fish solubles by two different methods. One of these methods is by proteolysis. That is, the insoluble proteins are solubilized by the action of an added proteolytic enzyme such as papain. The other method is to coagulate and remove the insoluble or readily coagulable proteins by the addition of a protein coagulant such as sulfuric acid.

Both of these methods have serious disadvantages. Neither of them solves the problem of gelling at low temperatures in a practical manner, and neither of them solves the problem posed by residual, easily oxidized oil. Also, the method which removes the solid proteins by coagulation has the disadvantage that the removal of these proteins detracts from the nutritive value of the fish solubles. Thus a large part of the riboflavin content of condensed fish solubles is removed, and the amino acid pattern is impoverished.

H.N. Brocklesby and J.R. Patrick; U.S. Patent 2,934,433; April 26, 1960; assigned to H.N. Brocklesby and Associates solved these problems by mixing fish stickwater and comminuted raw fish viscera, the stickwater containing discrete particles of insoluble proteins and also dissolved proteins and other nutritional materials; heating the mixture of stickwater and viscera at a pH of 4 to 8.5, at a temperature of about 170° to 200°F, and for a period of time sufficient to peptize and disperse the insoluble proteins; then cooling the treated liquor and separating nonprotein solid particles; then evaporating the liquor to a substantially higher solids content.

The process is preferably started with fish stickwater, i.e., with fish press water from which most of the oil has been removed by heating and centrifuging. Raw fish viscera contains substances which will act on fish stickwater at high temperatures and in a very short period of time to peptize the insoluble proteins, to disperse such proteins stably in the aqueous phase and to render the fish stickwater (and also condensed fish solubles prepared from the fish stickwater) nongelling at low temperatures.

It was found that an antioxidant can be more effectively incorporated in fish stickwater processed with raw fish viscera in accordance with this method. By fish viscera is meant the waste from fish cannery operations which comprises all or some combination of the fish organs and material such as bone, muscle and connective tissue. The viscera may come from a variety of fish such as tuna, sardines, mackerel, anchovies, and salmon. It may be fresh or thawed frozen viscera or it may be raw viscera that has been preserved by added alcohol, glycerin, propylene glycol or other suitable preservative. The viscera should be ground sufficiently fine to disperse and act readily.

Fish viscera itself presents a major disposal problem in the fish canning industry. It is known to contain many vitamins and other growth factors in substantial amount, e.g., vitamins A, D, B_1, B_2, B_6, B_{12}, pantothenic acid, niacin, biotin and choline. Fish viscera also contains a complex variety of enzymes and/or enzyme precursors such as fish pepsin (which differs from swine and other pepsin), erepsin or enterokinase and trypsinogen. The enzymes in fish viscera are more heat labile than those from mammalian sources.

Preferably the temperature is about 170° to 200°F. The pH may vary widely, e.g., from a pH of 4 to a pH of 8.5. The time of contact need not exceed a few minutes, e.g., 5 minutes, usually not in excess of about 10 minutes. Under these conditions of high temperature and short contact time, factors present in the raw fish viscera act in some way on the insoluble protein of the stickwater to peptize it, and they act also to impart nongelling properties. At the same time the fish viscera becomes uniformly and stably dispersed in the stickwater and makes a very important contribution to the nutritive value of the end product.

The raw fish viscera may be employed in widely varying proportions, e.g., from 1 to 200% of the weight of solids in the stickwater, preferably 50 to 100% of the weight of solids in the stickwater. Since the solids in fish stickwater are usually 6 to 10% of the weight of stickwater, it is best to add, in most cases, about 3 to 10% of raw fish viscera based on the weight of stickwater. The following specific examples will serve to illustrate the process.

Example: 15 parts by weight of fish stickwater from the canning of tuna were taken. The bulk of the oil had been removed from this stickwater by centrifuging. This stickwater was divided into three equal lots (lots No. 1, No. 2 and No. 3). Each lot was heated to 200°F. To lot No. 1 was added raw, ground tuna viscera in the amount of 4% of the weight of stickwater.

Lot No. 1 was then stirred thoroughly at 200°F for 5 minutes, after which the temperature was raised to boiling and the mixture was passed through a 40 mesh screen to remove connective tissue and bones contributed by the raw viscera. Lot No. 1 was then cooled to room temperature and its pH was adjusted to 4.5. It was then set aside overnight in a refrigerator at -4°C.

Lot No. 2 was treated in the same manner except that, instead of raw fish viscera, a proteolytic enzyme (papain) was added. The pH was 6.5 (optimum for papain) and the papain was added in the amount of 0.1% of the weight of stickwater, which is far in excess of that usually employed for proteolysis. After cooling, as in the case of lot No. 1, the pH was adjusted to 4.5.

Lot No. 3 was a control to which nothing was added but which was heated, stirred, etc. as in the case of lots Nos. 1 and 2. Results were as shown in the table below.

	Lot No. 1 (Viscera Treated)	**Lot No. 2 (Papain Treated)**	**Lot No. 3 (Control)**
Gel formation after -4°C storage overnight.	No gel formed	Set to stiff gel	Set to stiff gel
Separation of solid particles.	No separation	Solid particles settled to a paste-like layer with gel above.	Solid particles settled to a paste-like layer with gel above.

Two-Stage Digestion

R.T. Erickson; U.S. Patent 3,003,880; October 10, 1961; assigned to Star-Kist Foods, Inc. has developed a process for preparing an especially nutritious concentrate from fish stickwater by effecting digestion of the suspended solids originally present in the stickwater without requiring the addition of an outside proteolytic enzyme. In accordance with the process, the proteolytic enzymes necessary for effecting such digestion are provided from the viscera of the fish from which such stickwater is expressed. Accordingly, the process of the method may be carried out at less cost than where an outside proteolytic enzyme such as papain is employed.

Figure 7.7 shows a flow diagram of the process. The numeral **10** designates a primary digester. This primary digester receives fish viscera which has been finely ground as by a hammermill. If desired, finely ground fish scraps may additionally be received by this primary digester. Sufficient water is added from a source **12** to provide a pumpable mixture. This mixture is then violently agitated, as by pumping, while being held at a temperature between 120° to 140°F for a period sufficient for most of the suspended solids in the mixture to be dissolved. Generally, from 3 to 10 hours are required to break down the fish solids.

At the conclusion of this primary digestion period, the resulting mixture will be passed through a screen **14** to remove bones and foreign materials and emptied into a secondary digester **16**. The secondary digester also receives a mixture of fish stickwater and fish juices. The stickwater and juices are first mixed, however, in a mixer designated **18** while heated at a temperature preferably between 180° to 200°F. The fish juices are recovered from various processes in a fish canning plant, as for example, the precooking operation.

After the mixing operation, the mixture of stickwater and fish juices is passed through a centrifuge **20** which mechanically removes the free fish oils present therein. The term free oils as used herein refers to those oils which may be fairly readily mechanically separated from the mixture of stickwater and fish juices. It is an important feature of the process that the fish oils removed by the prior art processes are permitted to remain in the stickwater throughout and thereby render the concentrate particularly nutritious.

After separation of the free oil, the mixture of stickwater and fish juices pass into the secondary digester. It should be particularly observed at this point that the viscera which has undergone primary digestion in the digester is rich in natural proteolytic enzymes. These proteolytic enzymes are employed in the secondary digester in the digestion of the suspended solids and gelatinous materials contained in the mixture of stickwater and fish juices, as well as the digestion of any suspended solids remaining in the mixture coming from the primary digester.

It has been determined that such digestion will most readily take place at a temperature of approximately 150°F. Additionally, the mixture within the secondary digester should be violently agitated as by pumping. A time varying between 4 to 24 hours is generally required for the secondary digestion step depending upon the amount of suspended solids which must be dissolved.

It has been found that the oils present within the material entering the secondary digester will be stabilized during the secondary digestion step. Thus, after undergoing such secondary digestion such oils will not separate from the final product of the process, i.e., the nutritional concentrate. Additionally, such secondary digestion reduces the susceptibility of these oils to rancidity.

At the conclusion of the secondary digestion step the pH of the solution within the secondary digester is adjusted so as to provide the concentrate with a pH of approximately 4.5 or slightly less. In general, this requires that the pH of the solution within the secondary digester be adjusted to approximately 4.52 to 3 in order to achieve a concentrate pH of 4.5. By so adjusting the pH the keeping qualities of the concentrate is improved.

FIGURE 7.7: TWO-STAGE DIGESTION FOR STICKWATER

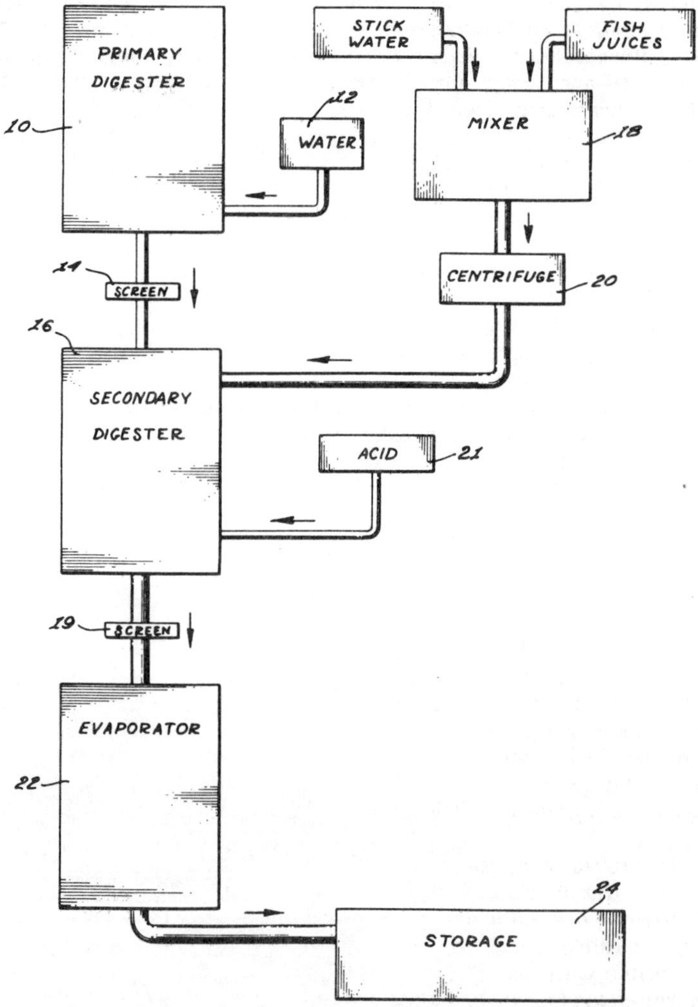

Source: R.T. Erickson; U.S. Patent 3,003,880; October 10, 1961

While it is not essential to adjust the pH at this point, and it is in fact possible to effect such adjustment after the solution from the secondary digester has been concentrated, such adjustment is most readily effected within the secondary digester. This is true since it is easier to disperse the adjusting medium in the thinner liquid and additionally the secondary digester contains means for agitating the solution. The adjusting medium may come from a source designated **18** in the drawing and can take several forms, as for example sulfuric acid, phosphoric acid, or even lime juice.

It should be particularly noted that despite a wide variation in the proportion of stickwater, fish juices and the material coming from the primary digester, the solution obtained by the secondary digestion step will be homogenous. The solution from the secondary digester **10** is preferably passed through a screen **20** to remove bones and foreign materials and passes to an evaporator **22**. The solution is then concentrated, preferably, until it has dissolved solids content of approximately 50%. After the concentrating step, the resulting concentrate may be transferred to a storage tank **24**.

The resulting concentrate includes in addition to regular fish oils, a percentage of valuable glandular oils of high vitamin content, the latter being provided by the viscera. These oils will not separate from the solution and additionally they will not turn rancid. As a result, the concentrate has very little fish odor and may therefore be used in higher concentrations in feeding poultry and animals without imparting a disagreeable taste to the flesh of the animals or poultry.

The solids originally present in the sources of the concentrate as suspended matter will remain dissolved in the concentrate even over extended periods of time and at ordinary temperatures. Accordingly, the concentrate will remain highly fluid and may therefore be readily handled both during shipping and use.

Simultaneous Fat Removal and Dehydration

J.C. Cavanagh; U.S. Patent 3,076,708; February 5, 1963 describes a process for producing a protein-rich material using the waste from fish processing plants and whaling stations. It comprises subjecting the protein-containing material having a water content of the order of 20% or greater to a simultaneous dehydration and fat removal stage with a solvent which is miscible with water and has a boiling point below that of water, separating the material from the liquid phase, treating the separated material with a defatting solvent, and separating the dehydrated and defatted material from the defatting solvent. A wide range of organic solvents may be used.

Acetone has proved to be particularly suitable and is capable of fulfilling both functions, i.e., dehydration and fat rendering as well as defatting or fat solubilization. In carrying out the process, the protein-containing material is comminuted to a suitable particle size of the order of one-eighth to one-quarter of an inch and then treated with the solvent either in a batch process or in countercurrent flow, the apparatus being designed so that the solvent loss is reduced to a minimum. The temperature at which the first stage is carried out is not critical but should be such that a considerable portion of the fat is rendered or melted. For most purposes a temperature within the range of 40° to 60°C is satisfactory.

Preferably the process is carried out at a temperature lower than the boiling point of the solvent and higher than the melting point of the fat. After the treatment has been completed, the mixture is filtered, pressed, squeezed, centrifuged or otherwise treated to separate the solid material from the liquid phase. The liquid phase is passed to a distillation and/or fractionation unit for the recovery of solvent while the solid material is then further treated with the defatting solvent to dissolve the residual fat.

FISH PROTEIN FOR USE IN CAKE MIXES

As is well-known to those skilled in the art, heat-coagulable proteins find wide use in industry and particularly in the food industry as a component of mixtures employed in baking, and for the preparation of toppings. One of the more common food uses of a heat-coagulable protein is in the production of an angel-food-type cake where the heat-coagulable protein may be egg white.

This material, like many other heat-coagulable proteins, may form an airy foam on whipping which in due course will fall. On heating this foam of heat-coagulable protein together with other ingredients, the protein coagulates to form a more-or-less rigid matrix or skeleton which provides the support for other ingredients such as sugar, flavor, etc.

Coagulation of protein under these conditions may result from various reactions including physical and/or chemical reactions which are initiated or energized by heat. Most commonly, however, coagulation is thought of in terms of its macroscopic features, particularly an apparent hardening of a protein foam or dispersion to yield a more-or-less rigid matrix.

To a large extent, most of the components of mixtures containing heat-coagulable protein are very economical. The main component which is characterized by its high price is the heat-coagulable protein, which is commonly egg white. Accordingly prior art investigators have long studied the problem of obtaining a relatively cheap heat-coagulable protein which would be characterized by substantially the same desirable heat-coagulation properties as those which characterize the more expensive egg white. Among these properties are the stability, as well as the ease and reproducibility of coagulation when the protein is subjected to heat.

A wide variety of substitutes have been tried, but many of these have proven to be unsatisfactory because of high price, low availability, poor stability, inability to hold any appreciable amount of other components, incompatability with other ingredients, or undesirable color. Accordingly, there are no low-cost, heat-coagulable proteins which can be employed as a substitute for egg white or the like.

W.I. Rogers; U.S. Patent 3,099,562; July 30, 1963; assigned to General Foods Corporation has developed a technique for obtaining a high-quality, low-cost, heat-coagulable protein from fish by a process which comprises subjecting whole fresh fish or portions of fresh fish containing substantial amounts of fish muscle to mild extraction conditions in the presence of an aqueous extraction liquor thereby solubilizing at least a portion of the heat-coagulable protein content thereof, and isolating the heat-coagulable protein.

The raw materials which may be treated in accordance with this process include fresh fish. More specifically, it is preferred to use fillets. Fish is said to be fresh, as the term is used herein, when it is free of undesirable fish odors, of any indications of rancidification of the natural oils, or of any evidence of undesirable bacterial degradation characterized by undue softness of the flesh and generally nonbland odor. More specifically, it may refer to fresh fish which has been stored at room temperature for not more than one day; fish refrigerated at 35°F for less than 5 to 9 days; or fish refrigerated at 0°F for less than 4 months and preferably less than 6 weeks.

The raw material may be either a low-oil fish such as cod or haddock (typically 1 to 2% oil) or a high-oil (typically 5 to 20%) fish such as menhaden, herring, mackerel, king salmon, sardine, or pilchard. However, the advantages of the process are particularly apparent when high-oil fish are employed.

The fish charge, whether in the form of whole fish or fish fillets, is gently comminuted (i.e., shredded) under conditions which minimize shearing, aerating, foaming, or other undesirable denaturing. This may be effected in a screw-type grinder such as a Hobart mill, to give a product most of which will pass through a 16 to 20 mesh screen on which the bulk of the bones and scale may be retained. This comminuted fish is then slurried with or dispersed in 1 to 10, preferably 5 volumes, of dispersing liquid, typically water.

The fish which has been mixed with water and which is in slurry form is then subjected to mild extraction conditions. The slurry is maintained at temperatures below 30°C and down to just above 0°C. A preferred extraction temperature is 5°C and, at this lower temperature, the amount of extract may be, e.g., 15 to 20% greater than that attained at about 30°C. The preferred time of extraction is 0.25 to 0.50 hours. The extraction may be effected in one or more steps, and if two steps be used, the yield may be increased by as much as 35%.

During the period when the fish in slurry form is being subjected to extraction conditions, it is believed that the main effects are a favorable denaturing and an extraction of solubles from the fish. More particularly, it is believed that proteinaceous nitrogen-containing moieties in the fish and especially those in the fish muscle, including water-soluble albumins and globulins, may be leached out or extracted.

The extract liquor comprises a solution of proteins including, e.g., albumins and globulins. More specifically, the liquor may have a protein content of 0.6 to 0.8%, say 0.7%, and the yield of soluble protein may typically be 3 to 4% of the wet weight of the fresh fish. The amount of protein in the aqueous solution may correspond to about 20% of the protein content of the fish.

When the extracting medium which is employed is substantially pure water, the extract liquor will contain primarily the albumins and the globulins, these being the two water-soluble components of the original fish muscle, and they compose typically as much as 20% of the fish muscle. The actomyosin fraction of the fish muscle, which comprises as much as 70% of the fish muscle, is substantially insoluble in water to which no salt has been added. If desired, the extract of globulins and albumins may be isolated as hereinafter disclosed in detail to obtain a heat-coagulable protein.

However, it is possible to increase the amount of heat-coagulable protein recovered by as much as 300% or more by use of a saline or salt-containing extracting medium. The preferred salt is sodium chloride, and such salt is present in the extracting medium in concentrations of 1 to 5%, preferably 2%. The operable temperature range of the saline extraction is the same as that for the pure water extraction and although the yield of heat-coagulable protein extracted at above 30°C, say 35°C, is increased somewhat over that obtained with pure water extraction, this yield is still too small to provide enough extraction product for further processing.

The saline extraction solution may contain up to 90% of the entire protein content of the treated fish and may have a protein content of about 9 to 45% protein based on the total weight of saline extraction solution. More typically, however, the protein content of the solution will be about 18%.

The saline extract liquor, containing albumin, globulin, and actomyosin may be subjected to enzymolysis to convert the actomyosin therein to (a) about 60% of heavy meromyosin, and (b) about 40% of light meromyosin. More specifically, the protein may be subjected to enzymatic digestion to give the desired product by adjusting the pH of the solution to 7 to 8, preferably 7, by addition of sodium hydroxide, typically in dilute solution of 0.1 N. Enzymolysis may be effected by use of a proteolytic enzyme such as trypsin, pancreatin, etc.

The enzyme, for example, trypsin, may be added to the solution of solubilized protein in amount of 0.01 up to 0.1%, preferably 0.02 to 0.05% of the protein, i.e., of the solids in the extract liquor. This may correspond to about 0.06% up to about 0.6%, say 0.12 to 0.3% by weight of the fresh fish which has a protein content of about one-sixth of its weight. The temperature of the enzymolysis will be below about 40°C to minimize coagulation of native protein, usually 25° to 0°C, preferably 5°C. Enzymolysis may be continued for 60 minutes.

As the digestion continues, the viscosity of the solution decreases considerably. For example, in one case when trypsin was employed at 25°C to digest a solubilized cod fillet, the viscosity (measured in a Brookfield viscometer at 15°C) dropped from about 980 to 40 cp after about 10 minutes. The course of the reaction may be determined by observing this fall to a low viscosity. After the digestion-enzymolysis has been continued for the desired time, the reaction mixture is modified to inhibit further enzyme activity by cooling or by changing the pH so that it falls outside of the range 7 to 8.

Treatment of this mixture to give a salt-free, heat-coagulable protein may be effected by dialysis. During dialysis, the saline liquor may be placed in a dialysis apparatus and dialyzed against preferably tap water. This removes the salt from the liquor. Dialysis may be effected preferably at a pH which may be 4 to 9, but preferably 4.5 to 6.0, say 5.6.

It is found that dialysis in the preferred range permits attainment of good yields of product of high quality, specifically product which is substantially odorless. Dialysis may be continued for as long as 20 to 120, say 72 hours; a purer product can be obtained by a long dialysis or a highly-efficient dialysis for a shorter time.

Further treatment of the dialyzed sol may include isolation of the soluble protein which is contained therein if the product be desired in dry form. This may be effected by drying, e.g., freeze drying, typically at pressure less than about 40 microns Hg, say 20 microns Hg. Spray drying or vacuum drum drying may be employed. Preferably, however, isolation of the desired product may be effected by solvent precipitation, preferably with ethanol.

95% ethanol may be added to the sol until the ethanol concentration rises to 40 to 50%, the mixture being maintained at less than 5°C. The precipitated heat-coagulable protein may be separated as by filtration and washed several times with e.g., 50% ethanol.

The so-prepared heat-coagulable protein may be characterized by its light color and by its freedom from strong offensive odor, etc. More particularly the product is characterized by its ease of heat-coagulation. The product of this process may have superior whipping properties and it will typically have an overrun which is at least about equal to, and frequently 25 to 30% greater than a standard egg albumin. Furthermore, this maximum overrun may frequently be obtained in a time which may be no longer than the time to obtain the maximum (but lower) overrun with egg albumin.

This newly-prepared, heat-coagulable protein may be characterized by its ease of coagulation when subjected to heat and by its low cost. It may be noted that it will effect heat-coagulation when present in amount of as low as 1% of a mixture, whereas egg white will not function similarly at concentrations below about 5%. The heat-coagulable protein of this process may be used in any one of a wide variety of food uses and the preferred use may be in cake mixes of the angel-food-type. The technique of this process may be illustrated by reference to the following example.

Example: In this example, 60 grams of fresh cod muscle was extracted with 300 ml water at 7°C in a Waring Blender. After 5 minutes of mixing, the slurry was centrifuged for 15 minutes at 24,000 rpm in a Sharples Supercentrifuge. Analysis of the supernatant liquor indicated a concentration of 0.69% protein.

The solution containing the desired product was freeze dried at 15 microns pressure for 16 hours to give a solid product weighing 2.46 grams (4.1% of the fresh fish muscle, about 20.5% of the total fish protein, and approaching 100% of the albumin content of the muscle).

The whipping properties of this material were tested by adding 1 gram thereof to 100 ml of water (which optionally may be buffered to pH 7 with a citrate-phosphate buffer). The mixture was allowed to stand for about 15 minutes before whipping; and it was whipped in a Sunbeam (#1406) mixmaster at highest speed. The whip reached its maximum overrun of 200% in 1 minute (about the same overrun as attained by egg albumin) and the whip became stable after about 5 minutes.

The fish foam was lighter in density and texture than a corresponding foam prepared from egg albumin; it was more stable on addition of fat than was the egg-foam. Heat-coagulability of the whipped protein was demonstrated by placing the whipped fish foam in a mold and placing this in an oven for 60 minutes at 70°C. The whip coagulated to produce a hard confection-like product which was of low density.

Production of a sugary confection can be readily effected by adding 120 grams of sugar and 0.1 gram of vanilla flavor to a 1 gram aliquot of the whipping agent in 100 ml of water. Preferably the sugar and the flavor may be folded in after the mixture has been whipped. The whip may be placed in an oven at, e.g., 70°C for 60 minutes to yield a white, light, airy, foamy cake or cookie which is pleasant to taste.

COMPANY INDEX

The company names listed below are given exactly as they appear in the patents, despite name changes, mergers and acquisitions which have, at times, resulted in the revision of a company name.

INVENTOR INDEX

U.S. PATENT NUMBER INDEX

NOTICE

POLLUTION CONTROL IN MEAT, POULTRY AND SEAFOOD PROCESSING 1974

by Harold R. Jones

Pollution Technology Review No. 6

Slaughtering and meat processing produce a highly loaded, but biologically degradable wastewater, containing dissolved, suspended and dispersed blood, bile, grease and fibrous animal matter with high nitrogen content, bringing with it also serious air pollution and odor removal problems.

Since animal slaughter is the primary factor in determing wastewater volumes and pollution loads, it was necessary to examine the size and structure of the meat packing industry. A meat packing plant usually is both a slaughterhouse and a meat processing plant, where edible and inedible fats are also processed.

In this book are condensed vital data from numerous government and other sources that are unusually difficult to pull together. Important pollution removal processes are interpreted and explained by examples from recent U.S. patents.

A partial and condensed table of contents follows. Chapter headings and some important subtitles are given.

ISBN 0-8155-0518-3

263 pages

FOOD ADDITIVES TO EXTEND SHELF LIFE 1974

by Nicholas D. Pintauro

Food Technology Review No. 17

Aside from freezing, canning and sophisticated methods of packaging, food is preserved by dehydration, salting, sugaring, smoking, curing, and certain types of fermentation.

A newer effective approach toward prevention of spoilage is by the use of chemical additives other than sugar, salt, vinegar, and spices.

Food additives, as defined by the National Academy of Sciences are those relatively nontoxic chemicals that may be incorporated into foodstuffs during the growing, processing, or storing periods. Every chemical added must serve one or more of these general purposes: Improve or maintain nutritional value, enhance quality, increase consumer acceptability, and facilitate preparation. In modern applications food additives are combined with established, classical methods of food preservation to maximize stability for extended shelf life. There is a great demand for additives to prevent or retard food deterioration. These additives include antioxidants, antibacterial agents, mold inhibitors, color stabilizers, anticaking agents, antibrowning agents, cloud stabilizers, metal scavengers, enzyme inhibitors.

This book describes over 140 processes involving the newest technology available in the U.S. patent literature using food additives. A partial and condensed table of contents follows here. Chapter headings are given, followed by examples of important subtitles.

ISBN 0-8155-0548-0

402 pages

FEEDS FOR LIVESTOCK, POULTRY AND PETS 1973

by M. H. Gutcho

Food Technology Review No. 6

Food fads, health claims and high prices notwithstanding, meat remains Western man's most preferred source of high quality protein.

The trend in modern livestock production is toward rearing animals with a rapid rate of growth. Feed efficiency in poultry and cattle has been increased 50%, and 40% in swine, with a corresponding weight gain, over what was considered normal 15 years ago.

Since estrogens and antibiotics are no longer accepted components of feeds, new accessory factors are of great importance. Many processes in this book are concerned with just such growth-promoting compositions, making the natural nutrients easier to digest and to assimilate.

Among the fastest selling animal feeds in a gaining market are those for pets. In the U.S. a pet population of 55 million dogs and cats eats 3 million net tons of food every year. Formulations for dog and cat food are therefore included in this book, as are some for birds, tropical fish, fur bearers, and laboratory animals.

The book describes 396 patent-based processes and formulations. A partial and condensed table of contents follows. Numbers in ( ) indicate the number of processes per topic.

ISBN 0-8155-0496 **389 pages**